AF692923

Sensors Network and Wearables for People Activities and Wellbeing Monitoring

Sensors Network and Wearables for People Activities and Wellbeing Monitoring

Guest Editors

Sara Casaccia
Gloria Cosoli

MDPI

Basel • Beijing • Wuhan • Barcelona • Belgrade • Novi Sad • Cluj • Manchester

Guest Editors
Sara Casaccia
Department of Industrial
Engineering and
Mathematical Sciences
Università Politecnica Delle
Marche
Ancona
Italy

Gloria Cosoli
Department of Theoretical
and Applied Sciences
eCampus University
Novedrate
Italy

Editorial Office
MDPI AG
Grosspeteranlage 5
4052 Basel, Switzerland

This is a reprint of the Special Issue, published open access by the journal *Sensors* (ISSN 1424-8220), freely accessible at: https://www.mdpi.com/journal/sensors/special_issues/41XQ0PVNXQ.

For citation purposes, cite each article independently as indicated on the article page online and as indicated below:

Lastname, A.A.; Lastname, B.B. Article Title. *Journal Name* **Year**, *Volume Number*, Page Range.

ISBN 978-3-7258-7915-1 (Hbk)
ISBN 978-3-7258-7916-8 (PDF)
https://doi.org/10.3390/books978-3-7258-7916-8

© 2026 by the authors. Articles in this reprint are Open Access and distributed under the Creative Commons Attribution (CC BY) license. The reprint as a whole is distributed by MDPI under the terms and conditions of the Creative Commons Attribution-NonCommercial-NoDerivs (CC BY-NC-ND) license (https://creativecommons.org/licenses/by-nc-nd/4.0/).

Contents

About the Editors vii

Preface ix

Sara Casaccia and Gloria Cosoli
Sensors Network and Wearables for People Activities and Wellbeing Monitoring
Reprinted from: *Sensors* **2026**, *26*, 2475, https://doi.org/10.3390/s26082475 1

Carlo Ricciardi, Noemi Pisani, Leandro Donisi, Filomena Abate, Marianna Amboni, Paolo Barone, et al.
Agreement between Optoelectronic System and Wearable Sensors for the Evaluation of Gait Spatiotemporal Parameters in Progressive Supranuclear Palsy
Reprinted from: *Sensors* **2023**, *23*, 9859, https://doi.org/10.3390/s23249859 7

Tommaso Di Libero, Chiara Carissimo, Gianni Cerro, Angela Marie Abbatecola, Alessandro Marino, Gianfranco Miele, et al.
An Overall Automated Architecture Based on the Tapping Test Measurement Protocol: Hand Dexterity Assessment through an Innovative Objective Method
Reprinted from: *Sensors* **2024**, *24*, 4133, https://doi.org/10.3390/s24134133 20

Jian-Hong Wang, Phuong Thi Le, Weng-Sheng Bee, Wenny Ramadha Putri, Ming-Hsiang Su, Kuo-Chen Li, et al.
Implementation of Sound Direction Detection and Mixed Source Separation in Embedded Systems
Reprinted from: *Sensors* **2024**, *24*, 4351, https://doi.org/10.3390/s24134351 38

Tania K. Hernández-Heredia, Cesar. F. Reyes-Manzano, Diego A. Flores-Hernández, Gabriel Ramos-Fernández, Lev Guzmán-Vargas
Proximity Sensor for Measuring Social Interaction in a School Environment
Reprinted from: *Sensors* **2024**, *24*, 4822, https://doi.org/10.3390/s24154822 55

Haruki Toda and Takaaki Chin
Physical Frailty Prediction Using Cane Usage Characteristics during Walking
Reprinted from: *Sensors* **2024**, *24*, 6910, https://doi.org/10.3390/s24216910 70

Grazia Iadarola, Alessandro Mengarelli, Paolo Crippa, Sandro Fioretti, Susanna Spinsante
A Review on Assisted Living Using Wearable Devices
Reprinted from: *Sensors* **2024**, *24*, 7439, https://doi.org/10.3390/s24237439 79

Alessia Fedele, Andrea Colantoni, Giuseppe Calabrò, Mauro Scungio, Stefano Rossi and Juri Taborri
Measuring CO_2 Concentration and Thermal Comfort in Italian University Classrooms: A Seasonal Analysis
Reprinted from: *Sensors* **2025**, *25*, 1970, https://doi.org/10.3390/s25071970 107

Nicole Morresi, Maite Puerta-Beldarrain, Diego López-de-Ipiña, Alex Barco, Oihane Gómez-Carmona, Carlos López-Gomollon, et al.
A Wearable Sensor Node for Measuring Air Quality Through Citizen Science Approach: Insights from the SOCIO-BEE Project
Reprinted from: *Sensors* **2025**, *25*, 2373, https://doi.org/10.3390/s2512373 124

Nouran Abdalazim, Leonardo Alchieri, Lidia Alecci, Pietro Barbiero and Silvia Santini
The Impact of Domain Shift on Predicting Perceived Sleep Quality from Wearables
Reprinted from: *Sensors* **2025**, *25*, 4012, https://doi.org/10.3390/s25134012 148

Francesco Scardulla, Gloria Cosoli, Cosmina Gnoffo, Luca Antognoli, Francesco Bongiorno, Gianluca Diana, et al.
Experimental Analysis on the Effect of Contact Pressure and Activity Level as Influencing Factors in PPG Sensor Performance
Reprinted from: *Sensors* **2025**, *25*, 4477, https://doi.org/10.3390/s25144477 . . . . **174**

Minakshi Debnath, Sana Alamgeer, Md Shariar Kabir and Anne H. Ngu
Enhancing Wearable Fall Detection System via Synthetic Data
Reprinted from: *Sensors* **2025**, *25*, 4639, https://doi.org/10.3390/s25154639 . . . . **190**

Céline Labie, Nils Runge, Zosia Goossens, Olivier Mairesse, Jo Nijs, Anneleen Malfliet, et al.
Can a Commercially Available Smartwatch Device Accurately Measure Nighttime Sleep Outcomes in Individuals with Knee Osteoarthritis and Comorbid Insomnia? A Comparison with Home-Based Polysomnography
Reprinted from: *Sensors* **2025**, *25*, 4813, https://doi.org/10.3390/s25154813 . . . . **211**

Su Young Kim and Yoon Sang Kim
A Study on Autonomous Driving Motion Sickness from the Perspective of Multimodal Human Signals
Reprinted from: *Sensors* **2026**, *26*, 1675, https://doi.org/10.3390/s26051675 . . . . **233**

About the Editors

Sara Casaccia

Sara Casaccia received her Ph.D. in Industrial Engineering from Università Politecnica delle Marche (UNIVPM), Ancona, Italy, in 2015, and is now an Assistant Professor at the Department of Industrial Engineering and Mathematical Sciences (DIISM) of UNIVPM. Her research activity focuses on mechanical and thermal measurements, with particular emphasis on measurement methodologies for well-being and comfort in the built environment, quality assessment, and non-invasive monitoring techniques. Her interests also include data analysis and artificial intelligence applications for smart environments, Ambient Assisted Living (AAL), and the monitoring of physiological and environmental parameters. She is the author of approximately 100 publications, including journal articles, conference proceedings, and book chapters, and is currently the Scientific Coordinator for Università Politecnica delle Marche (UNIVPM) of both a European HAAL project and a national PRIN project, WEPOP.

Gloria Cosoli

Gloria Cosoli received her B.S. in Biomedical Engineering (with honors) and M.S. in Electronic Engineering (with honors) from Università Politecnica delle Marche (UNIVPM), Ancona, Italy, in 2011 and 2013, respectively. She received her Ph.D. in Mechanical Engineering from the same university in 2017 with a thesis titled "Study and Development of a Novel Radio Frequency Electromedical Device for the Treatment of Peri-Implantitis: Experimental Performance Analysis, Modelling of the Electromagnetic Interaction with Tissues and In Vitro and In Vivo Evaluation". From 2023 to 2025, she was a Research Collaborator at DIISM; since May 2024, she has been an Associate Professor at the Department of Theoretical and Applied Sciences (DiSTA) of eCampus University, Novedrate, Italy. Additionally, since November 2024 she has been the Deputy Coordinator of the Bachelor's and Master's Degree courses in Information and Automation Engineering. She is the author of more than 45 articles, more than 55 conference proceedings, and two national patents. Her research interests include non-invasive physiological measurements, numerical modelling, mechanical measurements, signal processing, and NDT.

Preface

The aim of this Reprint is to provide a comprehensive and detailed overview of the high-quality scientific contributions published in the Special Issue "Sensors Network and Wearables for People Activities and Wellbeing Monitoring". By curating different studies, this volume explores cutting-edge advancements across several vital application fields. Key topics include the following: i) personalized comfort assessment in smart environments, ii) sleep monitoring, iii) Ambient Assisted Living (AAL) technologies for the elderly, and iv) human behavior analysis. This collection highlights how wireless sensor networks and unobtrusive wearable devices are revolutionizing the way we track physical activity and enhance long-term human wellbeing in daily life.

Sara Casaccia and Gloria Cosoli
Guest Editors

sensors

MDPI

Editorial

Sensors Network and Wearables for People Activities and Wellbeing Monitoring

Sara Casaccia [1] and Gloria Cosoli [2,*]

[1] Department of Industrial Engineering and Mathematical Sciences, Università Politecnica Delle Marche, v. Brecce Bianche 12, 60131 Ancona, Italy; s.casaccia@staff.univpm.it
[2] Department of Theoretical and Applied Sciences, eCampus University, v. Isimbardi 10, 22060 Novedrate, Italy
* Correspondence: gloria.cosoli@uniecampus.it

1. Introduction

In recent decades, distributed sensor network and wearable sensors have known an unprecedented development [1–3]. This has allowed researchers to study human activities [4–6] and behavioral parameters [7,8], exploiting low-cost devices preserving privacy and guaranteeing non-intrusivity. The definition of data processing pipelines plays a pivotal role in this context, and Artificial Intelligence (AI)-based algorithms have made great strides, supporting the effective analysis of multimodal data [9], both for classification and prediction purposes [10–13]. The human-in-the-loop paradigm [14] can now be considered essential, given that decision-making processes can benefit from AI, but humans must be always the final decision makers since the responsibility of decisions and, consequently, actions is fundamental and cannot be transferred to AI algorithms, whatever the application field.

At present, monitoring people's activities and behavior [15] is a widespread practice in several contexts, from personalized comfort assessment [16] to general health status monitoring [17] and wellbeing [15], and in diverse environments, both indoor and outdoor.

2. Overview of the Contributions

This Special Issue confirms what is stated above as well as the multidisciplinarity and wide applicability of distributed sensor networks and wearables for activity monitoring. The contributions present innovative solutions and advanced signal processing techniques as well as analyses and considerations on such systems' performance and characteristics.

Building on the general relevance of wearable and distributed sensor systems, the contributions in this Special Issue can be organized into four key application domains. The first domain addresses personalized comfort assessment, exploring how physiological, environmental, and perceptual data can be integrated through wearable devices to better understand individual comfort [18,19]. The second domain focuses on sleep monitoring, where continuous, non-intrusive sensing allows the analysis of sleep patterns and quality in real-life conditions [20,21]. The third domain concerns Ambient Assisted Living (AAL) and frailty monitoring [22], highlighting how sensors can support mobility assessment, fall detection [23], and independent living for older adults or individuals with neurodegenerative and mobility-impairing conditions [24]. Finally, the fourth domain investigates human behavior, perception, and functional abilities, showing how wearable sensors capture physiological responses, social interactions [25,26], and motor performance in everyday scenarios. The key contributions in each of these areas are summarized below.

Comfort assessment (and especially personalized comfort assessment) is one of the most interesting application fields for such techniques. In this context, multi-sensor approaches are of interest [27,28], since both environmental and physiological data, together with perceptual information, can contribute to a comprehensive assessment of personalized comfort. The following contributions address different aspects related to environmental monitoring and physiological sensing in this context.

In Contribution 1, the authors investigate how contact pressure and activity level affect PPG signal quality, showing that even subtle physiological and interaction factors can impact on the sensor performance. Hence, not only data processing techniques but also hardware design have a relevant impact on the reliability of data used in personalized comfort assessment. Optimizing the design since its early phase can significantly enhance measurement accuracy (mean absolute percentage errors—MAPEs—<7% are also obtained at a high activity intensity level), thus providing good quality data to Personal Comfort Models (PCMs).

Contribution 2 presents a wearable sensor node for measuring air quality through a citizen science approach, highlighting that integrating fine-grained environmental measurements with user behavior can provide deeper insights into individual comfort conditions. High correlation (R^2 = 0.75) and limited MAPE (0.54 $\mu g/m^3$) are reported, together with a high satisfaction level among participants (>70%, considering seniors, commuters, and students). Such a sensing node can provide reliable information, thus empowering citizens and supporting the definition of robust environmental policies.

In Contribution 3, the authors examine CO_2 concentration, temperature, and humidity in Italian university classrooms over different seasons, demonstrating that a comprehensive analysis of their interaction can provide relevant insights on the occupants' comfort. Indoor Air Quality (AIQ) and thermal comfort are related to these environmental quantities, which can significantly impact on overall wellbeing and performance.

Another relevant application field concerns sleep monitoring, where wearable devices and sensor networks enable continuous and non-intrusive assessment of sleep patterns and sleep quality in real-life conditions [29,30], also in the presence of pathologies.

The authors of Contribution 4 evaluate whether a commercially available smartwatch can accurately measure nighttime sleep outcomes in individuals with knee osteoarthritis and comorbid insomnia, comparing its performance to home-based polysomnography. The findings highlight the potential and limitations of wearable devices for continuous, non-intrusive sleep monitoring in real-life conditions; the device achieved high accuracy (85.76%) and sensitivity (95.95%), but low specificity (50.96%).

In Contribution 5, the authors investigate the impact of domain shift on predicting perceived sleep quality from wearable devices, showing how differences between populations or contexts can affect the accuracy of sleep quality predictions, and emphasizing the importance of personalized calibration for reliable sleep assessment, even if this presents some challenges (e.g., the need for user-provided labels). The authors also make a dataset available to the relevant research community.

Another important application domain treated in this Special Issue concerns the monitoring of older adults and individuals affected by neurodegenerative or mobility-impairing conditions. In these contexts, wearable sensors and distributed sensing systems enable the continuous assessment of mobility, fall risk, and functional status, supporting both clinical evaluation and independent living. Moreover, such technologies play a key role in Ambient Assisted Living (AAL) scenarios, where sensor-based monitoring can help detect early signs of frailty and support safer living conditions for people living independently at home [31,32].

https://doi.org/10.3390/s26082475

The authors of Contribution 6 propose methods to enhance wearable fall detection systems using synthetic data, underlining the importance of reliable fall monitoring for older adults and individuals with mobility impairments in AAL environments. The idea is to use such data to supplement the limited availability of real data; in particular, the researchers extract fall segments from public footage and also generate synthetic fall signals via diffusion models. Personalization techniques would be needed to fit the models to individual behaviors.

In Contribution 7, the authors assess the agreement between optoelectronic systems and wearable sensors for the evaluation of gait spatiotemporal parameters in individuals with Progressive Supranuclear Palsy, highlighting how wearable devices can support continuous, objective mobility assessment in clinical and home settings. Indeed, the authors found a perfect agreement for gait speed and a good agreement for both cadence and gait cycle duration; they also provide suggestions to deal with eventual misalignments between test and reference systems.

Contribution 8 is a review providing an overview of assisted living solutions using wearable devices. The authors emphasize how distributed sensor networks can enable the early detection of frailty and support independent living for older adults. Particular attention is paid to electromyographic sensors and inertial measurement units (IMUs) for activity recognition along with physiological sensors to provide an overview of the psycho-physiological conditions of the subject. These technologies can greatly contribute to clinical best practices and to care management policies.

The authors of Contribution 9 explore physical frailty prediction through cane usage patterns (considering accelerations and angular velocities acquired through an IMU) during walking in older people with and without physical frailty. The performance of ML models is evaluated to distinguish between people with and without physical frailty, and an accuracy of 78.6% is achieved, demonstrating how subtle behavioral and mobility data captured by sensors can inform the personalized assessment of frailty risk.

Sensor-based systems can also be employed to investigate different aspects of human behavior, perception, and functional abilities. By exploiting wearable and embedded sensing technologies, it is possible to explore physiological responses, social interactions, and motor performance in real-world scenarios.

In Contribution 10, the authors investigated motion sickness by analyzing multimodal human signals collected in an autonomous driving simulator. They found a multimodal correlation between human signals and the perceived motion sickness level, showing how wearable sensors can capture physiological and behavioral responses in real-world scenarios to support personalized assessment of discomfort and adaptation.

A proximity sensor integrated with a Bluetooth sensor, a global positioning system (GPS), and an accelerometer are employed in Contribution 11 to measure social interactions in a school environment. The authors evaluated interaction through the analysis of neighboring sensors proximity, highlighting the potential of wearable and embedded sensors for monitoring human behavior and social dynamics in real-world settings.

The authors of Contribution 12 propose an automated architecture based on the Tapping Test protocol to assess hand dexterity through the usage of inertial sensors worn at finger level. Data processing provides useful information on the tests, demonstrating how wearable-based measurements can provide an objective, fine-grained evaluation of motor skills for clinical or research applications. Upper limb coordination and neuroplasticity aspects are also evaluated in an assessment of the function of motor abilities.

Finally, Contribution 13 focuses on sound direction detection and mixed source separation in embedded systems. Wearables today can have a speech recognition capability, which

 https://doi.org/10.3390/s26082475

can be enhanced by an analysis of the auditory scene, going deeper into environmental perception in everyday contexts.

3. Conclusions

The contributions included in this Special Issue underline how wearable and distributed sensor systems enable continuous and non-intrusive monitoring while preserving user privacy, a key advantage over traditional approaches. By combining these sensors with AI-driven data analysis and human-in-the-loop paradigms, it is possible to extract meaningful insights across different domains, supporting more accurate, adaptive, and responsible interventions. These advancements highlight the potential of sensor technologies not only to improve wellbeing and safety but also to ensure ethical and privacy-conscious applications in real-world environments.

Author Contributions: Conceptualization, G.C. and S.C.; writing—original draft preparation, G.C.; writing—review and editing, G.C. and S.C.; visualization, G.C. All authors have read and agreed to the published version of the manuscript.

Data Availability Statement: Not applicable.

Conflicts of Interest: The authors declare no conflicts of interest.

List of Contributions

1. Scardulla, F.; Cosoli, G.; Gnoffo, C.; Antognoli, L.; Bongiorno, F.; Diana, G.; Scalise, L.; D'Acquisto, L.; Arnesano, M. Experimental Analysis on the Effect of Contact Pressure and Activity Level as Influencing Factors in PPG Sensor Performance. *Sensors* **2025**, *25*, 4477. https://doi.org/10.3390/s25144477
2. Morresi, N.; Puerta-Beldarrain, M.; López-de-Ipiña, D.; Barco, A.; Gómez-Carmona, O.; López-Gomollon, C.; Casado-Mansilla, D.; Kotzagianni, M.; Casaccia, S.; Udina, S.; et al. A Wearable Sensor Node for Measuring Air Quality Through Citizen Science Approach: Insights from the SOCIO-BEE Project. *Sensors* **2025**, *25*, 3739. https://doi.org/10.3390/s25123739
3. Fedele, A.; Colantoni, A.; Calabrò, G.; Scungio, M.; Rossi, S.; Taborri, J. Measuring CO_2 Concentration and Thermal Comfort in Italian University Classrooms: A Seasonal Analysis. *Sensors* **2025**, *25*, 1970. https://doi.org/10.3390/s25071970
4. Labie, C.; Runge, N.; Goossens, Z.; Mairesse, O.; Nijs, J.; Malfliet, A.; Van Assche, D.; de Vlam, K.; Menghini, L.; Verschueren, S.; et al. Can a Commercially Available Smartwatch Device Accurately Measure Nighttime Sleep Outcomes in Individuals with Knee Osteoarthritis and Comorbid Insomnia? A Comparison with Home-Based Polysomnography. *Sensors* **2025**, *25*, 4813. https://doi.org/10.3390/s25154813
5. Abdalazim, N.; Alchieri, L.; Alecci, L.; Barbiero, P.; Santini, S. The Impact of Domain Shift on Predicting Perceived Sleep Quality from Wearables. *Sensors* **2025**, *25*, 4012. https://doi.org/10.3390/s25134012
6. Debnath, M.; Alamgeer, S.; Kabir, M.S.; Ngu, A.H. Enhancing Wearable Fall Detection System via Synthetic Data. *Sensors* **2025**, *25*, 4639. https://doi.org/10.3390/s25154639
7. Ricciardi, C.; Pisani, N.; Donisi, L.; Abate, F.; Amboni, M.; Barone, P.; Picillo, M.; Cesarelli, M.; Amato, F. Agreement between Optoelectronic System and Wearable Sensors for the Evaluation of Gait Spatiotemporal Parameters in Progressive Supranuclear Palsy. *Sensors* **2023**, *23*, 9859. https://doi.org/10.3390/s23249859
8. Iadarola, G.; Mengarelli, A.; Crippa, P.; Fioretti, S.; Spinsante, S. A Review on Assisted Living Using Wearable Devices. *Sensors* **2024**, *24*, 7439. https://doi.org/10.3390/s24237439
9. Toda, H.; Chin, T. Physical Frailty Prediction Using Cane Usage Characteristics during Walking. *Sensors* **2024**, *24*, 6910. https://doi.org/10.3390/s24216910
10. Kim, S.Y.; Kim, Y.S. A Study on Autonomous Driving Motion Sickness from the Perspective of Multimodal Human Signals. *Sensors* **2026**, *26*, 1675. https://doi.org/10.3390/s26051675

https://doi.org/10.3390/s26082475

11. Hernández-Heredia, T.K.; Reyes-Manzano, C.F.; Flores-Hernández, D.A.; Ramos-Fernández, G.; Guzmán-Vargas, L. Proximity Sensor for Measuring Social Interaction in a School Environment. *Sensors* **2024**, *24*, 4822. https://doi.org/10.3390/s24154822
12. Di Libero, T.; Carissimo, C.; Cerro, G.; Abbatecola, A.M.; Marino, A.; Miele, G.; Ferrigno, L.; Rodio, A. An Overall Automated Architecture Based on the Tapping Test Measurement Protocol: Hand Dexterity Assessment through an Innovative Objective Method. *Sensors* **2024**, *24*, 4133. https://doi.org/10.3390/s24134133
13. Wang, J.-H.; Le, P.T.; Bee, W.-S.; Putri, W.R.; Su, M.-H.; Li, K.-C.; Chen, S.-L.; He, J.-L.; Pham, T.; Li, Y.-H.; et al. Implementation of Sound Direction Detection and Mixed Source Separation in Embedded Systems. *Sensors* **2024**, *24*, 4351. https://doi.org/10.3390/s24134351

References

1. Escobar-Linero, E.; Muñoz-Saavedra, L.; Luna-Perejón, F.; Sevillano, J.L.; Domínguez-Morales, M. Wearable Health Devices for Diagnosis Support: Evolution and Future Tendencies. *Sensors* **2023**, *23*, 1678. [CrossRef]
2. Kulkarni, M.B.; Rajagopal, S.; Prieto-Simón, B.; Pogue, B.W. Recent advances in smart wearable sensors for continuous human health monitoring. *Talanta* **2024**, *272*, 125817. [CrossRef] [PubMed]
3. Erdem, A.; Eksin, E.; Senturk, H.; Yildiz, E.; Maral, M. Recent developments in wearable biosensors for healthcare and biomedical applications. *TrAC Trends Anal. Chem.* **2024**, *171*, 117510. [CrossRef]
4. Shahid, A.R.; Imteaj, A.; Wu, P.Y.; Igoche, D.A.; Alam, T. Label Flipping Data Poisoning Attack Against Wearable Human Activity Recognition System. In Proceedings of the 2022 IEEE Symposium Series on Computational Intelligence (SSCI), Singapore, 4–7 December 2022; pp. 908–914. [CrossRef]
5. Miao, S.; Chen, L.; Hu, R. Spatial-Temporal Masked Autoencoder for Multi-Device Wearable Human Activity Recognition. In Proceedings of the ACM Interactive, Mobile, Wearable and Ubiquitous Technologies, Melbourne, Australia, 5–9 October 2024; Volume 7. [CrossRef]
6. Hu, R.; Chen, L.; Miao, S.; Tang, X. SWL-Adapt: An Unsupervised Domain Adaptation Model with Sample Weight Learning for Cross-User Wearable Human Activity Recognition. In Proceedings of the AAAI Conference on Artificial Intelligence, Washington, DC, USA, 7–14 February 2023; Volume 37, pp. 6012–6020. [CrossRef]
7. Su, J.; Wen, Z.; Lin, T.; Guan, Y. Learning disentangled behaviour patterns for wearable-based human activity recognition. In Proceedings of the ACM Interactive, Mobile, Wearable and Ubiquitous Technologies, New York, NY, USA, 11–15 September 2022; Volume 6, p. 19. [CrossRef]
8. Menassa, M.; Wilmont, I.; Beigrezaei, S.; Knobbe, A.; Arita, V.A.; Valderrama, J.F.; Bridge, L.; Verschuren, W.M.M.; Rennie, K.L.; Franco, O.H.; et al. The future of healthy ageing: Wearables in public health, disease prevention and healthcare. *Maturitas* **2025**, *196*, 108254. [CrossRef] [PubMed]
9. Cosoli, G.; Mansi, S.A.; Pigliautile, I.; Pisello, A.L.; Revel, G.M.; Arnesano, M. Enhancing personal comfort: A machine learning approach using physiological and environmental signals measurements. *Measurement* **2023**, *217*, 113047. [CrossRef]
10. Zhang, Z.; Wang, L.; Lee, C. Recent Advances in Artificial Intelligence Sensors. *Adv. Sens. Res.* **2023**, *2*, 2200072. [CrossRef]
11. Chen, S.; Fan, S.; Qiao, Z.; Wu, Z.; Lin, B.; Li, Z.; Riegler, M.A.; Wong, M.Y.H.; Opheim, A.; Korostynska, O.; et al. Transforming Healthcare: Intelligent Wearable Sensors Empowered by Smart Materials and Artificial Intelligence. *Adv. Mater.* **2025**, *37*, 2500412. [CrossRef]
12. Nong, H.; Jin, M.; Pan, C.; Zhou, H.; Zhang, C.; Pan, X.; Chen, Y.; Wei, X.; Lu, Y.; Zhao, K.; et al. Intelligent Sensing Technologies Based on Flexible Wearable Sensors: A Review. *IEEE Sens. J.* **2024**, *24*, 22197–22217. [CrossRef]
13. Ahmed, A.; Aziz, S.; Abd-Alrazaq, A.; Farooq, F.; Sheikh, J. Overview of Artificial Intelligence-Driven Wearable Devices for Diabetes: Scoping Review. *J. Med. Internet Res.* **2022**, *24*, e36010. [CrossRef]
14. Kumar, S.; Datta, S.; Singh, V.; Datta, D.; Singh, S.K.; Sharma, R. Applications, Challenges, and Future Directions of Human-in-the-Loop Learning. *IEEE Access* **2024**, *12*, 75735–75760. [CrossRef]
15. Naccarelli, R.; Casaccia, S.; Revel, G.M. The Problem of Monitoring Activities of Older People in Multi-Resident Scenarios: An Innovative and Non-Invasive Measurement System Based on Wearables and PIR Sensors. *Sensors* **2022**, *22*, 3472. [CrossRef]
16. Tartarini, F.; Schiavon, S.; Quintana, M.; Miller, C. Personal comfort models based on a 6-month experiment using environmental parameters and data from wearables. *Indoor Air* **2022**, *32*, e13160. [CrossRef] [PubMed]
17. Machorro-Cano, I.; Olmedo-Aguirre, J.O.; Alor-Hernández, G.; Rodríguez-Mazahua, L.; Sánchez-Morales, L.N.; Pérez-Castro, N. Cloud-Based Platforms for Health Monitoring: A Review. *Informatics* **2024**, *11*, 2. [CrossRef]

18. Ma, C.B.; Shang, X.; Sun, M.; Bo, X.; Bai, J.; Du, Y.; Zhou, M. Emerging Multifunctional Wearable Sensors: Integrating Multimodal Sweat Analysis and Advanced Material Technologies for Next-Generation Health Monitoring. *ACS Sens.* **2025**, *10*, 2388–2408. [CrossRef] [PubMed]
19. Sun, H.; Wang, B.; Du, Q. Work Comfort Evaluation in Virtual Reality Environment Based on Multimodal Signals. In Proceedings of the 2024 15th International Conference on Reliability, Maintenance and Safety (ICRMS), Guilin, China, 31 July–3 August 2024; pp. 1130–1136. [CrossRef]
20. Yoon, H.; Choi, S.H. Technologies for sleep monitoring at home: Wearables and nearables. *Biomed. Eng. Lett.* **2023**, *13*, 313–327. [CrossRef]
21. Yin, J.; Xu, J.; Ren, T.L. Recent Progress in Long-Term Sleep Monitoring Technology. *Biosensors* **2023**, *13*, 395. [CrossRef]
22. Guerra, B.M.V.; Torti, E.; Marenzi, E.; Schmid, M.; Ramat, S.; Leporati, F.; Danese, G. Ambient assisted living for frail people through human activity recognition: State-of-the-art, challenges and future directions. *Front. Neurosci.* **2023**, *17*, 1256682. [CrossRef]
23. Bustos, A.S.O.; Tramontano, M.; Morone, G.; Ciancarelli, I.; Panza, G.; Minnetti, A.; Picelli, A.; Smania, N.; Iosa, M.; Vannozzi, G. Ambient assisted living systems for falls monitoring at home. *Expert Rev. Med. Devices* **2023**, *20*, 821–828. [CrossRef]
24. Casaccia, S.; Revel, G.M.; Cosoli, G.; Scalise, L. Assessment of Domestic Well-Being: From Perception to Measurement. *IEEE Instrum. Meas. Mag.* **2021**, *24*, 58–67. [CrossRef]
25. Elmer, T.; Lodder, G. Modeling social interaction dynamics measured with smartphone sensors: An ambulatory assessment study on social interactions and loneliness. *J. Soc. Pers. Relat.* **2023**, *40*, 654–669. [CrossRef]
26. Wang, Z.; Larrazabal, M.A.; Rucker, M.; Toner, E.R.; Daniel, K.E.; Kumar, S.; Boukhechba, M.; Teachman, B.A.; Barnes, L.E. Detecting Social Contexts from Mobile Sensing Indicators in Virtual Interactions with Socially Anxious Individuals. In Proceedings of the ACM Interactive, Mobile, Wearable and Ubiquitous Technologies, Cancún, Mexico, 8–12 October 2023; Volume 7, p. 26. [CrossRef]
27. Chen, H.; Chen, W.; Bao, S.; Lu, C.; Wang, L.; Ma, J.; Wang, P.; Lu, H.; Shu, F.; Oetomo, S.B. Design of an integrated wearable multi-sensor platform based on flexible materials for neonatal monitoring. *IEEE Access* **2020**, *8*, 23732–23747. [CrossRef]
28. Booth, B.M.; Mundnich, K.; Feng, T.; Nadarajan, A.; Falk, T.H.; Villatte, J.L.; Ferrara, E.; Narayanan, S. Multimodal human and environmental sensing for longitudinal behavioral studies in naturalistic settings: Framework for sensor selection, deployment, and management. *J. Med. Internet Res.* **2019**, *21*, e12832. [CrossRef]
29. Zhai, B.; Duan, H.; Guan, Y.; Phan, H.; Woo, W.L. DSleepNet: Disentanglement Learning for Personal Attribute-Agnostic Three-Stage Sleep Classification Using Wearable Sensing Data. *IEEE J. Biomed. Health Inform.* **2025**, *29*, 4861–4872. [CrossRef]
30. Kwon, S.; Kim, H.; Yeo, W.H. Recent advances in wearable sensors and portable electronics for sleep monitoring. *iScience* **2021**, *24*, 102461. [CrossRef]
31. Rodrigues, M.J.; Postolache, O.; Cercas, F. Wearable Smart Sensing and UWB System for Fall Detection in AAL Environments. In Proceedings of the 2023 IEEE Sensors Applications Symposium, SAS, Ottawa, ON, Canada, 18–20 July 2023. [CrossRef]
32. Saragadam, A.; Saragadam, A.; Morita, P.; Kaur, J. A Comparative Study of Sensor Quantity and Data Granularity in AAL Systems for Activity Recognition. *Eur. J. Public Health* **2024**, *34*, 455. [CrossRef]

Disclaimer/Publisher's Note: The statements, opinions and data contained in all publications are solely those of the individual author(s) and contributor(s) and not of MDPI and/or the editor(s). MDPI and/or the editor(s) disclaim responsibility for any injury to people or property resulting from any ideas, methods, instructions or products referred to in the content.

https://doi.org/10.3390/s26082475

sensors

MDPI

Article

Agreement between Optoelectronic System and Wearable Sensors for the Evaluation of Gait Spatiotemporal Parameters in Progressive Supranuclear Palsy

Carlo Ricciardi [1,*,†], Noemi Pisani [2,†], Leandro Donisi [3], Filomena Abate [4], Marianna Amboni [4], Paolo Barone [4], Marina Picillo [4], Mario Cesarelli [5] and Francesco Amato [1,*]

1 Department of Electrical Engineering and Information Technology, University of Naples Federico II, 80125 Naples, Italy
2 Department of Advanced Biomedical Sciences, University of Naples Federico II, 80131 Naples, Italy
3 Department of Advanced Medical and Surgical Sciences, University of Campania Luigi Vanvitelli, 80138 Naples, Italy
4 Center for Neurodegenerative Diseases (CEMAND), Department of Medicine, Surgery and Dentistry "Scuola Medica Salernitana", University of Salerno, 84131 Salerno, Italy
5 Department of Engineering, University of Sannio, 82100 Benevento, Italy
* Correspondence: carloricciardi.93@gmail.com (C.R.); framato@unina.it (F.A.)
† These authors equally contributed to the work.

Abstract: The use of wearable sensors for calculating gait parameters has become increasingly popular as an alternative to optoelectronic systems, currently recognized as the gold standard. The objective of the study was to evaluate the agreement between the wearable Opal system and the optoelectronic BTS SMART DX system for assessing spatiotemporal gait parameters. Fifteen subjects with progressive supranuclear palsy walked at their self-selected speed on a straight path, and six spatiotemporal parameters were compared between the two measurement systems. The agreement was carried out through paired data test, Passing Bablok regression, and Bland-Altman Analysis. The results showed a perfect agreement for speed, a very close agreement for cadence and cycle duration, while, in the other cases, Opal system either under- or over-estimated the measurement of the BTS system. Some suggestions about these misalignments are proposed in the paper, considering that Opal system is widely used in the clinical context.

Keywords: gait analysis; optoelectronic system; wearable device; spatiotemporal parameters; progressive supranuclear palsy

1. Introduction

Progressive supranuclear palsy (PSP) is a rare neurodegenerative disease characterized by oculomotor gaze palsy, akinesia, postural instability, and cognitive dysfunction. Gait abnormalities represent one of the cornerstones of PSP and determine an increased risk of falls from the earliest stages of the disease [1–3]. The PSP Rating Scale (PSP-RS) is a semi-quantitative clinician-based scale considered the gold standard to evaluate disease severity in both clinical and research settings. However, only a few items of the PSP-RS are dedicated to gait and to axial impairment in general [4].

Beyond clinical scales, gait analysis is considered one of the main tools to measure quantitative gait-related parameters [5,6]. In detail, the instrumented gait analysis can assess an individual's gait patterns through the calculation of quantitative gait-related parameters, namely kinetic, kinematic, and spatiotemporal variables. In neurodegenerative diseases, gait analysis has been used for monitoring disease progression over time, quantifying parkinsonian symptoms [7], evaluating treatments' outcomes, implementing algorithms for diagnosis through the recognition of soft signs [8], and predicting the risk of falls [9–12]. As for PSP, recent evidence suggests that quantitative evaluation of gait may provide further insights in both the diagnostic process and in the evaluation of disease progression [1,13,14].

Optoelectronic motion capture systems are recognized as the gold standard in the evaluation of quantitative gait parameters [15,16]. These laboratory systems are based on passive light-reflecting markers, infrared cameras to track the trajectory of markers, and force platforms to extract kinetic data [17,18]. However, the limitations of optoelectronic systems consist of high equipment costs, the need for large capture volume, technical and clinical trained personnel, and time-consuming acquisitions [19]. These aspects have recently boosted the development of wearable sensors based on Inertial Measurement Units (IMUs) for calculating gait parameters [20–23]. IMU sensors can be used in several real-life environments, both indoor and outdoor, reducing the costs of lab-based systems and the complexity of the acquisitions. Additionally, these devices are easy to wear, require less patient preparation, and, consequently, can be used by a larger number of users [24,25].

Recently, inertial sensors and gold standard systems have been compared in gait parameters assessment (kinematic, spatiotemporal, or both) in healthy subjects [26–28]. González et al. studied the reliability of the Kinovea (version 0.8.15) software, a video annotation tool designed for movement analysis, and the agreement with a three-dimensional motion system, Vicon Motion System®, for extracting hip, knee, and ankle joint angles parameters [29]. The same kinematic parameters were extracted by Piche et al. to compare the iSen IMU sensor with the optoelectronic gold standard system OptiTrack [30]. Finally, Bartoszec et al. compared the IMU-based MyoMotion system and the optoelectronic BTS SMART DX system, simultaneously recording the upper and lower limb joint angles [31]. Furthermore, the comparison of the above-mentioned measurement systems is also carried out to develop an algorithm for the accurate calculation of spatiotemporal gait parameters from wearable IMUs for Parkinson disease (PD) rehabilitation [32]. The Vicon Motion System® was also used as gold standard reference against the system proposed in [33], based on a single RGB-D sensor to estimate the gait parameters in a limited space such as the domestic setting. Previous studies have also investigated the interoperability of IMUs sensors and optoelectronic systems in PD subjects [11,34].

In this context, the aim of the study is to evaluate the agreement between two systems for the measurement of the spatiotemporal gait parameters in patients with PSP. The BTS SMART DX optoelectronic system by BTS Bioengineering Inc. is used as reference device to the wearable Opal System by APDM Inc. To the best of our knowledge, no study has ever investigated the agreement between these two measurements systems in a population of patients with a rare disease such as the PSP. In this work, spatiotemporal parameters calculated by the two measurement systems were evaluated by means of different statistical techniques to assess their level of agreement. Since, as said above, a hard limit of the optoelectronic systems relies in the fact that they cannot be used outside the laboratorial context, the overall objective of this study is to prove that the Opal wearable system can guarantee a performance comparable to that one of the BTS SMART DX optoelectronic system.

2. Materials and Methods

2.1. Study Population and Clinical Assessment

Fifteen patients diagnosed with PSP, aged between 60 and 75 years, were recruited from the Center for Neurodegenerative diseases (CEMAND) at the University of Salerno, Italy, between January 2021 and January 2023. The diagnosis was established using the Movement Disorders Society clinical criteria [1], and all patients met the criteria for either probable or possible PSP [32]. Exclusion criteria included patients with gait requiring bilateral assistance, dementia according to the Diagnostic and Statistical Manual of Mental Disorders (DSM-V) criteria, significant comorbidities possibly impacting on gait (including other neurologic disorders, orthopedic diseases, or cardiovascular/respiratory diseases and/or brain surgery). All enrolled patients were treated with levodopa with a mean Levodopa Equivalent Daily Dose (LEDD) of 322.00 $\pm$ 177.13 [35]. The protocol was approved by the local ethics committee and all patients were enrolled upon signing a written informed consent form.

Disease severity was evaluated with the PSP-RS, administered by expert clinicians. The scale is composed of 28 items, distributed into 6 domains: history, mentation, bulbar, ocular motor, limb motor, gait/midline. The total score ranges from 0 to 100, where 100 represents the more severe outcome [4].

Demographic and clinical data are shown in Table 1.

Table 1. Patients' anthropometrics and clinical data (n = 15; 5 females and 10 males).

Clinical Parameters	Mean ± SD
Age (years)	68.13 ± 4.97
Disease duration (years)	3.13 ± 2.07
PSP-RS limb	3.33 ± 2.29
PSP-RS gait/midline	5.20 ± 4.78
LEDD (mg)	322.00 ± 177.13
BMI (kg/m^2)	27.48 ± 3.84

Abbreviations: BMI = Body Mass Index; LEDD = Levodopa Equivalent Daily Dose; PSP-RS = Progressive Supranuclear Palsy Rating Scale.

2.2. Gait Analysis Systems

Two different systems were compared for gait analysis in PSP patients: the BTS SMART DX-400 System (BTS Bioengineering Corp., Milan, Italy) and the Opal System (APDM Inc., Portland, OR, USA).

The BTS SMART DX-400 is an optoelectronic system made up of six infrared digital cameras that record at a sample frequency of 100 Hz, two dynamometric platforms, twenty-two retro-reflective passive markers (each one with a diameter of 14 mm), two video cameras, and an elaborator, as shown in Figure 1. For data acquisition, the Davis protocol [36] was applied, consisting of the following steps:

- anthropometric measurements,
- positioning of passive markers on patients' body (from the feet to the shoulders) according to the Davis protocol,
- standing phase,
- walking phases.

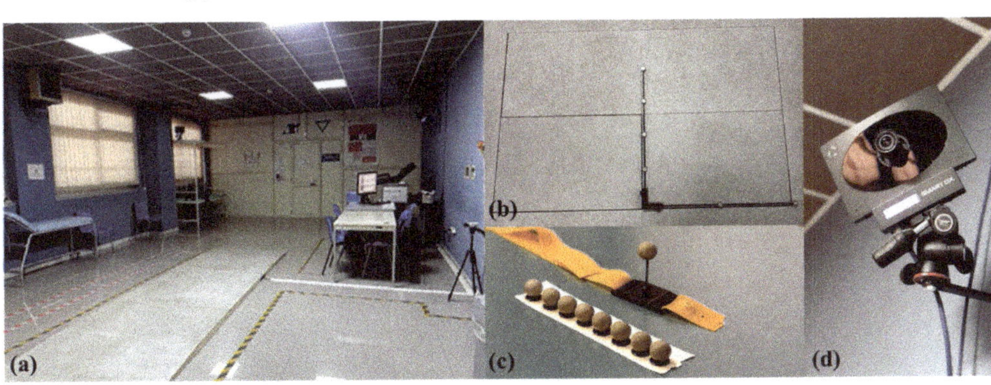

Figure 1. BTS Gait Lab: (**a**) Movement analysis laboratory at the University Hospital of Salerno; (**b**) Dynamometric platform; (**c**) Retro-reflective passive markers; (**d**) Infrared-camera.

Then, the acquired data were transferred to the elaborator using a dedicated proprietary software, able to store and compute spatiotemporal parameters, as reported in other studies [13,37,38].

The Opal System is a wearable inertial system made up of three IMU sensors, each one containing three-axes accelerometer with 14 bits resolution and selectable ±16 g or ±200 g ranges, three-axes gyroscope with 16 bits resolution and ±2000 deg/s, and three-axes magnetometer with 12 bits resolution and ±8 Gauss range. Opal sensors were placed on the patients' body by means of straps, including one sensor on each foot and one sensor on the lumbar region. Thanks to the Access Point, sensors were wirelessly connected by Bluetooth to a laptop equipped with the Mobility Lab (version 4) software capable of processing movement-related data. In addition, the Docking Station was used for charging and configuring the sensors. Figure 2 illustrates the Opal Systems parts.

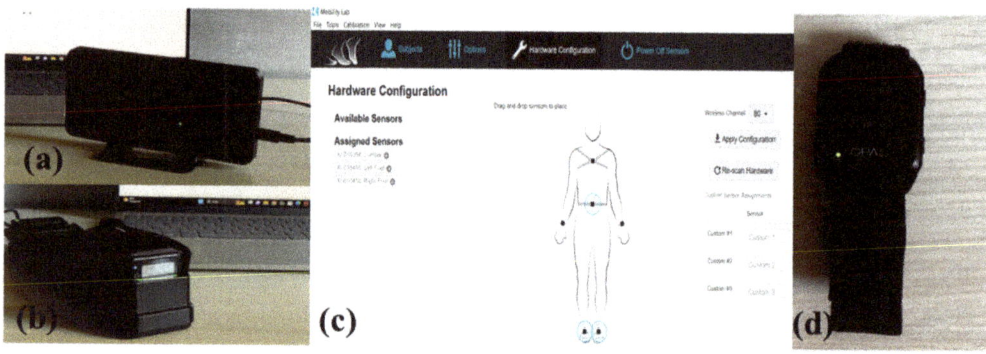

Figure 2. Opal System: (**a**) Access Point; (**b**) Docking Station; (**c**) Mobility Lab software; (**d**) Opal sensor.

2.3. *Study Protcol*

Gait analysis tasks were performed in the movement analysis laboratory at the University Hospital of Salerno. First, all patients were trained to walk at their normal speed on a 10 m straight pathway four times; overall, a straight path of 40 m has been considered. Then, the four walking phases were averaged. The trials were recorded by the optoelectronic system BTS SMART DX. Before conducting the analysis with the Opal sensors, patients rested for 15 min to allow them to perform the new test without feeling tired. After such resting time, patients repeated the trials on a straight path for 10 m, walking at their normal speed for two minutes while wearing the Opal sensors. All patients were evaluated more than once.

The spatiotemporal parameters analyzed were as follows:

- Cadence (steps/minute): stepping rate.
- Cycle duration (s): duration of a complete gait cycle.
- Speed (m/s): walking speed.
- Stance phase (% gait cycle time): average percentage of a gait cycle that either foot is on the ground.
- Swing phase (% gait cycle time): average percentage of a gait cycle that either foot is off the ground.
- Stride length (m): distance between two consecutive foot falls at the moments of initial contacts.

For the current analysis, the spatiotemporal parameters from the trials were averaged for each subject in order to obtain the mean value of both sides (right and left).

2.4. *Statistical Analyses*

All statistical analyses were performed using the SPSS® (version 28, IBM Corp., Armonk, NY, USA) and MATLAB (version 2021b, Mathworks Inc., Natick, MA, USA) software. In order to evaluate the agreement between the optoelectronic system (BTS SMART DX-100, BTS Bioengineering Inc., Milano, Italy) and the inertial sensors (Opal, APDM Inc., Portland, OR, USA), a two-tailed paired test was carried out. The paired data test, in its parametric and non-parametric form according to the Shapiro–Wilk normality test, was used to identify statistically significant differences between the means or the medians of the spatiotemporal variables measured by the two systems [39]. The absence of a statistically significant difference implies an overall agreement between the variables.

Passing–Bablok (PB) linear regression analysis was employed to evaluate the presence of systematic errors in the measurements. In particular, the slope of the regression line indicates a proportional systematic error, while the intercept of the regression line with the y-axis indicates a constant systematic error; such errors have a physical meaning whenever the slope and the intercept 95% confidence intervals contain the 1 and 0 values, respectively. The PB method was applied after verifying the linear relationship between the measurements [34,40].

The Bland–Altman (BA) method is a visual representation of the agreement between the two systems by means of a scatterplot in which the mean of the two measurements is plotted on the x-axis and the difference between the measurements is represented on the y-axis. To obtain the limits of agreement, defined as the mean of the differences ± 1.96 times the standard deviation (i.e., 95% of the confidence interval), the mean of differences (bias) and the standard deviation of differences were calculated. In this context, the presence of errors can be assessed through the bias distance from the 0 value, which implies a constant systematic error, the width of the limits of agreement and any trends in the distribution of data around the bias [29,41]. In particular, possible errors could be a fan-shaped distribution within the limits of agreement, known as heteroscedasticity of variance, a non-causal but systemic distribution in absolute value, or a linear distribution of points which determine a systematic proportional error [42,43].

In all the performed analyses, the uncertainty level was set at $\alpha = 0.05$.

3. Results

Table 2 shows the paired data test results in terms of p-value. Additionally, Table 2 reports the mean and standard deviation values for each spatiotemporal variable for both systems and test types, parametric or non-parametric, depending on the normality test.

Table 2. Two-tailed paired test analysis to test the overall agreement between each couple of parameters of the systems.

Spatiotemporal Parameters	BTS SMART DX System	Opal System	p-Value	Test Type
Cadence (steps/min)	91.72 ± 14.23	99.14 ± 14.28	0.002	Wilcoxon test
Cycle duration (s)	1.36 ± 0.33	1.21 ± 0.30	0.003	Wilcoxon test
Speed (m/s)	0.61 ± 0.20	0.65 ± 0.19	0.192	Wilcoxon test
Stance phase (% gait cycle time)	63.96 ± 2.92	62.12 ± 2.11	0.037	Wilcoxon test
Swing phase (% gait cycle time)	36.83 ± 3.92	36.88 ± 2.11	0.159	Wilcoxon test
Stride length (m)	0.41 ± 0.10	0.78 ± 0.20	<0.001	*T*-Student test

The PB linear regression results of cadence, cycle duration, and speed, illustrated in Table 3, exhibited a confidence interval containing 1 for the slope and 0 for the intercept. Other parameters deviated from such values.

Table 3. PB linear regression analysis.

Spatiotemporal Parameters	m *	LB_m	UB_m	q **	LB_q	UB_q
Cadence (steps/min)	1.09	0.74	1.81	−3.02	−67.89	30.17
Cycle duration (s)	1.03	0.63	1.74	−0.10	−1.04	0.40
Speed (m/s)	1.02	0.71	1.45	0.05	−0.17	0.22
Stance phase (% gait cycle time)	0.63	0.46	0.85	23.09	8.78	33.34
Swing phase (% gait cycle time)	0.58	0.32	0.74	15.96	10.13	25.68
Stride length (m)	2.10	1.64	2.69	−0.06	−0.28	0.13

* m is the slope of Passing-Bablok regression line; ** q is the intercept of Passing-Bablok regression line. Abbreviations: LB = Lower Bound of confidence interval; UB = Upper Bound of confidence interval.

The results of the BA analysis for each spatiotemporal parameter included in the study are shown in Table 4 in terms of bias (the difference between two measurement systems), the 95% lower and the upper bounds of bias, and the BA lower and upper bounds, known as limits of agreement. The bias approached zero, and the bias limits containing the zero value demonstrate a good agreement between the measured parameters.

Table 4. BA analysis applied per each couple of parameters of the systems.

Spatiotemporal Parameters	Bias	LB_b	UB_b	LB_LA	UB_LA
Cadence (steps/min)	−7.43	−11.36	−3.50	−28.95	14.10
Cycle duration (s)	0.15	0.05	0.25	−0.40	0.70
Speed (m/s)	−0.03	−0.08	0.02	−0.31	0.25
Stance phase (% gait cycle time)	0.84	0.11	1.57	−3.17	4.84
Swing phase (% gait cycle time)	−0.05	−1.37	1.27	−7.30	7.20
Stride length (m)	−0.37	−0.42	−0.32	−0.63	−0.11

Abbreviations: LA = Limit of Agreement; LB = Lower Bound; UB = Upper Bound.

Figures 3–8 show, in pairs, the PB plot and the BA graph for all parameters (i.e., cadence, cycle duration, speed, stance phase, swing phase, and stride length).

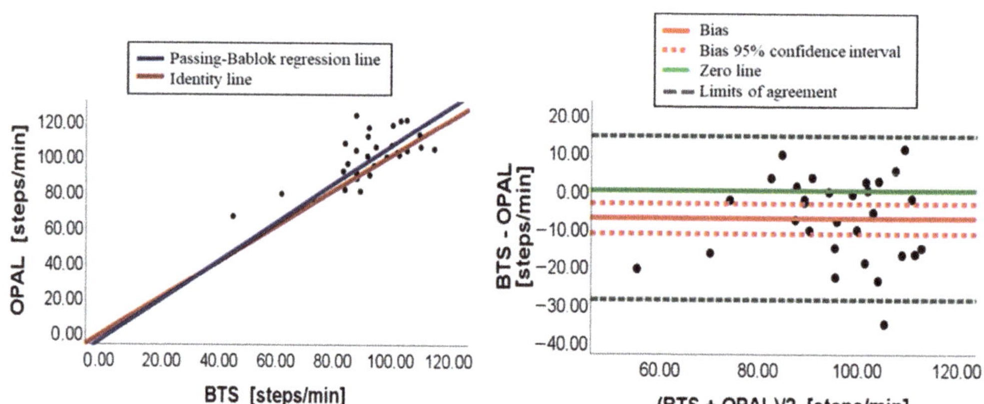

Figure 3. Cadence. On the left, scatter plot with PB regression line and identity line (BTS = OPAL). On the right, BA plot (average vs. difference).

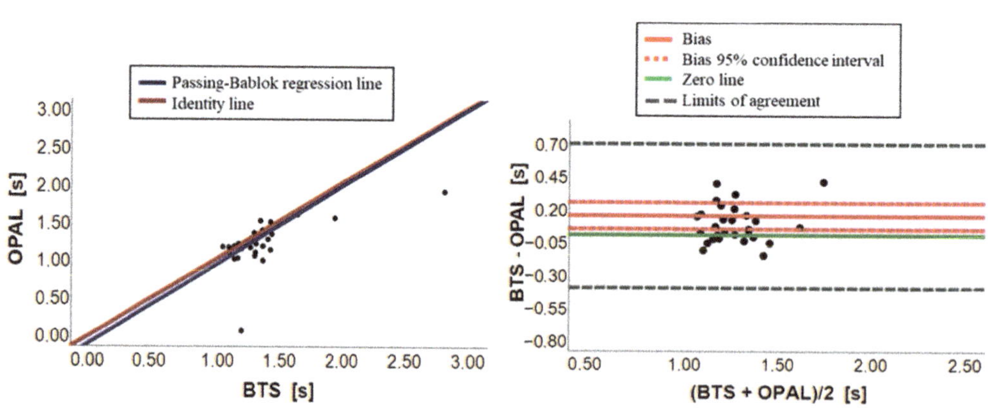

Figure 4. Cycle duration. On the left, scatter plot with PB regression line and identity line (BTS = OPAL). On the right, BA plot (average vs. difference).

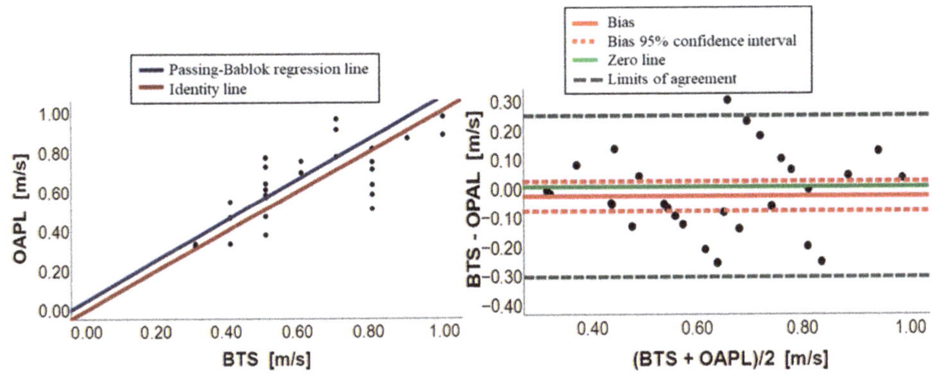

Figure 5. Speed. On the left, scatter plot with PB regression line and identity line (BTS = OPAL). On the right, BA plot (average vs. difference).

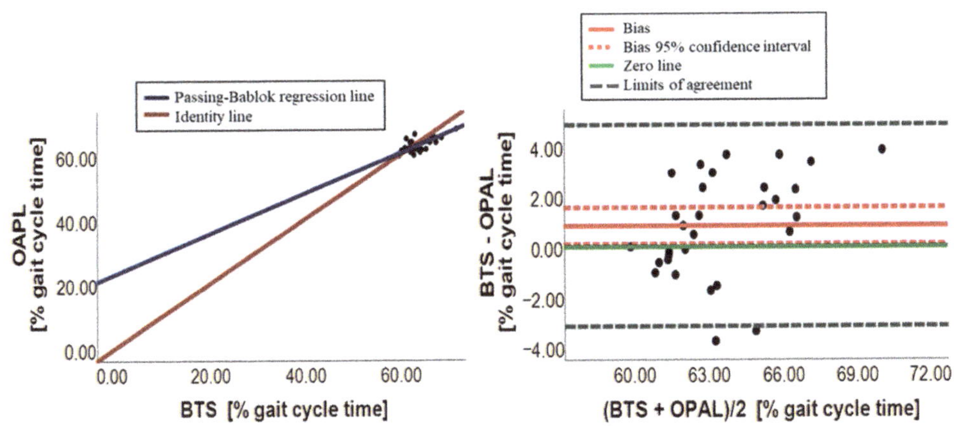

Figure 6. Stance phase. On the left, scatter plot with PB regression line and identity line (BTS = OPAL). On the right, BA plot (average vs. difference).

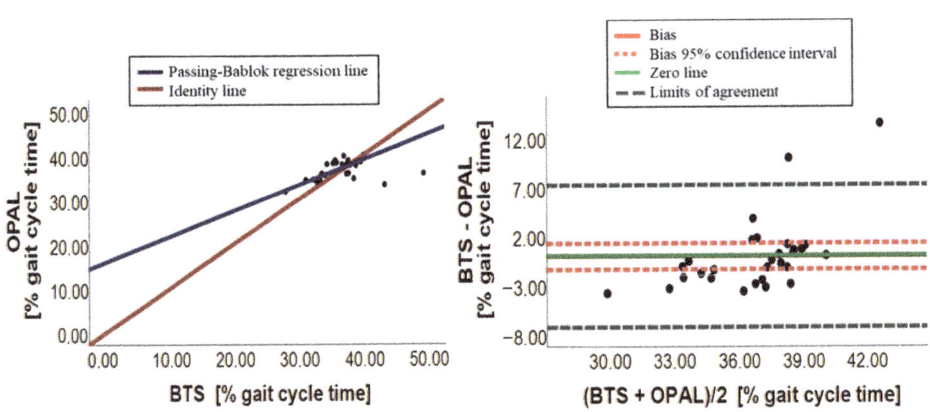

Figure 7. Swing phase. On the left, scatter plot with PB regression line and identity line (BTS = OPAL). On the right, BA plot (average vs. difference).

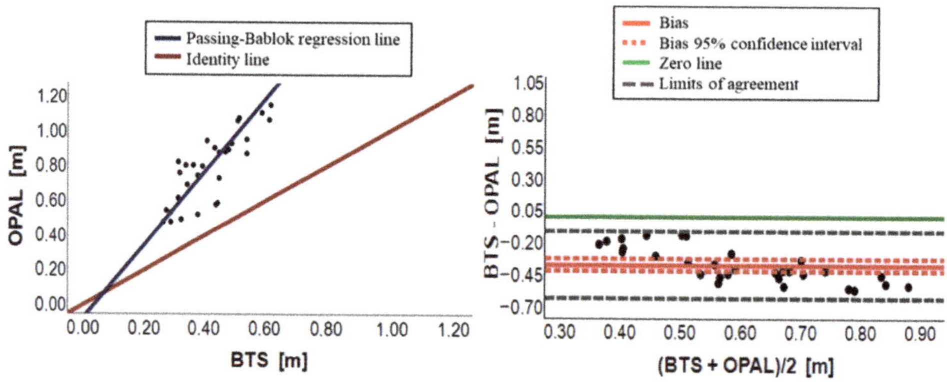

Figure 8. Stride length. On the left, scatter plot with PB regression line and identity line (BTS = OPAL). On the right, BA plot (average vs. difference).

4. Discussion

The current study aimed to compare the performance of Opal wearable sensors by APDM with the BTS SMART DX system in assessing spatiotemporal parameters in PSP patients during walking. Among all the spatiotemporal parameters evaluated, gait speed showed a perfect agreement between the two measurement systems, meaning it can be interchangeably evaluated with either of them. Cadence and cycle duration showed a very close agreement due to the presence of a constant systematic error, which could be removed by zeroing the bias. On the other hand, constant and proportional systematic errors were detected for the remaining parameters, meaning those cannot be interchangeably evaluated by the two systems.

Regarding speed, first, the overall agreement was demonstrated by the Wilcoxon test with no significance between the groups (p-value = 0.192). The agreement was then confirmed by the PB analysis (Table 3 and Figure 5) with the slope (m = 1.02) and the intercept (q = 0.05) values of regression line close to 1 and 0, respectively, and their 95% confidence intervals containing such values. Moreover, points were equally distributed around the identity line (OPAL = BTS), as shown in Figure 5 on the left. The BA analysis in Table 4 showed the presence of a small bias equal to −0.03 and its 95% confidence interval contained the 0 value. In the BA plot, points were randomly distributed around the 0 line (Figure 5). Therefore, even though there is a very small bias, there is an absolute agreement between the two measurement systems regarding speed.

Concerning cadence, the paired data test showed a significant overall difference (p-value equal to 0.002), with Opal overestimating the parameter. This result was confirmed by the PB plot (Figure 3) in which the majority of points were over the identity line (OPAL = BTS). Additionally, the slope of the PB regression line was 1.09 and its 95% confidence interval (0.74–1.81) included the 1 value, assuring the absence of a proportional systematic error, while the intercept value was equal to −3.02 with a quite large 95% confidence interval (−67.89–30.17) containing 0. Nevertheless, there may be a constant systematic error due to the width of the confidence interval. Indeed, the BA analysis, illustrated in Table 4, confirmed the presence of a constant systematic error for the presence of a bias (bias = −7.43), which was quite different from 0, and its 95% confidence interval did not include 0, as shown in Figure 3.

Similarly, as for the cadence parameter, Wilcoxon test for cycle duration showed a statistical overall difference (p-value equal to 0.003). In this case, the Opal system underestimated the evaluation of the cycle duration variable (Table 2 and Figure 4). The slope and the intercept of the PB regression line were close to 1 and 0, respectively (m = 1.03; q = −0.10), and their 95% confidence intervals contained the 1 and 0 values, which allows to suppose the presence of agreement. Despite this, the presence of a constant systematic error has been detected by the BA analysis (Table 4 and Figure 4). Indeed, there was a small bias, equal to 0.15 with a confidence interval, ranging from 0.05 to 0.25. The BA plots for cadence and cycle duration illustrated a random distribution of points around the zero-line, confirming the presence of only a constant systematic error. In conclusion, due to a constant systematic error, there is no agreement between the BTS systems and wearable sensors for both cadence and cycle duration parameters, which could be achieved by zeroing the bias.

Concerning the stance phase, the paired data test exhibited a significant overall difference (p-value = 0.037), implying the absence of an overall agreement. The slope of the PB regression line was 0.63 and its 95% confidence interval ranging from 0.46 to 0.85 did not include the 1 value, showing the presence of a proportional systematic error; the intercept was 23.09 with a 95% confidence interval that did not contain the 0, indicating the presence of a constant systematic error; the presence of such errors has emerged also from the analysis of BA (Table 4 and Figure 6). The bias of the stance phase was close to 0 (bias = 0.84) but its 95% confidence interval did not include the 0 value. Furthermore, the points distribution followed a linear trend around the bias, as shown in Figure 6. In conclusion, there is no agreement between the two measurement systems due to the presence of a constant systematic error and a proportional systematic error.

Regarding the swing phase, no significance between groups was found (p-value = 0.159) from Wilcoxon test, suggesting the presence of an overall agreement. Nevertheless, the slope and the intercept of the PB regression line (m = 0.58; q = 15.96) were far from the identity line values and their 95% confidence intervals did not include 1 and 0, respectively, showing the presence of a proportional and a constant systematic error (Table 3). The BA analysis, illustrated in Table 4, confirmed the presence of both errors for the presence of bias (bias = −0.05), whose 95% confidence interval contained 0, and for the linear distribution of points around the bias (Figure 7). Therefore, due to a constant systematic error, we can conclude that there is no agreement between BTS and Opal systems for the swing phase.

Regarding stride length, the T-Student paired test showed a statistical overall difference (p-value < 0.001) with Opal overestimating the parameter (Table 2). This result was confirmed by the PB scatter plot in Figure 8. The PB analysis, in Table 3, illustrated the presence of a proportional systematic error with a slope value far from 1 and its 95% confidence interval, ranging from 1.64 to 2.69, which did not include it. The intercept of the PB regression line was −0.05 and its 95% confidence interval (−0.28–0.13) contained the 0 value, assuring the absence of a constant systematic error. The absence of a constant systematic error was not confirmed by the BA analysis due to the presence of a small bias equal to −0.37, and its 95% confidence interval did not include the 0, as shown in Table 4. In addition, the BA plot in Figure 8, where the linear trend of points around the bias is shown, implies the presence of a proportional systematic error. In conclusion, there is no agreement between the two measurement systems due to the presence of a constant systematic error and a proportional systematic error. Results are summarized in Table 5.

Table 5. Agreement results.

Spatiotemporal Parameters	Level of Agreement	Error Types
Cadence (steps/min)	Very close agreement	Constant systematic error
Cycle duration (s)	Very close agreement	Constant systematic error
Speed (m/s)	Agreement	/
Stance phase (% gait cycle time)	No agreement	Constant and proportional systematic errors
Swing phase (% gait cycle time)	No agreement	Constant and proportional systematic errors
Stride length (m)	No agreement	Constant and proportional systematic errors

Recently, Zago et al. compared the BTS SMART DX optoelectronic system to the G-Sensor wearable system, assessing spatiotemporal parameters in patients' diagnosis with PD [11]. The variables cadence, stride length, stride duration, step duration, stance duration, swing duration, and double support showed no statistical significance from Wilcoxon rank-sum test (p-value > 0.05) except for velocity showing a p-value equal to 0.01. The

Spearman correlation analysis obtained a good correlation between the two measurement systems, reaching high correlation coefficient values (from 0.64 to 0.91) except for step duration that achieved a correlation coefficient value equal to 0.41. High values, over 10% of the mean absolute error, were obtained for stride length, step duration, stance, and double support duration. The overall results reached by the IMUs system were sufficiently accurate for spatiotemporal parameters evaluations, showing a discrepancy with our results. The same optoelectronic system in comparison with an IMU-based system was used in another study for kinematic gait parameters assessment in healthy subjects [31]. A high relationship between the two measurement systems was obtained by the correlation coefficient r for shoulder flexion-extension, ankle dorsi-plantar flexion, elbow flexion-extension, knee flexion-extension, hip abduction-adduction, and hip flexion-extension. The BA analysis results showed that the IMU-based system overestimates or underestimates the joint angles measurements, reaching bias values greater than 10 degrees. Moreover, Vítecková et al. achieved results comparable to ours by means of different measurement systems (the GAITRite optoelectronic system and the wearable G-Walk system) [34]. Compared to our study, other parameters were additionally evaluated, namely stride duration, double, and single support. Cadence, speed, stride length, and stride duration demonstrated no statistical differences from the paired *t*-test. The PB analysis showed slope and intercept values very close to 1 and 0 for cadence and stride duration, whereas the interclass correlation coefficient exhibited high correlation value ($r > 0.75$) for speed, stride duration, cadence, and stride length. In particular, the agreement was reached for the speed variable in both studies, whereas parameters that could not be considered interchangeable for the two systems and for both studies were stance and swing phase.

The use of gait analysis systems in movement disorders is of utmost importance for regular outpatient follow-ups and for research purposes. Recent evidence demonstrated that Opal wearable sensors present a good correlation with PSP-RS and can effectively detect disease severity and progression in PSP [14,44,45]. Wearable sensors present several advantages compared to more sophisticated systems gathering spatiotemporal parameters of gait for both clinical and research settings. First, set-up and placement are swift and straightforward and only require a short training. Then, the report is promptly provided by the parent company software and normal and previous evaluation values are always shown on the output, making it easy to be interpreted even by non-expert clinicians. As a matter of fact, in just a few minutes, the clinician can have a complete overview of patients gait impairment. Moreover, since PSP patients usually suffer from a severe and rapidly progressive disability, it could be easier for them to use wearable sensors at home or in their outpatient clinic rather than reaching a specific movement analysis laboratory. From a research point of view, wearable sensors can be easily added to a protocol evaluating the efficacy of disease treatments and implemented even in research centers lacking a specific gait lab with more sophisticated instruments.

One limitation of the present study concerns the acquisitions that could not be conducted at the same time due to the incompatibility of the two systems (positions of sensors and markers in the same points of the body and possibility of infrared ray reflection), but they were conducted in the same conditions to reduce the risk of bias at the minimum. Other studies in the literature emphasized the small sample size [11,28,31,32,34] in addition to the fact that patients walk straight paths in lab-based or clinical settings; nevertheless, our study focused on a rare disease, which means that our sample size is consistent with other studies investigating PSP [11,13]. It could also be useful for evaluating a cohort of patients with several gait-related conditions. Furthermore, the different algorithms implemented in the measurement systems can introduce discrepancies in the measurement of parameters.

5. Conclusions

This study proposed a comparison between the wearable Opal system, and the opto-electronic lab-based system, BTS SMART DX, for assessing gait-related parameters in PSP patients. The agreement study conducted between the two measurement systems does not allow the conclusion that the two devices can be used with a complete interchangeability, due to the presence of two types of errors: a constant systematic error detectable with bias zeroing (cadence and cycle duration), and a proportional error (stance phase, swing phase and stride length). This result does not apply to the speed variable that showed perfect agreement between the two measurement systems. This is a first step in defining the reliability of a wearable sensor system compared to the actual gait analysis gold standard. Given the increasing importance of implementing wearable technologies in both clinical and research settings, the feasibility for non-expert clinicians and the known sensitivity of the Opal system, the present data represent a further step forward the implementation of quantitative evaluation of gait in PSP patients. A further development could be focused on the need to study the algorithm to detect gait events underlying systems by reducing the offset.

Author Contributions: Conceptualization, M.C. and F.A. (Francesco Amato); methodology, N.P., C.R., L.D., M.C. and F.A. (Francesco Amato); software, N.P., C.R., M.C. and F.A. (Francesco Amato); validation, M.A., P.B., F.A. (Filomena Abate) and M.P.; formal analysis, N.P., C.R. and L.D.; investigation, M.A., P.B., F.A. (Filomena Abate) and M.P.; data curation, N.P., C.R., M.A., P.B., F.A. (Filomena Abate) and M.P.; writing—original draft preparation, N.P., C.R. and F.A. (Filomena Abate); writing—review and editing, C.R., L.D., P.B., M.P., M.C. and F.A. (Francesco Amato) and supervision, M.C. and F.A. (Francesco Amato); funding acquisition, M.P. All authors have read and agreed to the published version of the manuscript.

Funding: The present study was funded by the Italian Ministry of Health—GR-2019-12370133, Principal Investigator: Marina Picillo.

Institutional Review Board Statement: This study was performed in accordance with the 1964 Declaration of Helsinki and was approved by Campania Sud, the reference ethics committee of the Center for Neurodegenerative Diseases of the University of Salerno (approval date 2 December 2022).

Informed Consent Statement: Informed consent was obtained from all subjects involved in the study.

Data Availability Statement: The data presented in this study are available on request from the corresponding author. The data are not publicly available due to the privacy policy.

Conflicts of Interest: The authors declare no conflict of interest.

References

1. Hatanaka, N.; Sato, K.; Hishikawa, N.; Takemoto, M.; Ohta, Y.; Yamashita, T.; Abe, K. Comparative Gait Analysis in Progressive Supranuclear Palsy and Parkinson's Disease. *Eur. Neurol.* **2016**, *75*, 282–289. [CrossRef] [PubMed]
2. Nieuwboer, A.; Dom, R.; De Weerdt, W.; Desloovere, K.; Fieuws, S.; Broens-Kaucsik, E. Abnormalities of the Spatiotemporal Characteristics of Gait at the Onset of Freezing in Parkinson's Disease. *Mov. Disord.* **2001**, *16*, 1066–1075. [CrossRef] [PubMed]
3. Rogers, M.W. Disorders of Posture, Balance, and Gait in Parkinson's Disease. *Clin. Geriatr. Med.* **1996**, *12*, 825–845. [CrossRef] [PubMed]
4. Golbe, L.I.; Ohman-Strickland, P.A. A Clinical Rating Scale for Progressive Supranuclear Palsy. *Brain* **2007**, *130*, 1552–1565. [CrossRef] [PubMed]
5. Nonnekes, J.; Goselink, R.J.M.; Růžička, E.; Fasano, A.; Nutt, J.G.; Bloem, B.R. Neurological Disorders of Gait, Balance and Posture: A Sign-Based Approach. *Nat. Rev. Neurol.* **2018**, *14*, 183–189. [CrossRef] [PubMed]
6. Salarian, A.; Russmann, H.; Vingerhoets, F.J.G.; Dehollain, C.; Blanc, Y.; Burkhard, P.R.; Aminian, K. Gait Assessment in Parkinson's Disease: Toward an Ambulatory System for Long-Term Monitoring. *IEEE Trans. Biomed. Eng.* **2004**, *51*, 1434–1443. [CrossRef]
7. Sveinbjornsdottir, S. The Clinical Symptoms of Parkinson's Disease. *J. Neurochem.* **2016**, *139*, 318–324. [CrossRef]
8. Cesarelli, G.; Donisi, L.; Amato, F.; Romano, M.; Cesarelli, M.; D'Addio, G.; Ponsiglione, A.M.; Ricciardi, C. Using Features Extracted from Upper Limb Reaching Tasks to Detect Parkinson's Disease by Means of Machine Learning Models. *IEEE Trans. Neural Syst. Rehabil. Eng.* **2023**, *31*, 1056–1063. [CrossRef]

9. Morris, R.; Stuart, S.; McBarron, G.; Fino, P.C.; Mancini, M.; Curtze, C. Validity of Mobility Lab (Version 2) for Gait Assessment in Young Adults, Older Adults and Parkinson's Disease. *Physiol. Meas.* **2019**, *40*, 095003. [CrossRef]
10. Fasano, A.; Plotnik, M.; Bove, F.; Berardelli, A. The Neurobiology of Falls. *Neurol. Sci. Off. J. Ital. Neurol. Soc. Ital. Soc. Clin. Neurophysiol.* **2012**, *33*, 1215–1223. [CrossRef]
11. Zago, M.; Sforza, C.; Pacifici, I.; Cimolin, V.; Camerota, F.; Celletti, C.; Condoluci, C.; De Pandis, M.F.; Galli, M. Gait Evaluation Using Inertial Measurement Units in Subjects with Parkinson's Disease. *J. Electromyogr. Kinesiol.* **2018**, *42*, 44–48. [CrossRef] [PubMed]
12. Warmerdam, E.; Schumacher, M.; Beyer, T.; Nerdal, P.T.; Schebesta, L.; Stürner, K.H.; Zeuner, K.E.; Hansen, C.; Maetzler, W. Postural Sway in Parkinson's Disease and Multiple Sclerosis Patients During Tasks with Different Complexity. *Front. Neurol.* **2022**, *13*, 857406. [CrossRef] [PubMed]
13. Picillo, M.; Ricciardi, C.; Tepedino, M.F.; Abate, F.; Cuoco, S.; Carotenuto, I.; Erro, R.; Ricciardelli, G.; Russo, M.; Cesarelli, M.; et al. Gait Analysis in Progressive Supranuclear Palsy Phenotypes. *Front. Neurol.* **2021**, *12*, 674495. [CrossRef] [PubMed]
14. Abate, F.; Russo, M.; Ricciardi, C.; Tepedino, M.F.; Romano, M.; Erro, R.; Pellecchia, M.T.; Amboni, M.; Barone, P.; Picillo, M. Wearable Sensors for Assessing Disease Severity and Progression in Progressive Supranuclear Palsy. *Park. Relat. Disord.* **2023**, *109*, 105345. [CrossRef] [PubMed]
15. Stamatakis, J.; Crémers, J.; Maquet, D.; Macq, B.; Garraux, G. Gait Feature Extraction in Parkinson's Disease Using Low-Cost Accelerometers. In Proceedings of the 2011 Annual International Conference of the IEEE Engineering in Medicine and Biology Society, Boston, MA, USA, 30 August–3 September 2011; pp. 7900–7903.
16. Zhou, H.; Hu, H. Human Motion Tracking for Rehabilitation—A Survey. *Biomed. Signal Process. Control* **2008**, *3*, 1–18. [CrossRef]
17. Hreljac, A.; Marshall, R.N. Algorithms to Determine Event Timing during Normal Walking Using Kinematic Data. *J. Biomech.* **2000**, *33*, 783–786. [CrossRef] [PubMed]
18. McGinley, J.L.; Baker, R.; Wolfe, R.; Morris, M.E. The Reliability of Three-Dimensional Kinematic Gait Measurements: A Systematic Review. *Gait Posture* **2009**, *29*, 360–369. [CrossRef]
19. Saggio, G.; Tombolini, F.; Ruggiero, A. Technology-Based Complex Motor Tasks Assessment: A 6-DOF Inertial-Based System Versus a Gold-Standard Optoelectronic-Based One. *IEEE Sens. J.* **2021**, *21*, 1616–1624. [CrossRef]
20. Santos, T.M.O.; Barroso, M.F.S.; Ricco, R.A.; Nepomuceno, E.G.; Alvarenga, É.L.F.C.; Penoni, Á.C.O.; Santos, A.F. A Low-Cost Wireless System of Inertial Sensors to Postural Analysis during Human Movement. *Measurement* **2019**, *148*, 106933. [CrossRef]
21. Wu, Y.; Su, Y.; Feng, R.; Yu, N.; Zang, X. Wearable-Sensor-Based Pre-Impact Fall Detection System with a Hierarchical Classifier. *Measurement* **2019**, *140*, 283–292. [CrossRef]
22. Muro-de-la-Herran, A.; Garcia-Zapirain, B.; Mendez-Zorrilla, A. Gait Analysis Methods: An Overview of Wearable and Non-Wearable Systems, Highlighting Clinical Applications. *Sensors* **2014**, *14*, 3362–3394. [CrossRef] [PubMed]
23. Donisi, L.; Pagano, G.; Cesarelli, G.; Coccia, A.; Amitrano, F.; D'Addio, G. Benchmarking between Two Wearable Inertial Systems for Gait Analysis Based on a Different Sensor Placement Using Several Statistical Approaches. *Measurement* **2021**, *173*, 108642. [CrossRef]
24. Mancini, M.; Smulders, K.; Harker, G.; Stuart, S.; Nutt, J.G. Assessment of the Ability of Open- and Closed-Loop Cueing to Improve Turning and Freezing in People with Parkinson's Disease. *Sci. Rep.* **2018**, *8*, 12773. [CrossRef] [PubMed]
25. Muthukrishnan, N.; Abbas, J.J.; Shill, H.A.; Krishnamurthi, N. Cueing Paradigms to Improve Gait and Posture in Parkinson's Disease: A Narrative Review. *Sensors* **2019**, *19*, 5468. [CrossRef] [PubMed]
26. Bugané, F.; Benedetti, M.G.; Casadio, G.; Attala, S.; Biagi, F.; Manca, M.; Leardini, A. Estimation of Spatial-Temporal Gait Parameters in Level Walking Based on a Single Accelerometer: Validation on Normal Subjects by Standard Gait Analysis. *Comput. Methods Programs Biomed.* **2012**, *108*, 129–137. [CrossRef] [PubMed]
27. Mariani, B.; Hoskovec, C.; Rochat, S.; Büla, C.; Penders, J.; Aminian, K. 3D Gait Assessment in Young and Elderly Subjects Using Foot-Worn Inertial Sensors. *J. Biomech.* **2010**, *43*, 2999–3006. [CrossRef]
28. Webster, K.E.; Wittwer, J.E.; Feller, J.A. Validity of the GAITRite® Walkway System for the Measurement of Averaged and Individual Step Parameters of Gait. *Gait Posture* **2005**, *22*, 317–321. [CrossRef]
29. Fernández-González, P.; Koutsou, A.; Cuesta-Gómez, A.; Carratalá-Tejada, M.; Miangolarra-Page, J.C.; Molina-Rueda, F. Reliability of Kinovea® Software and Agreement with a Three-Dimensional Motion System for Gait Analysis in Healthy Subjects. *Sensors* **2020**, *20*, 3154. [CrossRef]
30. Piche, E.; Guilbot, M.; Chorin, F.; Guerin, O.; Zory, R.; Gerus, P. Validity and Repeatability of a New Inertial Measurement Unit System for Gait Analysis on Kinematic Parameters: Comparison with an Optoelectronic System. *Measurement* **2022**, *198*, 111442. [CrossRef]
31. Bartoszek, A.; Struzik, A.; Jaroszczuk, S.; Wozniewski, M.; Pietraszewski, B. Comparison of the Optoelectronic BTS Smart System and IMU-Based MyoMotion System for the Assessment of Gait Variables. *Acta Bioeng. Biomech. Wroclaw Univ. Technol.* **2022**, *24*, 103–116. [CrossRef]
32. Muthukrishnan, N.; Abbas, J.J.; Krishnamurthi, N. A Wearable Sensor System to Measure Step-Based Gait Parameters for Parkinson's Disease Rehabilitation. *Sensors* **2020**, *20*, 6417. [CrossRef] [PubMed]
33. Cimolin, V.; Vismara, L.; Ferraris, C.; Amprimo, G.; Pettiti, G.; Lopez, R.; Galli, M.; Cremascoli, R.; Sinagra, S.; Mauro, A.; et al. Computation of Gait Parameters in Post Stroke and Parkinson's Disease: A Comparative Study Using RGB-D Sensors and Optoelectronic Systems. *Sensors* **2022**, *22*, 824. [CrossRef] [PubMed]

34. Vítečková, S.; Horáková, H.; Poláková, K.; Krupička, R.; Růžička, E.; Brožová, H. Agreement between the GAITRite® System and the Wearable Sensor BTS G-Walk® for Measurement of Gait Parameters in Healthy Adults and Parkinson's Disease Patients. *PeerJ* **2020**, *8*, e8835. [CrossRef] [PubMed]
35. Tomlinson, C.L.; Stowe, R.; Patel, S.; Rick, C.; Gray, R.; Clarke, C.E. Systematic Review of Levodopa Dose Equivalency Reporting in Parkinson's Disease. *Mov. Disord.* **2010**, *25*, 2649–2653. [CrossRef] [PubMed]
36. Davis, R.B.; Õunpuu, S.; Tyburski, D.; Gage, J.R. A Gait Analysis Data Collection and Reduction Technique. *Hum. Mov. Sci.* **1991**, *10*, 575–587. [CrossRef]
37. Ricciardi, C.; Amboni, M.; De Santis, C.; Ricciardelli, G.; Improta, G.; D'Addio, G.; Cuoco, S.; Picillo, M.; Barone, P.; Cesarelli, M. Machine Learning Can Detect the Presence of Mild Cognitive Impairment in Patients Affected by Parkinson's Disease. In Proceedings of the 2020 IEEE International Symposium on Medical Measurements and Applications (MeMeA), Bari, Italy, 1 June–1 July 2020; pp. 1–6.
38. Landolfi, A.; Ricciardi, C.; Donisi, L.; Cesarelli, G.; Troisi, J.; Vitale, C.; Barone, P.; Amboni, M. Machine Learning Approaches in Parkinson's Disease. *Curr. Med. Chem.* **2021**, *28*, 6548–6568. [CrossRef]
39. Blair, R.C.; Cole, S. Two-Sided Equivalence Testing of the Difference Between Two Means. *J. Mod. Appl. Stat. Methods* **2002**, *1*, 18. [CrossRef]
40. Passing, H.; Bablok, W. A New Biometrical Procedure for Testing the Equality of Measurements from Two Different Analytical Methods. Application of linear regression procedures for method comparison studies in Clinical Chemistry, Part I. *Clin. Chem. Lab. Med. (CCLM)* **1983**, *21*, 709–720. [CrossRef]
41. Benchoufi, M.; Matzner-Lober, E.; Molinari, N.; Jannot, A.-S.; Soyer, P. Interobserver Agreement Issues in Radiology. *Diagn. Interv. Imaging* **2020**, *101*, 639–641. [CrossRef]
42. Ludbrook, J. Confidence in Altman–Bland Plots: A Critical Review of the Method of Differences. *Clin. Exp. Pharmacol. Physiol.* **2010**, *37*, 143–149. [CrossRef]
43. Mansournia, M.A.; Waters, R.; Nazemipour, M.; Bland, M.; Altman, D.G. Bland-Altman Methods for Comparing Methods of Measurement and Response to Criticisms. *Glob. Epidemiol.* **2021**, *3*, 100045. [CrossRef] [PubMed]
44. Sotirakis, C.; Conway, N.; Su, Z.; Villarroel, M.; Tarassenko, L.; FitzGerald, J.J.; Antoniades, C.A. Longitudinal Monitoring of Progressive Supranuclear Palsy Using Body-Worn Movement Sensors. *Mov. Disord.* **2022**, *37*, 2263–2271. [CrossRef] [PubMed]
45. Dale, M.L.; Prewitt, A.L.; Harker, G.R.; McBarron, G.E.; Mancini, M. Perspective: Balance Assessments in Progressive Supranuclear Palsy: Lessons Learned. *Front. Neurol.* **2022**, *13*, 801291. [CrossRef] [PubMed]

Disclaimer/Publisher's Note: The statements, opinions and data contained in all publications are solely those of the individual author(s) and contributor(s) and not of MDPI and/or the editor(s). MDPI and/or the editor(s) disclaim responsibility for any injury to people or property resulting from any ideas, methods, instructions or products referred to in the content.

sensors

MDPI

Article

An Overall Automated Architecture Based on the Tapping Test Measurement Protocol: Hand Dexterity Assessment through an Innovative Objective Method

Tommaso Di Libero [1,†], Chiara Carissimo [2,*,†], Gianni Cerro [2], Angela Marie Abbatecola [1,3], Alessandro Marino [4], Gianfranco Miele [4], Luigi Ferrigno [4] and Angelo Rodio [1]

[1] Department of Human, Social and Health Sciences, University of Cassino and Southern Lazio, 03043 Cassino, Italy; tommaso.dilibero@unicas.it (T.D.L.); angelamarie.abbatecola@unicas.it (A.M.A.); a.rodio@unicas.it (A.R.)

[2] Department of Medicine and Health Sciences "Vincenzo Tiberio", University of Molise, 86100 Campobasso, Italy; gianni.cerro@unimol.it

[3] Alzheimer's Disease Day Clinics, Azienda ria Locale, 03100 Frosinone, Italy

[4] Department of Electrical and Information Engineering, University of Cassino and Southern Lazio, 03043 Cassino, Italy; al.marino@unicas.it (A.M.); g.miele@unicas.it (G.M.); ferrigno@unicas.it (L.F.)

* Correspondence: chiara.carissimo@unimol.it

† These authors contributed equally to this work.

Abstract: The present work focuses on the tapping test, which is a method that is commonly used in the literature to assess dexterity, speed, and motor coordination by repeatedly moving fingers, performing a tapping action on a flat surface. During the test, the activation of specific brain regions enhances fine motor abilities, improving motor control. The research also explores neuromuscular and biomechanical factors related to finger dexterity, revealing neuroplastic adaptation to repetitive movements. To give an objective evaluation of all cited physiological aspects, this work proposes a measurement architecture consisting of the following: (i) a novel measurement protocol to assess the coordinative and conditional capabilities of a population of participants; (ii) a suitable measurement platform, consisting of synchronized and non-invasive inertial sensors to be worn at finger level; (iii) a data analysis processing stage, able to provide the final user (medical doctor or training coach) with a plethora of useful information about the carried-out tests, going far beyond state-of-the-art results from classical tapping test examinations. Particularly, the proposed study underscores the importance interdigital autonomy for complex finger motions, despite the challenges posed by anatomical connections; this deepens our understanding of upper limb coordination and the impact of neuroplasticity, holding significance for motor abilities assessment, improvement, and therapeutic strategies to enhance finger precision. The proof-of-concept test is performed by considering a population of college students. The obtained results allow us to consider the proposed architecture to be valuable for many application scenarios, such as the ones related to neurodegenerative disease evolution monitoring.

Keywords: neuroplasticity; hand dexterity; tapping test; IMU sensors; coordinative abilities; measurement platform

1. Introduction

The evaluation of motor abilities to assess human aptitude is a challenging task from a measurement point of view, as it involves complex interactions between neurophysiological systems [1]. Specifically, coordinative abilities encompass a multifaceted array of skills regulating our ability to manage, coordinate, and integrate movements to accomplish tasks with accuracy and refinement [2]. The development of manual dexterity, which plays a significant role in several areas of human activity, is an essential component of coordinative abilities [3]. Manual dexterity assessment is crucial in various fields such as health care, rehabilitation [4], occupational therapy, sports, and some professions [5]. Understanding and assessing hand function makes it possible to provide higher-quality care, support, and interventions, ultimately leading to better outcomes and a higher quality of life for those facing hand-related challenges [6]. Researchers and clinicians have turned to tapping tests (TTs), tests in the Halstead–Reitan battery [7], as valuable assessment tools to understand and improve this essential aspect of human motor performance. Gaining a comprehensive understanding of the intricate aspects of manual dexterity and assessing its progression and potential diversities can yield significant and invaluable knowledge regarding an individual's motor abilities and functional abilities [2,8]. There are different methods of conducting TTs. The original way involves the subject tapping with their index finger as many times as possible in 10 s [7]. However, there are more intricate versions that use both the hands and the feet [9], or multiple fingers or hands alternately or simultaneously [10,11]. Several solutions converge towards wearable motion-capture technologies to obtain an objective assessment of this motion and improve evaluation performance. These devices are small, low-cost, non-invasive, and easy to use; these features leading to an increase in their use in various fields, such as sports [12] and medicine [13]. In this scenario, the authors propose an architecture for performing a novel measurement method based on the tapping test by adopting multiple synchronized inertial sensors and providing a wide data-analysis framework; this will provide the final user (medical doctor or training coach) with a comprehensive information panel going beyond classical indexes that could be extracted from traditional approaches. The core of the work is the measurement protocol, as it deals with an enhanced use of the TT, employing multi-finger movements. Participants must move two fingers simultaneously or alternately, considering the dominant hand only (in this case, index and middle fingers) or a two-hands case (both index fingers). Such choices are motivated by medical and physiological reasons, such as the possibility of assessing coordinative and conditional abilities by exploiting both the cerebral hemisphere and the efficiency of the corpus callosum structures, as well as the participant's abilities. The joint use of measurements and suitable processing algorithms, specifically designed for TTs evaluation, has already been exploited by the authors in [14]: in addition to recording tapping counts, the intertemps, amplitudes, speed, and finger–hand coordination are all computable. Stemming from the TT evaluation according to the gender and age parameters presented in [15,16], this work provides extensions in two main directions: (i) provide an insight evaluation of bi-manual vs. uni-manual coordination performance; (ii) assess dexterity capabilities adopting simultaneous and alternate double-finger TT either on one single hand, or requiring movements from both hands. The impact of the study goes beyond academia and finds practical applications in sports training, motor rehabilitation, and cognitive assessment. The measurement protocol involves a population of 30 voluntary students who have been suitably informed about the procedure and it deals with an extensive experimental campaign, from which significant statistical analyses and neurophysiological-related outcomes are derived. The paper is organized into five main parts: Section 2 describes the motivation and the specific contribution of the authors to the state of the art; Section 3 reports the system's architecture, the adopted measurement materials and methods, the participants' description, and the proposed protocol.

2. Motivation and Contribution of the Work

The standard TT has proven to be a widely used method for assessing coordination and cognitive abilities in the scientific literature [17]. Several alternative devices for obtaining objective measures of movement have emerged over the years. The most popular system is a lever equipped with a mechanical counter [18]. However, electronic and software versions have also been developed, such as the Digital Tapping Test offered by the Western Psychological Service, which uses a device with an electrical switch and a digital counter [19,20]. In addition, even the zero on the keypad has been exploited as a tapping device in a software-only variant of the WPS. Some recent innovations involve using keyboards, smartphone and tablet applications, traditional cameras, and cameras built into mobile devices to perform TT efficiently and accurately [21]. These approaches are generally limited by the low number of monitored parameters, such as number of taps and velocity.

In our research, we have introduced an innovative automated measurement protocol that combines IMU sensors with various versions of the TT found in the existing literature [22,23]. This specific measurement protocol (uni-manual and bi-manual tests) has the potential to become an excellent method of assessing motor efficiency by analyzing the movement of one or more fingers of the hand. These tests allow assessment of the ability to perform alternating and simultaneous sequential movements, thus providing insight into the synchronization between the central and peripheral nervous systems [24]. Using these tests to assess performance and compare the two sides of the body can help monitor the integrity of brain functions and motor learning abilities. The proposed system can become a valuable tool for screening and monitoring diseases, such as Alzheimer's and Parkinson's, which involve a progressive loss of motor and cognitive abilities with gradual deterioration [25]. In addition, by analyzing finger movements, the system can provide the physician with detailed information about the patient's condition. As an example, it can help determine whether medications produce the desired effects or whether drug treatment needs adjustment. In addition, it could play a key role in disease prevention by serving as an early warning for potential health problems. This approach could not only offer true prevention, but could also help to significantly reduce the costs faced by national healthcare systems by anticipating the early stage of diseases and enabling timely and targeted interventions. Leveraging the capabilities of powerful data processing, real-time analysis, and direct transmission of results to physicians, this automated system offers a promising solution for continuous patient monitoring and rapid response to medical needs. In detail, based on previous experience in this field [14,16], the authors extend the present state-of-the-art approaches in the following directions:

- Proposal of an automated measurement protocol to assess coordinative abilities through tapping test-based exercises;
- Computation of a large number of features to give a quantitative, objective, and exhaustive movement assessment;
- Comparative analysis of the proposed tasks to obtain general outcomes about the analyzed population samples.

3. The Proposed Measurement Architecture

To assess the performance of the proposed approach, a population of young students was recruited on a voluntary basis and a specific measurement protocol was designed.

3.1. The Measurement Protocol

A protocol with four different TTs is proposed to obtain more information on hand dexterity. Specifically, the tests are codified and described below:

- Alternate uni-manual (UniALT): tapping is performed by alternating the index and middle fingers of the dominant hand.
- Simultaneous uni-manual (UniSIM): tapping is performed by simultaneously moving the index and middle fingers of the dominant hand.
- Alternate bi-manual (BimALT): tapping is performed by alternately moving the index fingers of the right and left hands.
- Simultaneous bi-manual (BimSIM): tapping is performed by simultaneously moving the index fingers of both hands.

The management of fingers' fine movements represents the manifestation of human motor abilities, involving a complex network of highly specialized neural pathways. These movements enable individuals to perform various everyday activities that require precision and coordination, such as writing, playing musical instruments, or manipulating objects. The hand's fingers are extraordinarily dexterous tools, capable of performing a wide range of precise and complex movements. However, behind this apparent simplicity lies an intricate neural network that coordinates and controls every finger action. This network is responsible for fine hand movements and involves several key brain regions [26].

These regions include the primary motor cortex (M1), the supplementary motor cortex (SMA), the parietal cortex, the cerebellum, the basal ganglia, the associative cerebral cortex, and the limbic system [27]. Each area has a specific role in planning, executing, and controlling finger movements. Good brain plasticity efficiency is critical, as the brain constantly adapts neural pathways to optimize coordination [28]. The ability to perform alternating and simultaneous movements in sequence could provide valuable insights into the synchronization between the central and peripheral nervous systems. These motor tests pose considerable challenges, requiring rapid attention shifts between different fingers and timely motor accuracy. Several factors, including the nature of the task, the complexity of movements, the level of motor control, and practice, can affect finger coordination [29]. Understanding how the brain manages fine finger movements is a fascinating field of research. It has important implications for rehabilitation, occupational therapy, and the development of new technologies to assist people with motor disabilities.

3.1.1. Uni-Manual

This test requires coordination of index and middle finger movements (UniSIM–UniALT) and a great amount of neural complexity. The brain is activated through a series of neural pathways orchestrating these intricate movements. To precisely execute the movements of these fingers, the brain activates specific regions in M1 [30]. Each finger has a dedicated area in M1 that controls the muscles involved in these movements, ensuring a firm basis for action. The SMA is central in planning and sequencing movements [31]. When we want the index and middle fingers to move, the SMA coordinates the order and timing of these movements, ensuring that they occur smoothly. The brain sends specific signals to each muscle involved in the movements. The index and middle finger muscles contract synchronously or sequentially, depending on what we intend to do, reflecting detailed communication between the brain and the muscular system [32]. The brain constantly receives sensory feedback from finger movement. This feedback comes from receptors in the fingers' muscles, joints, and skin. This information helps the brain monitor movements' position and progress, helping maintain control. The cerebellum plays a crucial role in regulating the precision of finger movements [33]. It ensures that movements are smooth and precise, intervening to correct any deviations. The basal ganglia comes into play to regulate muscle activity, helping to inhibit unwanted or excessive movements and ensuring that action is appropriate and controlled [34]. The brain is an amazing learning machine. When we learn to coordinate the index and middle fingers in different activities, the brain modifies neural connections to improve our abilities. This process of constant adaptation allows us to refine our coordination more and more [35].

3.1.2. Bi-Manual

This test requires simultaneous or alternating movement of the index fingers of both hands (BimSIM–BimALT), necessitating the intervention of specific motor regions in the cerebral cortex to control the left and right hands [36]. The control of simultaneous or alternating movements involves the coordination of different brain regions, muscle sequencing, and sensory feedback. This requires greater intermanual coordination and synchronization. The coordination between the two hands may involve the interhemispheric commissure, the corpus callosum, which enables the communication between the cerebral hemispheres cite [37]. The accuracy and fluidity of such movements are achieved through a complex interaction of neural pathways and brain adaptation. The left motor area controls the right hand, while the right motor area controls the left hand [38]. The SMA and premotor areas are involved in planning and sequencing hand movements, helping to coordinate the order and timing of movements. To perform simultaneous or alternating movements between the index fingers, the brain sends specific signals to each hand; the muscles of the two index fingers are activated synchronously or sequentially, depending on the movement's goal [10].

The cerebellum plays a crucial role in regulating the precision of these movements, ensuring that they are smooth and accurate in both hands. The basal ganglia help regulate muscle activity, ensuring both hands perform the desired movements. The brain can adapt and learn new movement patterns when performing simultaneous or alternating movements. Neural connections can be modified to improve coordination between the limbs [11].

3.1.3. Participants

Participants are required to sit near a table with their hands resting on a flat surface. They are asked to perform each test 20 times over a period of 30 s at the highest possible speed. A population of 30 students from the University of Cassino and Southern Lazio were involved in the TT study. Specifically, 16 females and 14 males aged between 21 and 30 were considered. All participants were informed about how the protocol would have been carried out and provided their consent before participating in the study. Ethical guidelines were followed throughout the study. This work was approved by the Institutional Review Board of the University of Cassino and Southern Lazio (no. 24777.2022.12.12). The informed consent and approval on benefits and risks are derived from the 1964 Declaration of Helsinki for research on humans. After the description of the test phase, all general data of the participants were collected, in particular age and the lengths of the index and middle fingers. The collected data are summarized and reported in Table 1. The resulting data prove the very low variability in the collected information, thus allowing the removal of possible additional influence factors when analyzing the test outcomes.

Table 1. Age and finger length of participants: mean and standard deviation.

	Mean ± St.Dev
Age	25.67 ± 0.33
Index Finger Length [cm]	10.03 ± 0.12
Middle Finger Length [cm]	11.07 ± 0.12

3.2. The Adopted Measurement Platform

Four Xsens IMUs produced by the MOVELLA company, Henderson, NV, USA [39] were adopted in the test phase. These are wearable devices with a weight of 11.2 g, compact dimensions (36.30 × 30.35 × 10.80 mm) and a latency of 30 ms. Each IMU has three triaxial sensors: an accelerometer with a maximum range of ±16 g (g = 9.81 m/s^2, gravity acceleration constant), a gyroscope with a maximum range of ±2000 deg/s, and a magnetometer with a full scale of ±8 G. In recording mode, the maximum acquisition frequency is 120 Hz; meanwhile, in the real-time mode, it is 60 Hz. This study considers only accelerometric measurements: the chosen range is ±8 g , while the sampling frequency is set at 120 Hz.

Before starting the test, the sensors are connected to a smartphone via Bluetooth Low-Energy 5.0 Smart® module and are configured using a proprietary MOVELLA DOT app. 2023.6.1 The operator sets the accelerometer range and the sampling frequency. A synchronization operation is then carried out between the sensors so that they share the same time reference. At the end of the test, the data are saved on the internal memory, which has a capacity of 65 MB, and subsequently downloaded to a PC via the APP. A diagram of the measurement setup used is summarized in Figure 1. The optimal sensor position was defined after an experimental characterization, which will be described in Section 4.1. Figure 2 shows how the adopted sensors were anchored on the participant's hand: the index and middle finger of the dominant hand (Figure 2a); the right and left index fingers (Figure 2b).

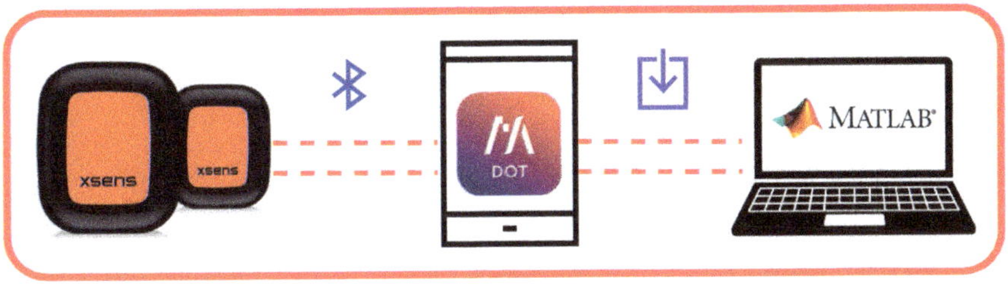

Figure 1. The measurement setup: data are acquired through a couple of IMU sensors, driven by a proprietary MOVELLA DOT App, which communicates with a PC where data processing is carried out in a MATLAB® environment R2023b.

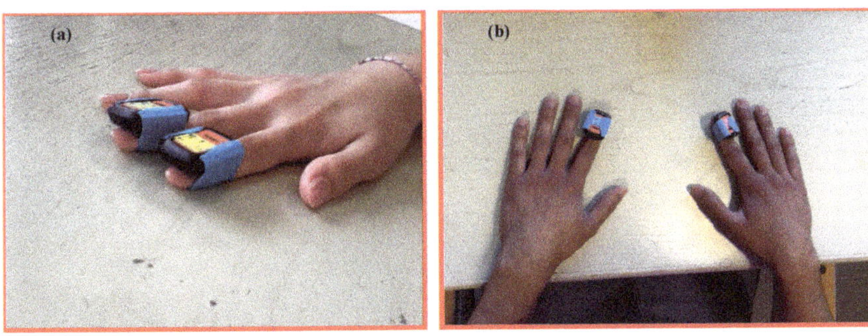

Figure 2. Two different configurations: (**a**) IMUs are placed on the index and middle of the dominant hand; (**b**) IMUs are placed on the index fingers of the right and left hand.

3.3. The Data Analysis

An algorithm was developed and implemented in MATLAB® R2022a to analyze the data acquired during the experimental tests. The block diagram in Figure 3 shows the logical process used by the algorithm to acquire the data, analyze it, and extract metric characteristics. The algorithm receives raw accelerometric data as input from the IMU sensors placed on selected fingers. The z-axis is selected as the axis of highest sensitivity, by evaluating the orientation of the sensor and the main direction of motion, i.e., the direction in which the greatest acceleration excursion is observed. On each trace, the tap finding is achieved by getting the peaks through MATLAB® *findpeaks* [40] routine, after a tuning phase to avoid getting spurious noise spikes. The found peaks are then used in three different metrics: calculation of the number of taps, evaluation of the inter-tap times for each single finger, and estimation of the time intervals between adjacent peaks acquired by the sensors placed on different fingers. In the last case, the first peak is taken on one trace and the second on the other (for a more in-depth understanding, see step 4 in Figure 3). Time intervals have different meanings according to test typology: in simultaneous cases (UniSIM, BimSIM), they should approach zero, as they estimate the time mismatch between taps of the considered fingers that should be perfectly time-aligned; in the alternate case (UniALT, BimALT), time intervals should be as much regular as possible, as they represent the rhythm each participant takes when alternate movement is required. To assess them in a statistical way, the means and standard deviations for each participant are computed (see step 5 in Figure 3). The Algorithm 1 pseudocode shows the metric characteristic computation, the main logical steps for the calculation of the inter-times of the single-finger analysis, and the alternate and simultaneous times in the case of the two-finger analysis.

Algorithm 1 Data analysis: Computation of metrics features.

```
Choose the test: UniALT–UniSIM–BimALT–BimSIM
NID= Total number of participants
NTest= Total number of tests
for k = 1, ..., NID do
  for y = 1, ..., NTest do
    load Acc_finger1(k, y)
    load Acc_finger2(k, y)
    [A1, B1] = findpeaks(Acc_finger1(k, y))
    [A2, B2] = findpeaks(Acc_finger2(k, y))
    for i = 1, ..., length(B1 − 1) do
      InterTimeFinger1 = abs(B1(i) − B1(i − 1))
      InterTimeFinger2 = abs(B2(i) − B2(i − 1))
      SATime12(i) = abs(B2(i) − B1(i))
    end for
    SATime12 = mean(SATime12)
    σ(SATime12) = std(SATime12)
  end for
end for
```

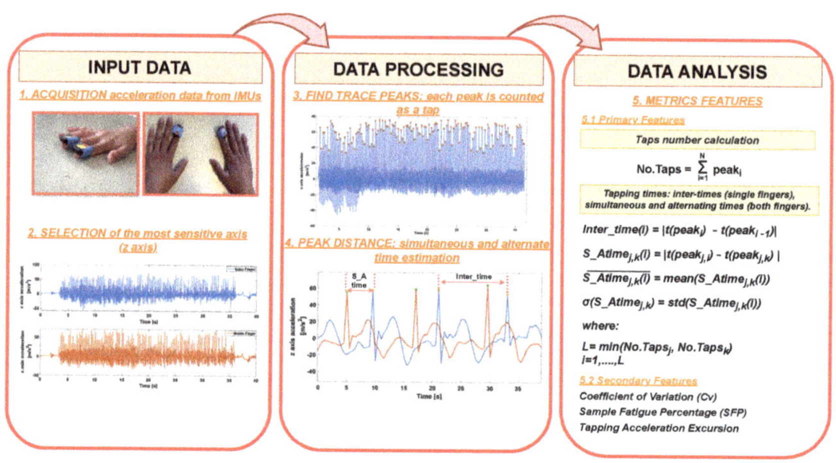

Figure 3. Algorithm block diagram: description of the main steps for acquiring and selecting the most sensitive axis (INPUT DATA), processing (DATA PROCESSING) and analyzing (DATA ANALYSIS) IMU inertial data.

4. Results

This section reports the obtained results, particularly focusing on two main aspects: (i) the investigation of the effect of the device positioning on the finger; (ii) a quantitative analysis of the computed features in the single-finger and dual-finger modes.

4.1. Experimental Characterization

An experimental characterization has been necessary to define the sensor position that would have the minimum possible influence on the measurement results. In our previous work [16], the influence of the sensor's weight was studied, which is shown to have a negligible effect on performance; for this reason, this aspect is not investigated in this subsection. The experimental characterization considers two different tasks, each repeated five times in a row, which were performed by 10 students. Two configurations were considered (Figure 4):

- Position A: a fixed position from the metacarpal joint of the index finger (2 cm) is chosen to place the IMU.
- Position B: IMU placed on the distal phalanx of an index finger, without considering joint distance.

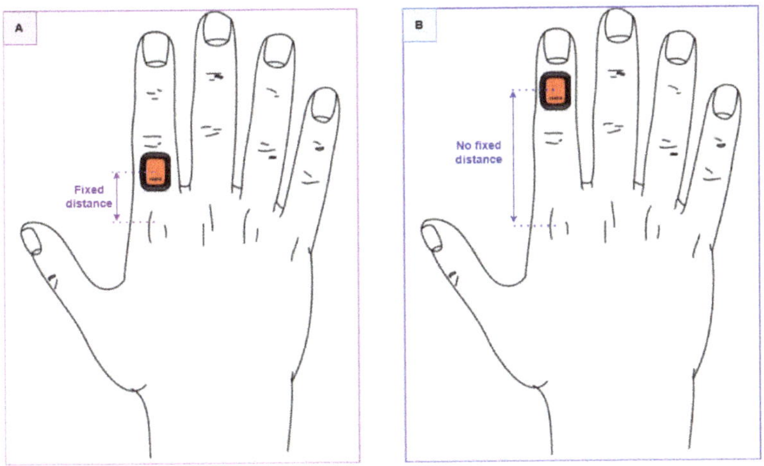

Figure 4. In configuration (**A**), the sensor is placed at a fixed distance from the metacarpal joint of 2 cm. In configuration (**B**), the sensor is placed on the distal phalanx with an unfixed distance.

Participants were asked to tap the index finger of the dominant hand as fast as possible in both conditions. A variation coefficient (Cv) was computed to check the best position of the sensor. Equation (1) shows the described mathematical procedure:

$$Cv = \frac{std(NTap)}{mean(NTap)} * 100 \quad (1)$$

where $NTap$ is the number of taps that have been performed. It is therefore important to assess how much the standard deviation weighs against the mean value . The best repeatability and configuration are obtained at a low (Cv).

Figure 5 and Table 2 show that the lowest variability of Cv is obtained in the case of position B. Evaluating the average value of the CVs obtained in the two different configurations shows that the mean value of A is 5.57%, while position B obtains 4.23%. These results emphasize how the sensor positioned on the distal phalanx affects the measurement process less than it does in case A. Therefore, the protocol presented in this thesis envisages positioning the sensor in configuration B.

Table 2. Coefficient of variation calculated in configurations A and B.

	Cv [%] Configuration A	Cv [%] Configuration B
ID 1	10.60	7.40
ID 2	3.31	4.54
ID 3	6.16	5.05
ID 4	8.58	5.03
ID 5	3.27	1.71
ID 6	7.40	5.53
ID 7	4.54	4.91
ID 8	5.05	4.61
ID 9	5.03	2.76
ID 10	1.71	0.78
Mean value	5.57	4.23

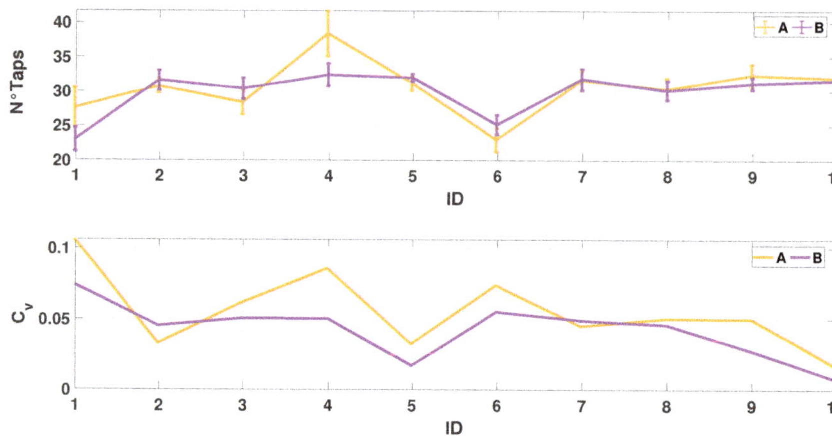

Figure 5. The top figure shows the average number of taps obtained in the two different configurations. The second plot compares the coefficient of variation calculated under conditions A and B.

4.2. Measurement Protocol Results

4.2.1. Single-Finger Analysis

The first analysis focuses on the behavior of individual fingers in the various administered tests. In particular, once the test and the finger of interest are fixed, the inter-times obtained during the test are calculated, i.e., the time difference between two consecutive peaks. A linear interpolation was performed to check fatigue once inter-times had been calculated for each participant, and the angular coefficient of the straight line was derived from this. Figure 6 shows, as an example, an interpolation of the inter-times obtained from the index finger of ID 19 during the UniALT test. In this example, the angular coefficient is 0.026 and the obtained R^2 is equal to 94%. The angular coefficients are computed for each participant and then the mean and the standard deviation are derived. These metrics were used to calculate Gaussian probability density functions (pdf) obtained from the central limit theorem, to compare the fingers distribution during the same test.

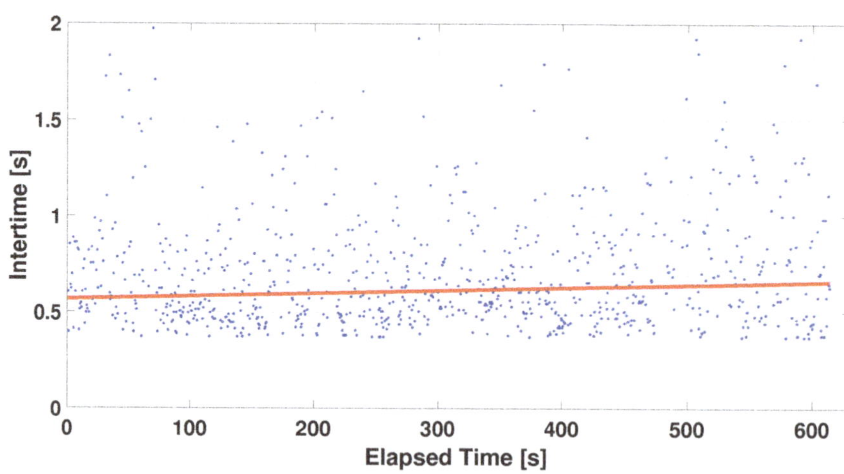

Figure 6. Example of linear inter-times fitting versus execution time—participant ID: 19; test: UniALT (index finger). The blue points are the raw intertime evaluation data, the red line is the linear fitting curve.

Table 3 shows the mean and standard deviation values for the index finger and middle finger in the case of the simultaneous and alternating uni-manual tests.

Table 3. The mean and standard deviation of the estimated line slopes of the inter-times (uni-manual test).

Test UniSIM	Mean	Dev.Std
Index	−0.006	0.010
Middle	−0.005	0.010
Test UniALT	**Mean**	**Dev.Std**
Index	0.006	0.060
Middle	0.000	0.033

In Table 4, the same parameters are calculated; but, in this case, it is performed by considering the bi-manual test.

Table 4. The mean and standard deviation of the estimated line slopes of the inter-times (bi-manual test).

Test BimSIM	Mean	Dev.Std
Index R	−0.0022	0.0075
Index L	−0.0033	0.0062
Test BimALT	**Mean**	**Dev.Std**
Index R	0.000	0.013
Index L	−0.001	0.017

This first analysis shows that, in the UniALT test, the participants showed higher difficulty in achieving the same response and rhythm between the two fingers. This is also confirmed by the results shown in Table 5: the complementary normal cumulative distribution function is considered to check whether there was fatigue or training during the exercise. In particular, the line slope can define a fatigue behavior if positive, since it reports the inter-time versus the elapsed time. Indeed, if a positive slope is experienced, this means that the inter-times are gradually increasing with respect to the test elapsed time. On the other hand, if a negative slope is observed, then the inter-times are gradually decreasing with respect to the elapsed test time; thus, the exercise produces a training effect for the participant.

Considering all participants, the line slopes define a sample from which an estimation of the probability density function of a Gaussian random variable can be carried out. Therefore, adopting the above-cited definition for fatigue or training, the probability that the random variable is greater than zero is an estimation of the portion of the sample experiencing fatigue during the exercise. On the contrary, the probability that the same variable is less than zero quantifies the participants' portion that reports a training effect.

To express it in an equation, let X be the random variable expressing the behavior of a specific test on a particular finger (e.g., UniSIM test, index finger). The sample fatigue percentage (SFP) computation can be achieved as in Equation (2).

$$SFP = \int_0^{\infty} f_X(x)dx = CCDF_X(0) \quad (2)$$

where $f_X()$ is the probability density function of the random variable X and $CCDF_X()$ expresses the complementary cumulative distribution function of X.

In the case of the UniSIM, BimSIM, and BimALT tests, the SFP is less than 0.5. This implies that most of the area of the Gaussian curves is concentrated in the negative semi-axis, corresponding the idea that one is more likely to experience a training session than they are to experience a fatigue-inducing exercise . In the case of UniALT, the difficulty of the test is once again confirmed; the SFP is above 0.5. In this case, the exercise execution, performed for both the index and middle finger, leads the participant not only to lack good coordination between the two fingers, but also to obtain a fatigue effect (as in the case shown in Table 6).

Table 5. Finger fatigue assessment: complementary normal cumulative distribution function.

	Test UniSIM	Test UniALT		Test BimSIM	Test BimALT
Index	0.29	0.54	**Index R**	0.38	0.48
Middle	0.31	0.50	**Index L**	0.30	0.47

In this case, it can be seen that this exercise, performed for both the index and middle finger, leads the participant not only to lack good coordination between the two fingers, but also to have fatigue defined by positive angular coefficients. The acceleration excursion has been evaluated to provide more information on the finger movement. Specifically, the signal was divided into 5 s windows and the average peak amplitude excursion was calculated for each. Subsequently, the mean and standard deviation of these amplitudes were derived for each participant. Figure 7 briefly shows the average of the acceleration excursions obtained in the different tests.

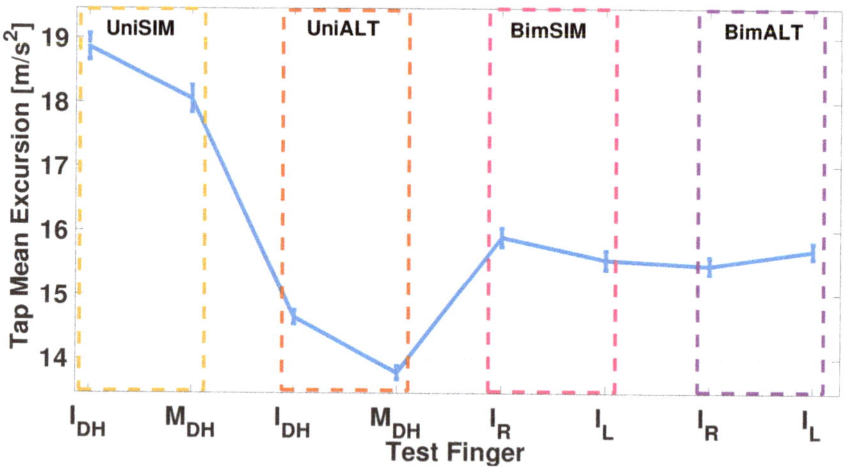

Figure 7. Mean excursion of tap movement acceleration calculated for each finger for the different case studies.

In the different scenarios, the index finger exhibits different behaviors in terms of the acceleration excursion. Table 6 shows the percentage differences between the excursions gained in the UniALT, BimSIM, and BimALT tests compared to the UniSIM test. The worst difference was obtained in the UniALT condition, with a 22% difference. In the bi-manual condition, the loss amplitude was 18% for BimALT and 15% for BimSIM.

Table 6. Percentage differences in the acceleration of the excursion of the index finger between the UniALT, BimSIM, and BimALT tests compared to the UniSIM reference test.

	UniSIM vs. UniALT	UniSIM vs. BimSIM	UniSIM vs. BimALT
Right Index	22.33%	15.70%	18.06%

4.2.2. Dual-Finger Analysis—A Simultaneous Case

To assess coordination abilities, it is necessary to perform an analysis on both fingers. Figures 8 and 9 show the error-bar of the obtained results for the uni-manual and the bi-manual tests considering simultaneous case. For the simultaneity exercises, the inter-times, in the ideal case, should be 0 s. The results show that, in the UniSIM test (considered as the reference test in this work for its easiness to accomplish), the worst case obtained an average inter-time value of 0.1 s. Starting from this experimental result, a threshold is defined to make comparisons with the other tests. In particular, for each test, the time difference between the adjacent peaks was calculated; for the uni-manual test, Figure 8 highlights how all participants kept their tapping time (simultaneity error) under 0.1 s. As the bi-manual case shows (Figure 9), the results are analogous to the single-hand situation . Particularly, a slight increase in the average value is observed, but most participants (28 out of 30) maintained their simultaneity error under 0.1 s.

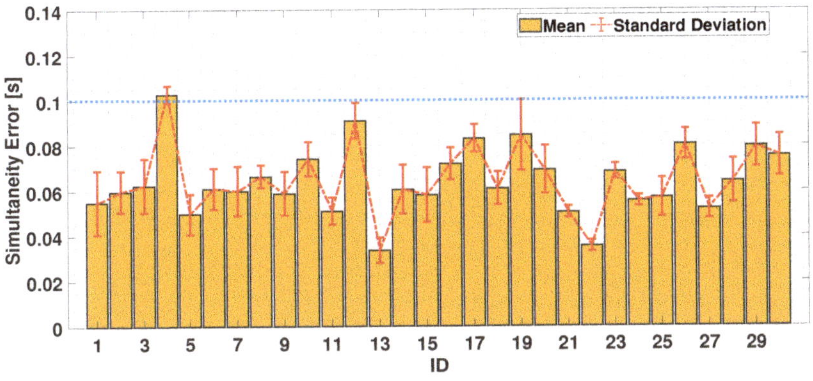

Figure 8. Bar chart of average trends of simultaneity times during UniSIM test. The horizontal dashed blue line is the experimental threshold for the simultaneity check.

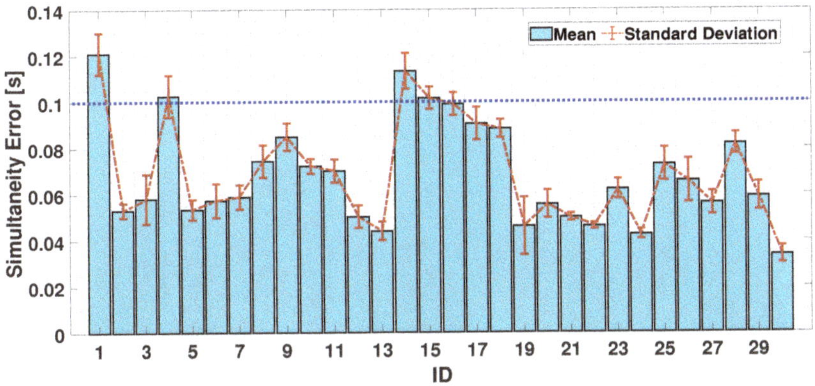

Figure 9. Bar chart of average trends of alternation and simultaneity times during BimSIM test. The horizontal dashed blue line is the experimental threshold for the simultaneity check.

A comparison of the mean and standard deviation of the UniSIM and BimSIM tests is shown in Figure10: it can be seen that the mean behavior is given by a value of 0.0734 s (UniSIM) and 0.0821 s (BimSIM), whereas the standard deviation is 0.0060 s in the UniSIM case and 0.0058 s in BimSIM case.

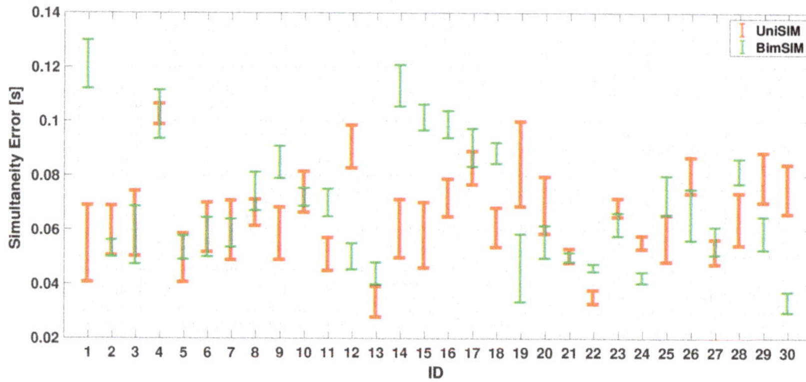

Figure 10. Comparison between the mean and standard deviation of UniSIM and BimSIM tests.

4.2.3. Dual-Finger Analysis—An Alternate Case

The alternating finger movement case is now considered. In Figure 11, it can be seen that the alternate tapping time is bounded in [0.02, 0.3]; most participants (21 out of 30) exceeded a tapping time higher than 0.1 s (the exercise could be considered as correctly executed if a participant has an increased value of inter-tap time, which exceeds the upper limit of the interval (mean $\pm$ standard deviation) obtained in the simultaneous case). In the bi-manual exercise, participants are required to perform the same movements, but using the index fingers belonging to both hands. The coordination could change as the movement involves the left and right sides of the body; the cerebral stimulus needed to achieve this may differ. It can be seen from Figure 12 that 28 out of 30 participants achieved an alternating tap time of more than 0.1 s, with a reduction in standard deviation compared to the uni-manual case. It can be seen that, in this case, the ability to achieve an alternating tap movement is more evident. The described finding is a significant result, as most participants clearly follow the required task and provide significantly higher tapping times in cases of alternate movement.

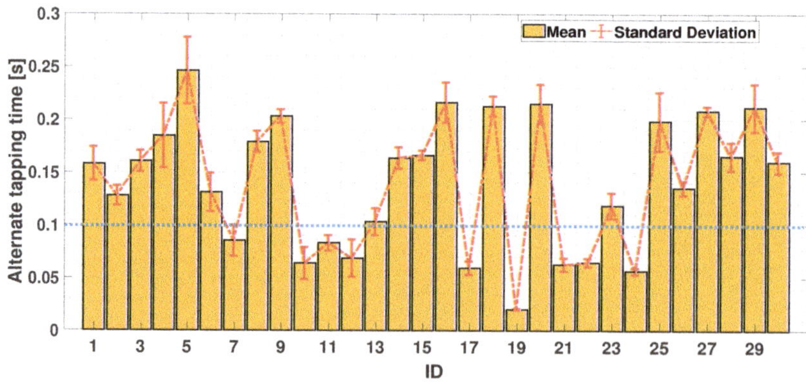

Figure 11. Bar chart of average trends of simultaneity times during UniALT test. The horizontal dashed blue line is the experimental threshold for the simultaneity check.

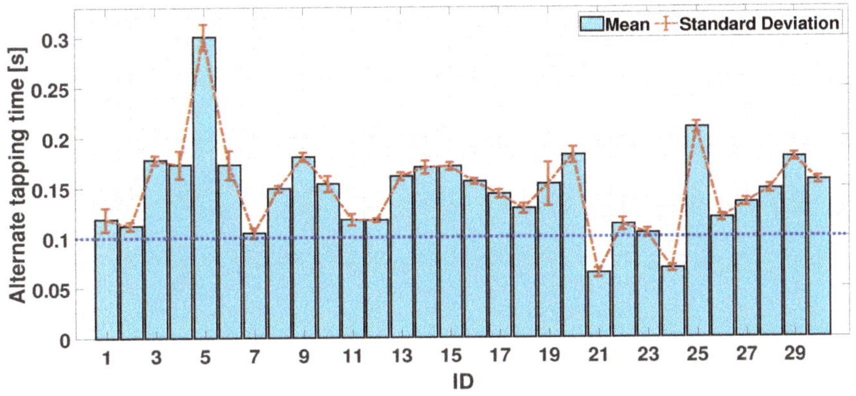

Figure 12. Bar chart of average trends of alternation and simultaneity times during BimALT test. The horizontal dashed blue line is the experimental threshold for the simultaneity check.

Finally, Figure 13 report a comparison of the uni-manual and bi-manual alternate cases. This gave a mean of 0.167s with a standard deviation of 0.012 s (UniALT) and a mean of 0.1457 s with a standard deviation of 0.0053 s (BimALT).

As shown above, the UniALT exercise proves to be more complex, with 33% of the participants failing to complete the task correctly, whereas in the BimAlt case, only 6% failed to complete the task correctly.

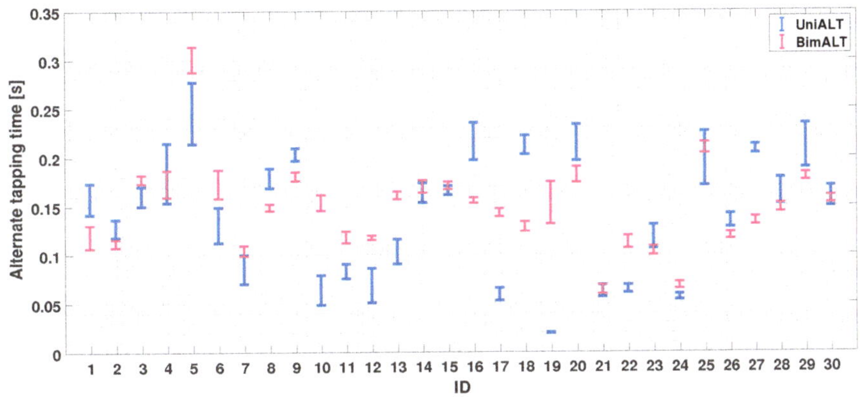

Figure 13. Comparison between the means and standard deviations of UniALT and BimALT tests.

5. Discussion

The performance of a simple test, such as the TT, has often been associated with the relationship between manual dexterity and the brain's ability to adapt and learn, as indicated in the literature [41]. This measurement methodology can be helpful in assessing movements involving oculo–manual abilities, as confirmed by studies highlighting individuals' ability to learn and adapt [42]. Information management software could be a potential tool to provide predictive information or indications of the early stages of a neurodegenerative disease [43]. This could be paramount, as knowing the predictors and intervening in the early stages can reduce intervention costs by using a platform that provides such information. It highlights the essential role of brain plasticity in refining manual dexterity through practice and learning, using the TT as an investigative tool [44]. The test results provide valuable insights into participants' fingers coordination abilities and motor dexterity in different scenarios. From the results obtained from the single-hand test, we could see that the UniALT test presented more difficulty in terms of performance, causing a more significant loss in performance acceleration (Figure 7). This may be related to the demand for the performance of a more difficult task in the UniALT test. This difficulty could be attributed to brain resource allocation, where the index finger tends to be dominant due to its superior motor and sensory control [45] This dominance creates difficulty in achieving accurate coordination. The index finger, therefore, served as a pacesetter in both simultaneous and alternating movements. Table 5 shows a more in-depth analysis of the measurements, focusing on fatigue dynamics between the index and middle fingers. In the UniALT case, the index finger tends to experience a more significant fatiguing effect than the middle finger. From the graph analysis, clear disparities emerge in the evolution of fatigue between the index and middle finger within the considered task. The observed differences might be influenced by various factors, including the movement of the fingers or the specific nature of the actions carried out by each of the two fingers [46].

In a further analysis, in both the UniSIM and UniALT cases, the participants failed to coordinate the finger movements; in fact, it is possible to observe a difference in the balance in the results. In Figure 7, we can observe similar balance differences in the two uni-manual tests. Despite the coordination between the index and middle finger shows similar values in both the UniSIM and UniALT tests, a notable difference in the average outcomes is observed between the two tests. As previously assumed, there is a greater demand for work in the UniALT test as it involves a significant amount of engagement of the motor processes that are required to perform the movements as required [47]. On the other hand, the results of the BimSIM and BimALT tests could highlight the brain's ability to differentiate and coordinate index finger movements when tasks require simultaneous and alternating movements [48]. Indeed, although coordination of two hands is required in this test, the participants achieved good synchronization of the index finger movements of both hands and in both TT variants, as shown in Figure 7. A possible explanation behind these findings could lie in the adaptive nature and plasticity of the human brain areas [49]. The brain learn and adapts new abilities through experience and training, and this adaptability may be particularly pronounced in regions involved in motor control, including the corpus callosum [50]. This structure plays a crucial role in bi-manual coordination, enabling effective communication between the hemispheres and facilitating the transfer of sensory information and motor commands necessary for hand coordination [51]. In Figures 10 and 13, after analyzing the results of both simultaneous and alternating tests, it is noticeable that, in simultaneous exercises, a majority of participants were successful in performing the exercise correctly in both the UniSIM and BimSIM modes. However, in the case of alternating exercises, there was more variability among subjects in UniALT, with 10 out of 30 participants performing the exercise as if it were simultaneous. On the other hand, in BimALT, most participants achieved an average alternation time of 0.15 s.

One hypothesis could be related to the effect of regular practice of bi-manual tasks, such as those presented in BimSIM and BimALT, justifying greater motor task efficiency [52]. In addition, the human brain can separate activities between the two hands so that each hand can focus on its specific task, which may help improve the synchronization of movements. This activity division could reflect the brain's ability to process sensory information, enabling better management of different tasks [53]. This phenomenon can be explained through functional and neuroplastic brain characteristics, the communication between the hemispheres, and the division of activities between the hands, which enable the good synchronization of bi-manual movements [54]. Understanding these mechanisms can inform the design of interventions to refine finger coordination, particularly in contexts where flawless motor precision is needed, such as rehabilitation programs and sports training regimens [55]. It also holds promise for developing technologies to improve manual accuracy, benefiting individuals in certain professions, such as musicians and surgeons. Moreover, this simple test, supplemented by the accurate analysis of measured vestments, could be a helpful tool for predicting or monitoring the course of neurodegenerative diseases.

6. Conclusions

An overall architecture exploiting the capabilities of an automatic measurement protocol, IMU-related quantities, and suitable data processing techniques has been presented in the context of sports-related objective performance evaluation. The approach could be implemented through more in-depth studies and testing as a valuable tool to provide objective data on motor signals that characterize neurodegenerative diseases. Several tapping tests have been performed (UniSIM, BimSIM, UniALT, and BimALT) to compare and analyze the effects on fatigue and hand coordination. This study found that the simultaneous tapping test produced better coordination and perceived fatigue results than other exercises. These findings could serve as a foundation for future research into coordinative abilities. The next step will be to adapt this measurement protocol to assess the motor abilities of individuals with neurodegenerative diseases. This further campaign will require some customizations of the software part to better characterize the pathological part, but the whole setup will remain unchanged, demonstrating the generalization property of the developed system. In addition, it could serve as an early intervention; it could ensure prevention and promise substantial cost savings for national healthcare systems by identifying the early stages of the disease [56]. On the other hand, the use of inertial sensors will provide objective support for movement measurement to improve even classical clinical evaluations.

Author Contributions: Conceptualization, G.C., T.D.L., C.C., L.F., A.M. and A.R.; methodology, T.D.L., C.C. and G.C.; software, C.C. and G.C.; validation, C.C. and G.C.; formal analysis, T.D.L., C.C. and G.C.; investigation, T.D.L., C.C. and G.C.; data curation, T.D.L., C.C. and G.C.; writing—original draft preparation, T.D.L., C.C. and G.C.; writing—review and editing, all authors; visualization, L.F., A.M. and A.R.; supervision, L.F., A.M., A.R., G.M. and A.M.A.; project administration, L.F., A.M. and A.R. All authors have read and agreed to the published version of the manuscript.

Funding: This research received no external funding.

Institutional Review Board Statement: This work was approved by the Institutional Review Board of the University of Cassino and Southern Lazio (no. 24777.2022.12.12).

Informed Consent Statement: The informed consent and approval on benefits and risks are derived from the 1964 Declaration of Helsinki for research on humans. Written informed consent has been obtained from the participants to publish this paper.

Data Availability Statement: Data are contained within the article.

Conflicts of Interest: The authors declare no conflicts of interest.

References

1. Stănciulescu, R. Dexterity, Mobility and Flexibility, Essential Components in the Development of the Military Students' Motor Skills. *Int. Conf. Knowl.-Based Organ.* **2023**, *29*, 242–246. [CrossRef]
2. Stergiou, N.; Decker, L.M. Human movement variability, nonlinear dynamics, and pathology: Is there a connection? *Hum. Mov. Sci.* **2011**, *30*, 869–888. [CrossRef] [PubMed]
3. Chen, W.; Xiong, C.; Huang, X.; Sun, R.; Xiong, Y. Kinematic analysis and dexterity evaluation of upper extremity in activities of daily living. *Gait Posture* **2010**, *32*, 475–481. [PubMed]
4. Alamri, A.; Eid, M.; Iglesias, R.; Shirmohammadi, S.; El Saddik, A. Haptic virtual rehabilitation exercises for poststroke diagnosis. *IEEE Trans. Instrum. Meas.* **2008**, *57*, 1876–1884. [CrossRef]
5. Morris, M.E.; Bilney, B.; Miller, K.J.; Nijkrake, M.J.; Munneke, M.; McGinley, J.L. Hand dexterity assessment in Parkinson's disease: Construct validity of the 9-Hole peg test for the more affected hand. *Disabil. Rehabil.* **2021**, *43*, 3834–3838.
6. Voola, S.I.; Hema, S. Effectiveness of Tele Rehabilitation on Manual Dexterity and its Impact on Quality of Life in Patients with Parkinson's Disease: A Pilot Study. *Indian J. Physiother. Occup. Ther.* **2020**, *14*. 186–193. [CrossRef]
7. Reed, J.C.; Reed, H.B. The halstead—Reitan neuropsychological battery. In *Contemporary Approaches to Neuropsychological Assessment*; Springer: Boston, MA, USA, 1997; pp. 93–129.
8. Axelrod, B.N.; Meyers, J.E.; Davis, J.J. Finger tapping test performance as a measure of performance validity. *Clin. Neuropsychol.* **2014**, *28*, 876–888. [CrossRef] [PubMed]
9. Nguyen, K.D.; Pathirana, P.N.; Horne, M.; Power, L.; Szmulewicz, D.J. Entropy-based analysis of rhythmic tapping for the quantitative assessment of cerebellar ataxia. *Biomed. Signal Process. Control* **2020**, *59*, 101916. [CrossRef]

10. Suzumura, S.; Kanada, Y.; Osawa, A.; Sugioka, J.; Maeda, N.; Nagahama, T.; Shiramoto, K.; Kuno, K.; Kizuka, S.; Sano, Y.; et al. Assessment of finger motor function that reflects the severity of cognitive function. *Fujita Med. J.* **2021**, *7*, 122–129.
11. Suzumura, S.; Osawa, A.; Maeda, N.; Sano, Y.; Kandori, A.; Mizuguchi, T.; Yin, Y.; Kondo, I. Differences among patients with Alzheimer's disease, older adults with mild cognitive impairment and healthy older adults in finger dexterity. *Geriatr. Gerontol. Int.* **2018**, *18*, 907–914. [CrossRef]
12. Waqar, A.; Ahmad, I.; Habibi, D.; Hart, N.; Phung, Q.V. Enhancing Athlete Tracking Using Data Fusion in Wearable Technologies. *IEEE Trans. Instrum. Meas.* **2021**, *70*, 4004013. [CrossRef]
13. Talitckii, A.; Kovalenko, E.; Shcherbak, A.; Anikina, A.; Bril, E.; Zimniakova, O.; Semenov, M.; Dylov, D.V.; Somov, A. Comparative Study of Wearable Sensors, Video, and Handwriting to Detect Parkinson's Disease. *IEEE Trans. Instrum. Meas.* **2022**, *71*, 2509910. [CrossRef]
14. Carissimo, C.; Cerro, G.; Di Libero, T.; Ferrigno, L.; Marino, A.; Rodio, A. Objective evaluation of coordinative abilities and training effectiveness in sports scenarios: An automated measurement protocol. *IEEE Access* **2023**, *11*, 76996–77008. [CrossRef]
15. Di Libero, T.; Carissimo, C.; Cerro, G.; Abbatecola, A.M.; Marino, A.; Miele, G.; Ferrigno, L.; Rodio, A. Motor abilities analysis using a standardized tapping test enhanced by a detailed processing stage: Gender and age comparison. In Proceedings of the 2023 IEEE International Symposium on Medical Measurements and Applications (MeMeA), Jeju, Republic of Korea, 14–16 June 2023; pp. 1–6. [CrossRef]
16. Di Libero, T.; Carissimo, C.; Zagaglia, A.; Cerro, G.; Ferrigno, L.; Rodio, A. Assessment of coordinative abilities through upper extremity wearable device technology. In Proceedings of the 2022 IEEE International Workshop on Sport, Technology and Research, STAR 2022-Proceedings, Trento, Italy, 6–8 July 2022; pp. 175–179. [CrossRef]
17. Schmitt, L. Finger-tapping test. In *Encyclopedia of Autism Spectrum Disorders*; Springer: Cham, Switzerland, 2021; p. 2030.
18. Mitrushina, M.; Boone, K.B.; Razani, J.; D'Elia, L.F. *Handbook of Normative Data for Neuropsychological Assessment*; Oxford University Press: Oxford, UK, 2005.
19. Akram, N.; Li, H.; Ben-Joseph, A.; Budu, C.; Gallagher, D.A.; Bestwick, J.P.; Schrag, A.; Noyce, A.J.; Simonet, C. Developing and assessing a new web-based tapping test for measuring distal movement in Parkinson's disease: A Distal Finger Tapping test. *Sci. Rep.* **2022**, *12*, 386. [PubMed]
20. Coleman, A.R.; Moberg, P.J. Comparison of the Halstead-Reitan and infrared light beam finger tappers. *Assessment* **1997**, *4*, 277–286. [CrossRef] [PubMed]
21. Guo, Z.; Zeng, W.; Yu, T.; Xu, Y.; Xiao, Y.; Cao, X.; Cao, Z. Vision-based finger tapping test in patients with Parkinson's disease via spatial-temporal 3D hand pose estimation. *IEEE J. Biomed. Health Inform.* **2022**, *26*, 3848–3859. [CrossRef] [PubMed]
22. Chettouf, S.; Rueda-Delgado, L.M.; de Vries, R.; Ritter, P.; Daffertshofer, A. Are unimanual movements bilateral? *Neurosci. Biobehav. Rev.* **2020**, *113*, 39–50. [CrossRef] [PubMed]
23. Koeneke, S.; Lutz, K.; Wüstenberg, T.; Jäncke, L. Bimanual versus unimanual coordination: What makes the difference? *Neuroimage* **2004**, *22*, 1336–1350. [CrossRef] [PubMed]
24. Georgopoulos, A.P.; Carpenter, A.F. Coding of movements in the motor cortex. *Curr. Opin. Neurobiol.* **2015**, *33*, 34–39. [CrossRef]
25. Di Libero, T.; Langiano, E.; Carissimo, C.; Ferrara, M.; Diotaiuti, P.; Rodio, A. Technological support for people with Parkinson's disease: A narrative review. *J. Gerontol. Geriatr.* **2023**, *71*, 87–101. [CrossRef]
26. Serrien, D.J. Coordination constraints during bimanual versus unimanual performance conditions. *Neuropsychologia* **2008**, *46*, 419–425. [CrossRef]
27. Lang, C.E.; Schieber, M.H. Differential impairment of individuated finger movements in humans after damage to the motor cortex or the corticospinal tract. *J. Neurophysiol.* **2003**, *90*, 1160–1170. [CrossRef]
28. Sainburg, R.L.; Schaefer, S.Y. Interlimb differences in control of movement extent. *J. Neurophysiol.* **2004**, *92*, 1374–1383. [CrossRef]
29. Shim, J.K.; Lay, B.S.; Zatsiorsky, V.M.; Latash, M.L. Age-related changes in finger coordination in static prehension tasks. *J. Appl. Physiol.* **2004**, *97*, 213–224. [CrossRef]
30. Zhao, M.; Marino, M.; Samogin, J.; Swinnen, S.P.; Mantini, D. Hand, foot and lip representations in primary sensorimotor cortex: A high-density electroencephalography study. *Sci. Rep.* **2019**, *9*, 19464. [CrossRef]
31. Yokoi, A.; Arbuckle, S.A.; Diedrichsen, J. The role of human primary motor cortex in the production of skilled finger sequences. *J. Neurosci.* **2018**, *38*, 1430–1442. [CrossRef]
32. Schieber, M.H.; Santello, M. Hand function: Peripheral and central constraints on performance. *J. Appl. Physiol.* **2004**, *96*, 2293–2300. [CrossRef]
33. Wiegel, P.; Kurz, A.; Leukel, C. Evidence that distinct human primary motor cortex circuits control discrete and rhythmic movements. *J. Physiol.* **2020**, *598*, 1235–1251. [CrossRef]
34. François-Brosseau, F.E.; Martinu, K.; Strafella, A.P.; Petrides, M.; Simard, F.; Monchi, O. Basal ganglia and frontal involvement in self-generated and externally-triggered finger movements in the dominant and non-dominant hand. *Eur. J. Neurosci.* **2009**, *29*, 1277–1286. [CrossRef]
35. Kornhuber, H.H.; Deecke, L. Brain potential changes in voluntary and passive movements in humans: Readiness potential and reafferent potentials. *Pflügers Arch.-Eur. J. Physiol.* **2016**, *468*, 1115–1124. [CrossRef]
36. Van Vugt, F.T.; Altenmüller, E. On the one hand or on the other: Trade-off in timing precision in bimanual musical scale playing. *Adv. Cogn. Psychol.* **2019**, *15*, 216. [CrossRef]

37. Saito, Y.; Maezawa, T.; Kawahara, J.I. Beat patterns determine inter-hand differences in synchronization error in a bimanual coordination tapping task. *i-Perception* **2021**, *12*, 20416695211053882. [CrossRef]
38. Sisti, H.M.; Beebe, A.; Bishop, M.; Gabrielsson, E. A brief review of motor imagery and bimanual coordination. *Front. Hum. Neurosci.* **2022**, *16*, 1037410. [CrossRef]
39. Movella. Xsens Products. Available online: https://www.movella.com/products/xsens (accessed on 24 August 2023).
40. MATLAB^{TM}. Find Abrupt Changes. Available online: https://it.mathworks.com/help/signal/ref/findchangepts.html (accessed on 1 March 2024).
41. Mawase, F.; Uehara, S.; Bastian, A.J.; Celnik, P. Motor learning enhances use-dependent plasticity. *J. Neurosci.* **2017**, *37*, 2673–2685. [CrossRef]
42. Mollà-Casanova, S.; Llorens, R.; Borrego, A.; Salinas-Martínez, B.; Serra-Añó, P. Validity, reliability, and sensitivity to motor impairment severity of a multi-touch app designed to assess hand mobility, coordination, and function after stroke. *J. Neuroeng. Rehabil.* **2021**, *18*, 70. [CrossRef]
43. Roalf, D.R.; Rupert, P.; Mechanic-Hamilton, D.; Brennan, L.; Duda, J.E.; Weintraub, D.; Trojanowski, J.Q.; Wolk, D.; Moberg, P.J. Quantitative assessment of finger tapping characteristics in mild cognitive impairment, Alzheimer's disease, and Parkinson's disease. *J. Neurol.* **2018**, *265*, 1365–1375. [CrossRef]
44. Lövdén, M.; Garzón, B.; Lindenberger, U. Human skill learning: Expansion, exploration, selection, and refinement. *Curr. Opin. Behav. Sci.* **2020**, *36*, 163–168. [CrossRef]
45. Crichton, P. Penfield's homunculus. *J. Neurol. Neurosurg. Psychiatry* **1994**, *57*, 525; Erratum in: *J. Neurol. Neurosurg. Psychiatry* **1994**, *57*, 772. [CrossRef]
46. Makai-Bölöni, S.; Thijssen, E.; van Brummelen, E.M.; Groeneveld, G.J.; Doll, R.J. Touchscreen-based finger tapping: Repeatability and configuration effects on tapping performance. *PLoS ONE* **2021**, *16*, e0260783. [CrossRef]
47. Naito, E.; Morita, T.; Asada, M. Importance of the primary motor cortex in development of human hand/finger dexterity. *Cereb. Cortex Commun.* **2020**, *1*, tgaa085. [CrossRef]
48. Zhou, G.; Chen, Y.; Wang, X.; Wei, H.; Huang, Q.; Li, L. The correlations between kinematic profiles and cerebral hemodynamics suggest changes of motor coordination in single and bilateral finger movement. *Front. Hum. Neurosci.* **2022**, *16*, 957364. [CrossRef]
49. Afreen, S.; Mazhar, K.; Malik, R.; Asif, I. Role of Neuroplasticity in Neurorehabilitation. *Asia Pac. J. Allied Health Sci. Vol.* **2021**, *4*, 46–51.
50. Machan, T.; Krupps, K. The Neuroplastic Adaptation Trident Model: A suggested novel framework for ACL rehabilitation. *Int. J. Sport. Phys. Ther.* **2021**, *16*, 896. [CrossRef]
51. Pauwels, L.; Gooijers, J. The Role of the Corpus Callosum (Micro) Structure in Bimanual Coordination: A Literature Review Update. *J. Mot. Behav.* **2023**, *55*, 525–537. [CrossRef]
52. Tomita, Y.; Tanaka, S.; Takahashi, S.; Takeuchi, N. Detecting cognitive decline in community-dwelling older adults using simple cognitive and motor performance tests. *Geriatr. Gerontol. Int.* **2020**, *20*, 212–217. [CrossRef]
53. Carey, J.R.; Bhatt, E.; Nagpal, A. Neuroplasticity promoted by task complexity. *Exerc. Sport Sci. Rev.* **2005**, *33*, 24–31.
54. Ivanenko, Y.P.; Cappellini, G.; Solopova, I.; Grishin, A.; MacLellan, M.J.; Poppele, R.; Lacquaniti, F. Plasticity and modular control of locomotor patterns in neurological disorders with motor deficits. *Front. Comput. Neurosci.* **2013**, *7*, 123. [CrossRef]
55. Yue, Z.; Zhang, X.; Wang, J. Hand rehabilitation robotics on poststroke motor recovery. *Behav. Neurol.* **2017**, *2017*, 3908135. [CrossRef]
56. Kotseva, K.; De Backer, G.; De Bacquer, D.; Rydén, L.; Hoes, A.; Grobbee, D.; Maggioni, A.; Marques-Vidal, P.; Jennings, C.; Abreu, A.; et al. Primary prevention efforts are poorly developed in people at high cardiovascular risk: A report from the European Society of Cardiology EURObservational Research Programme EUROASPIRE V survey in 16 European countries. *Eur. J. Prev. Cardiol.* **2021**, *28*, 370–379.

Disclaimer/Publisher's Note: The statements, opinions and data contained in all publications are solely those of the individual author(s) and contributor(s) and not of MDPI and/or the editor(s). MDPI and/or the editor(s) disclaim responsibility for any injury to people or property resulting from any ideas, methods, instructions or products referred to in the content.

sensors

MDPI

Article

Implementation of Sound Direction Detection and Mixed Source Separation in Embedded Systems

Jian-Hong Wang [1,†], Phuong Thi Le [2,†], Weng-Sheng Bee [3], Wenny Ramadha Putri [3], Ming-Hsiang Su [4,*], Kuo-Chen Li [5,*], Shih-Lun Chen [6], Ji-Long He [1], Tuan Pham [7], Yung-Hui Li [8] and Jia-Ching Wang [3]

1 School of Computer Science and Technology, Shandong University of Technology, Zibo 255000, China; jhwang@sdut.edu.cn (J.-H.W.); 23505020692@stumail.sdut.edu.cn (J.-L.H.)
2 Department of Computer Science and Information Engineering, Fu Jen Catholic University, New Taipei City 242062, Taiwan; ptle@csie.fju.edu.tw
3 Department of Computer Science and Information Engineering, National Central University, Taoyuan City 320314, Taiwan; wengshengbee2@hotmail.com (W.-S.B.); wenny@g.ncu.edu.tw (W.R.P.); jcw@csie.ncu.edu.tw (J.-C.W.)
4 Department of Data Science, Soochow University, Taipei City 10048, Taiwan
5 Department of Information Management, Chung Yuan Christian University, Taoyuan City 320317, Taiwan
6 Department of Electronic Engineering, Chung Yuan Christian University, Taoyuan City 320314, Taiwan; chrischen@cycu.edu.tw
7 Faculty of Digital Technology, University of Technology and Education—University of Đà Nẵng, Danang 550000, Vietnam; ptuan@ute.udn.vn
8 AI Research Center, Hon Hai Research Institute, New Taipei City 207236, Taiwan; yunghui.li@foxconn.com
* Correspondence: huntfox.su@gmail.com (M.-H.S.); kuochen@cycu.edu.tw (K.-C.L.)
† These authors contributed equally to this work.

Abstract: In recent years, embedded system technologies and products for sensor networks and wearable devices used for monitoring people's activities and health have become the focus of the global IT industry. In order to enhance the speech recognition capabilities of wearable devices, this article discusses the implementation of audio positioning and enhancement in embedded systems using embedded algorithms for direction detection and mixed source separation. The two algorithms are implemented using different embedded systems: direction detection developed using TI TMS320C6713 DSK and mixed source separation developed using Raspberry Pi 2. For mixed source separation, in the first experiment, the average signal-to-interference ratio (SIR) at 1 m and 2 m distances was 16.72 and 15.76, respectively. In the second experiment, when evaluated using speech recognition, the algorithm improved speech recognition accuracy to 95%.

Keywords: embedded systems; position detection; hybrid sound source separation; signal-to-interference ratio (SIR); speech recognition

1. Introduction

In recent years, embedded system technology and products have become the focus of the global IT industry. As people pursue a more convenient and comfortable lifestyle, the information industry and smart homes are booming. Embedded systems are increasingly integrated into our daily lives in various forms, such as sensor networks and wearable devices for monitoring people's activities and health. Although many people still do not fully understand what embedded systems are, they are closely related to our daily lives and have already permeated various fields, such as home applications [1–7], wireless communications [4,8], network applications [9,10], medical devices [4,11,12], consumer electronics, etc. Embedded systems encompass many applications, including smart homes, gaming consoles, electronic stethoscopes, automated teller machines (ATMs), and car-mounted Global Positioning Systems (GPSs).

This paper discusses the implementation of audio localization and enhancement in embedded systems, focusing on embedded algorithms for direction detection and mixed sound source separation. These two algorithms are implemented using different embedded systems: the TI TMS320C6713 DSK [13–20] for direction detection development and the Raspberry Pi 2 [21–25] for mixed sound source separation. The objective is to develop audio localization and noise reduction techniques applicable to intelligent living to bring convenience and comfort to users' lives.

Direction detection entails capturing audio from a microphone array and determining the direction of the sound source through a specialized algorithm. Azimuth detection is utilized for audio tracking, with the TDE method [26] employed for direction detection. By utilizing Cross-Power Spectral Density (XPSD) based on the Generalized Cross-Correlation with Phase Transform (GCC-PHAT) estimate and detecting the peak of cross-correlation, we can accurately identify the azimuthal relationship between the sound signal and the microphone array.

The current research on direction detection, domestically and internationally, can be divided into two categories. The first category utilizes beamforming or subspace theory in conjunction with microphone arrays to determine the angle of the sound source. The most widely used method in this category is Multiple Signal Classification (MUSIC) [27]. The second category employs Time Delay of Arrival (TDOA) to estimate the angle of the sound source based on the time delay between the arrival of the sound at different microphones [28,29]. Among these, the Generalized Cross-Correlation (GCC) method proposed by Kanpp and Carter [26] is considered one of the most common TDOA methods. The first research category requires prior measurement of the impulse frequency response corresponding to each microphone in the array, resulting in significant computational complexity. Considering the real-time nature of the system, this paper will adopt the Generalized Cross-Correlation PHAT [26,30] method.

Mixed sound source separation involves extracting multiple individual source signals from the mixed signal captured by the microphone. Since the inception of this problem, it has garnered significant attention from researchers. We aim to extract the desired sounds embedded within the observed signal using mixed sound source separation techniques. Blind source separation of mixed signals can be classified in two ways based on the type of mixing model: the instantaneous mixing model [31,32] and the convolution mixing model [33–36]. Our method is as follows [37]. First, the Fourier transform transfers the received mixed signal to the frequency domain, and then features are extracted and input into K-Means to cluster the signal, where the k value is set to 2, as there are two mixed sound sources. Next, the mixed signal is subjected to Binary Masking to reconstruct the source in the frequency domain. The signal is finally converted back to the time domain using the inverse Fourier transform. This paper primarily explores the convolution mixing model as it finds application in real-world environments.

Due to the limited processing power and memory capacity of embedded system processors compared to PCs, efficient utilization of memory, computational resources, and program storage space become critical in the embedded system environment. Therefore, we need to further optimize the computational load in the algorithm and streamline the code to ensure smooth execution in embedded systems. Code Composer Studio (CCS) is an integrated development environment developed by TI. It provides an optimized compiler that compiles program code into efficient executable programs. CCS also provides a real-time operating system, DSP/BIOS, which can provide simple and effective management of programs. This study will utilize CCS to optimize the computational load and ensure smooth execution in embedded systems. To enhance the speech recognition capabilities of wearable devices, the contribution of this article is to realize audio positioning and enhancement in embedded systems, use embedded algorithms to perform direction detection and mixed source separation, and ultimately increase the speech recognition accuracy, further improving the practicality of wearable devices.

2. Embedded System Design for Direction Detection

2.1. Algorithm Flow and Overview

Firstly, we conduct voice activity detection (VAD) preprocessing on the received audio signal to identify segments containing speech [38]. Subsequently, we utilize the spectral subtraction method [39,40] to remove noise from the audio. Finally, the denoised audio is forwarded to the DOA (Direction of Arrival) recognizer for direction detection [26,30]. Figure 1 illustrates the architecture of the embedded system proposed in this paper for direction detection.

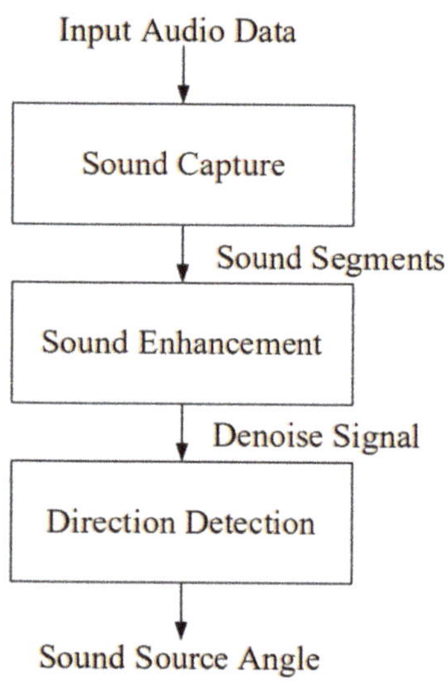

Figure 1. Architecture diagram of the embedded system for direction detection.

2.1.1. Voice Activity Detection (VAD)

We use conventional energy-based Voice Activity Detection (VAD) [38] to extract sound events. Let $x_t(n)$ represent the received audio, where t indicates the audio frame and n ranges from 1 to N samples. M_t denotes the average value of audio x_t. E_t is determined by a threshold value, T, resulting in either A = 1 for sound events or A = 0 for non-sound events. Since different microphones may have different threshold values, it is necessary to conduct testing to determine the appropriate threshold value.

$$E_t = \sum_{n=1}^{N} (x_t(n) - \mu_t)^2 \quad (1)$$

$$A = \begin{cases} 1, & E_t > T \\ 0, & \text{otherwise} \end{cases} \quad (2)$$

2.1.2. Sound Enhancement—Spectral Subtraction

Sound enhancement is achieved using the spectral subtraction method [39,40]. The benefit of the spectral subtraction method over some machine learning-based sound enhancement methods [41–43] is its lower computational complexity. This method involves subtracting the averaged noise spectrum from the spectrum of the noisy signal to eliminate environmental noise. The averaged noise spectrum is obtained from the signals received during non-sound events.

If the noise, $n(k)$, of one audio frame is added to the original signal, $s(k)$, of the same audio frame, resulting in a noisy signal $s(k)$ for that frame, we have the following equation:

$$y(k) = s(k) + n(k) \tag{3}$$

After performing the Fourier transform, we obtain the following:

$$Y\left(e^{j\omega}\right) = S\left(e^{j\omega}\right) + N\left(e^{j\omega}\right) \tag{4}$$

The general formula for the spectral subtraction method is as follows:

$$\begin{gathered} \left|S_{S(e^{j\omega})}\right|^2 = \left|Y\left(e^{j\omega}\right)\right|^2 - \alpha\left|\mu\left(e^{j\omega}\right)\right|^2, \\ \text{if } \left|Y\left(e^{j\omega}\right)\right|^2 > \alpha\left|\mu\left(e^{j\omega}\right)\right|^2 \beta\left|Y\left(e^{j\omega}\right)\right|^2,\ \left|\mu\left(e^{j\omega}\right)\right|^2 = E\left\{\left|N\left(e^{j\omega}\right)\right|^2\right\} \end{gathered} \tag{5}$$

where $\mu\left(e^{j\omega}\right)$ represents the average noise spectrum, α lies between 0 and 1, and β is either 0 or a minimum.

After subtracting the spectral energy, we obtain the denoised signal spectrum $\hat{S}\left(e^{j\omega}\right)$. $\theta_Y\left(e^{j\omega}\right)$ represents the phase of $Y\left(e^{j\omega}\right)$.

$$\hat{S}\left(e^{j\omega}\right) = \left|S_{S(e^{j\omega})}\right| e^{j\theta_Y\left(e^{j\omega}\right)} \tag{6}$$

Alternatively, by obtaining the ratio $H\left(e^{j\omega}\right)$ of the energy-subtracted spectrum $\left|S_{S(e^{j\omega})}\right|^2$ to the spectrum of the noisy signal $\left|Y\left(e^{j\omega}\right)\right|^2$, we multiply it with $Y\left(e^{j\omega}\right)$ to obtain the denoised signal spectrum $\hat{S}\left(e^{j\omega}\right)$.

$$\hat{S}\left(e^{j\omega}\right) = H\left(e^{j\omega}\right) Y\left(e^{j\omega}\right) \tag{7}$$

$$H\left(e^{j\omega}\right) = \frac{\left|S_{S(e^{j\omega})}\right|^2}{\left|Y\left(e^{j\omega}\right)\right|^2} \tag{8}$$

2.1.3. Direction Detection—TDE-to-DOA Method

We referred to related papers that utilize the GCC-PHAT [26,30] estimation for the XPSD (Cross-Power Spectral Density) and the peak detection of cross-correlation for direction detection using the Time Delay Estimation (TDE) method. In addition, we also refer to the research of Varma et al. [44], which uses cross-correlation-based time delay estimates (TDE) for direction-of-arrival (DOA) estimation of acoustic arrays in less reverberant environments. The TDE method determines the direction of a single sound source, and multiple sound sources cannot be differentiated simultaneously. However, its advantage lies in its simplicity, as it only requires two microphones and has a relatively straightforward hardware architecture, making it suitable for real-time applications.

Firstly, we assume the presence of a sound source in the space. Under ideal conditions, the signals received by the two microphones can be represented as follows:

$$x_1(t) = s_1(t) + n_1(t) \tag{9}$$

$$x_2(t) = \alpha s_1(t - D) + n_2(t) \tag{10}$$

$s_1(t)$ represents the sound source; $x_1(t)$ and $x_2(t)$ represent the signals received by the two microphones. $n_1(t)$ and $n_2(t)$ are the noises present.

We assume $s_1(t)$, $n_1(t)$, and $n_2(t)$ to be wide-sense stationary (WSS) and $s_1(t)$ and $n_1(t)$, as well as $n_2(t)$, to be uncorrelated. Here, D represents the actual delay, and α represents the scale value for changing the magnitude. Furthermore, the changes in D and α are slow, and at this stage, the cross-correlation between the microphones can be expressed as follows:

$$R_{x1,x2}(\tau) = E[x_1(t)x_2(t-\tau)] \quad (11)$$

where E represents the expectation value, and τ, which maximizes Equation (11), is the time delay between the two microphones. Since the actual observation time is finite, the estimation of cross-correlation can be expressed as follows:

$$\hat{R}_{x1,x2}(\tau) = \frac{1}{T-\tau}\int_{\tau}^{T} x_1(t)x_2(t-\tau)dt \quad (12)$$

where T represents the observation time interval, and the relationship between cross-correlation and cross-power spectrum can be expressed in the following Fourier representation:

$$R_{x1,x2}(\tau) = \int_{-\infty}^{\infty} G_{x1,x2}(f)e^{j2\pi f\tau}df \quad (13)$$

Now, let us consider the actual state of the physical space, where the sound signals received by the microphones undergo spatial transformations. Therefore, the actual cross-power spectrum between the microphones can be represented as follows:

$$G_{y1,y2}(f) = H_1(f)H_2^*(f)G_{x1,x2}(f) \quad (14)$$

where $H_1(f)$ and $H_2(f)$ represent the spatial transformation functions from the sound source to the first microphone and the second microphone, respectively. Therefore, we define the generalized correlation between the microphones as follows:

$$R_{x1,x2}^{(g)}(\tau) = \int_{-\infty}^{\infty} \Psi_g(f)G_{x1,x2}(f)e^{j2\pi f\tau}df \quad (15)$$

wherein

$$\Psi_g(f) = H_1(f)H_2^*(f) \quad (16)$$

In practice, due to the limited observation time, we can only use the estimated $\hat{G}_{x1,x2}(f)$ instead of $G_{x1,x2}(f)$. Therefore, Equation (16) is rewritten as follows:

$$\hat{R}_{x1,x2}^{(p)}(\tau) = \int_{-\infty}^{\infty} \Psi_p(f)\hat{G}_{x1,x2}(f)e^{j2\pi f\tau}df \quad (17)$$

Using Equation (17), we can estimate the time delay between the microphones. The choice of $\psi_p(f)$ also has an impact on the estimation of time delay. In this paper, we employ the PHAT (Phase Transform) method proposed by Carter et al. [30], which can be expressed as follows:

$$\Psi_p(f) = \frac{1}{|G_{x1,x2}(f)|} \quad (18)$$

This method works remarkably well when the noise distributions between the two microphones are independent. By employing the aforementioned approach, we can accurately detect the azimuth relationship between our sound signal and the microphones.

2.2. Embedded System Hardware Devices

The embedded system for azimuth detection in this study utilizes the TI TMS320C6713 DSK [16–23,37] as the development platform, as shown in Figure 2. In the following sections, we provide a detailed introduction to the specification for the TI TMS320C6713 DSK, which is divided into three parts: peripheral equipment, DSP core, and multi-channel audio input expansion card.

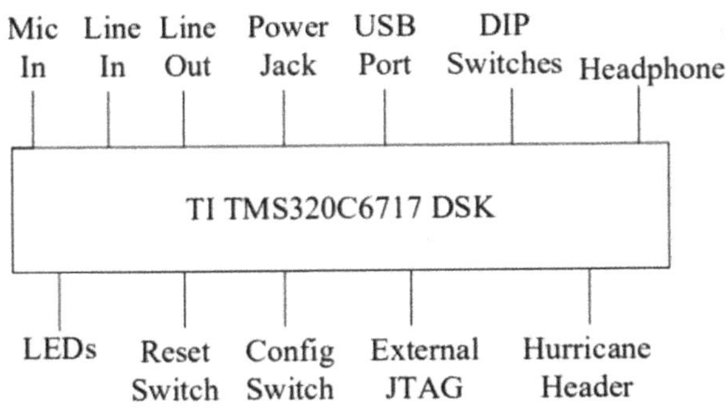

Figure 2. The physical image of the TI TMS320C6713 DSK.

3. Design of Embedded System for Separation of Mixed Audio Sources

3.1. Algorithm Flow and Introduction

We begin by applying the blind source separation algorithm [37] to the received audio signal for separating the mixed sources. Then, we upload the separated signals to the Google Speech API for recognition. Figure 3 illustrates the architecture diagram of the embedded system, proposed in this paper, for mixed source separation.

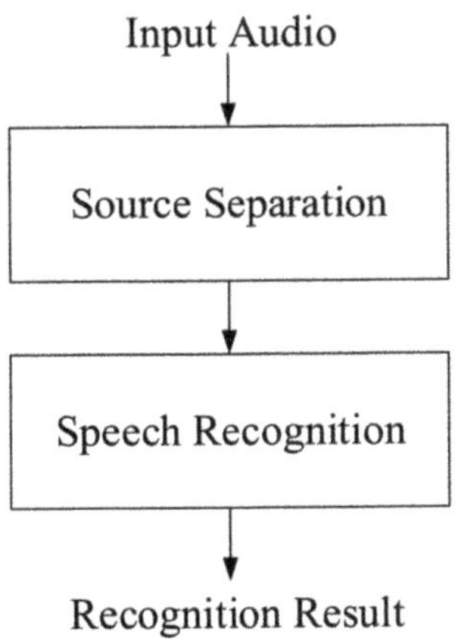

Figure 3. Architecture diagram of a hybrid audio source embedded system.

Hybrid Audio Source Separation

We set up the system with two receivers (microphones). Initially, the received mixed signal is transformed from the time domain to the frequency domain using Fourier transform to leverage its sparsity for further processing. Then, feature extraction is applied to the transformed signal and input into K-Means for clustering. During the clustering process, corresponding masks are generated, and a Binary Mask is adopted for implementation. Subsequently, Binary Masking is applied to the mixed signal to reconstruct the source signals in the frequency domain. Finally, the signals are transformed back to the time domain using the inverse Fourier transform. Figure 4 illustrates the flowchart of the hybrid audio source separation algorithm [37].

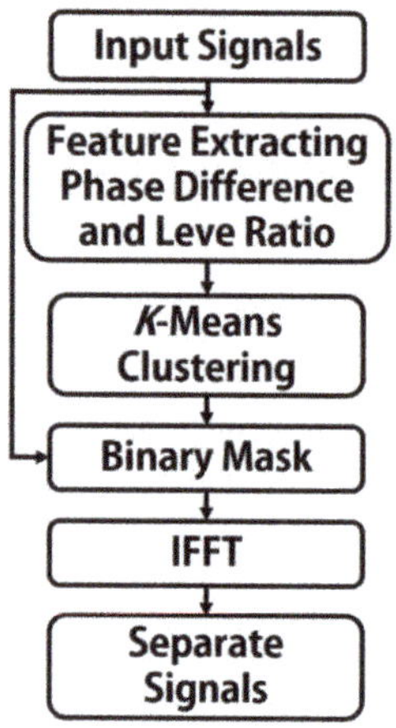

Figure 4. Flowchart of the hybrid audio source separation algorithm.

3.2. Embedded System Hardware Devices

In this paper, Raspberry Pi 2 [21–25] is the development platform for embedding mixed audio sources. The physical diagram is depicted in Figure 5. Here, we will provide a detailed introduction to the specifications for both the Raspberry Pi 2 and the Cirrus Logic Audio Card [45,46] audio module.

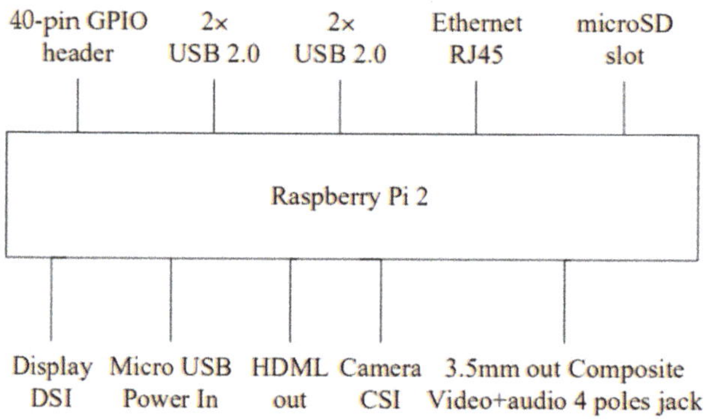

Figure 5. The physical image of the Raspberry Pi 2.

Cirrus Logic Audio Card Audio Module

Figure 6 illustrates the Cirrus Logic Audio Card [45,46], an audio expansion board designed for the Raspberry Pi. Compatible with the Raspberry Pi models A+ and B+, it features a 40-pin GPIO interface that seamlessly connects to the Raspberry Pi's 40-pin GPIO Header. The card supports high-definition audio (HD Audio) and incorporates two digital micro-electromechanical microphones (DMICs) and D-class power amplifiers for the direct driving of external speakers. The analog signals include line-level input/output and headphone output/headphone microphone input, while the digital signals encompass stereo headphone audio input/output (SPDIF). Moreover, it includes an Expansion Header, enabling connections to devices beyond the Raspberry Pi. Figure 7 depicts the Raspberry Pi 2 connected to the Cirrus Logic Audio Card.

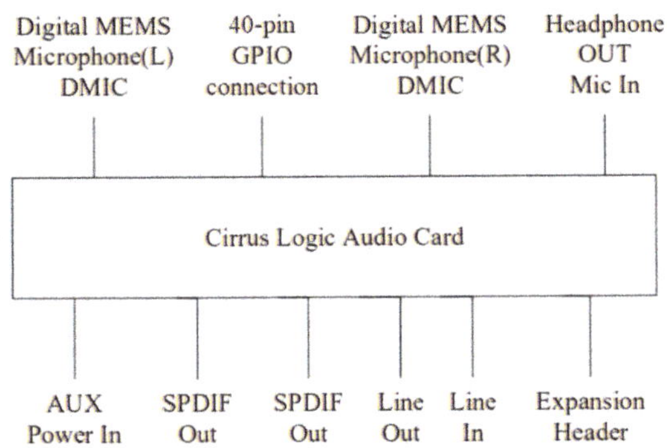

Figure 6. The physical image of the Cirrus Logic Audio Card.

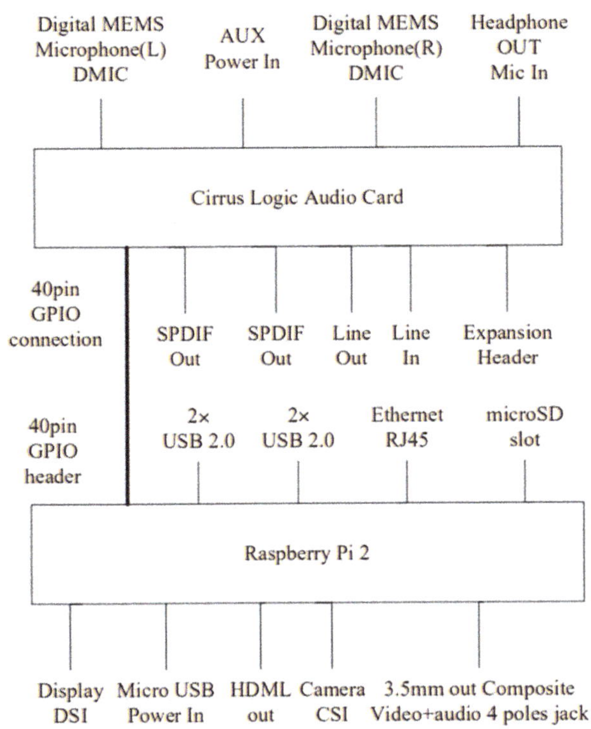

Figure 7. Connection between Raspberry Pi 2 and Cirrus Logic Audio Card.

4. Experimental Results

4.1. Direction Detection of the Embedded System

4.1.1. Experimental Environment Setup

We utilized a classroom measuring 15 m × 8.5 m × 3 m for the experiment. Four omnidirectional microphones were strategically placed within the classroom. The sound source was positioned 2 m away from the center of the microphone array. To assess the azimuth detection capability, we tested 18 angles, ranging from 5° to 175°, with a 10° interval between each test angle (see Table 1 for details). Figure 8 depicts the setup for the azimuth detection experiment.

Table 1. Orientation detection under different environment settings.

Parameters	Value
angle	5°, 15°, 25°, 35°, 45°, 55°, 65°, 75°, 85°, 95°, 105°, 115°, 125°, 135°, 145°, 155°, 165°, 175°
microphone distance	0.22 m
distance between source and microphone center	2 m
sampling frequency	16 kHz

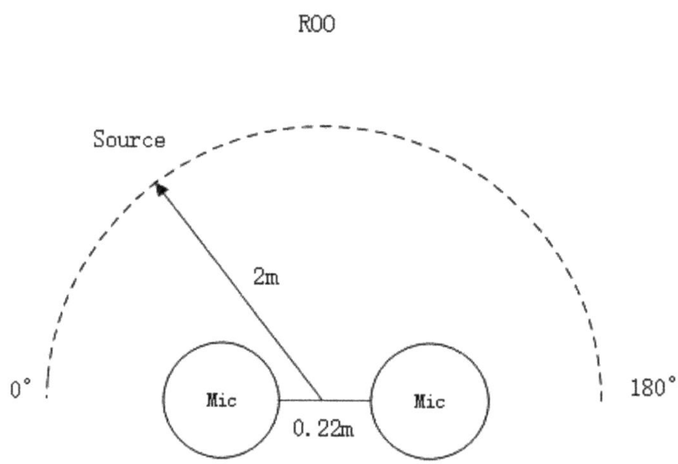

Figure 8. Setup for azimuth detection experiment.

4.1.2. Experimental Environment Equipment

We utilized the CM503N omnidirectional microphone (depicted in Figure 9). For the equipment setup, the microphone was initially connected to the phantom power supply, and then the phantom power supply was connected to DSK AUDIO 4.

CM503N	
Type	Capacitive Audio Head
Frequency Range	40 to 18,000Hz
Directionality	Omnidirectivity
Sensitivity	-60±3dB×(1mV)×0dB=1V/μbar
Attenuation Switch	-10db
Low-Frequency Attenuation	80Hz
Output Impedance	220Ω
Maximum Sound Pressure 1% THD	140dB
S/N Ratio	68dB
Current Loss	≤3mA
Power Supply	9-52VDC Phantom Power Supply
Output Connecto	XLR(M) Type (Power Module)

Figure 9. Microphone CM503N.

4.1.3. Experimental Results

The development version of the functionality was completed, and the measured angles yielded satisfactory results, with errors within 10 degrees (refer to Table 2). However, due to the limited memory and processor speed of the development version, achieving real-time measurements is currently not feasible. To address this limitation, a compromise was made by allocating the longest execution time program segments to the smaller internal memory, which offers the fastest execution speed. Meanwhile, the remaining parts were stored in the external memory. This approach ensures a reasonably fast execution speed. Figure 10 depicts the experimental scenario of direction detection.

Table 2. Test results for direction detection.

Parameters	Value								
actual angle measured angle	5° 8.53°	15° 8.53°	25° 19.13°	35° 28.54°	45° 41.77°	55° 52.17°	65° 64.12°	75° 75.60°	85° 89.68°
actual angle measured angle	95° 100.50°	105° 107.69°	115° 118.72°	125° 131.11°	135° 141.17°	145° 152.81°	155° 165.21°	165° 168.54°	175° 180.00°

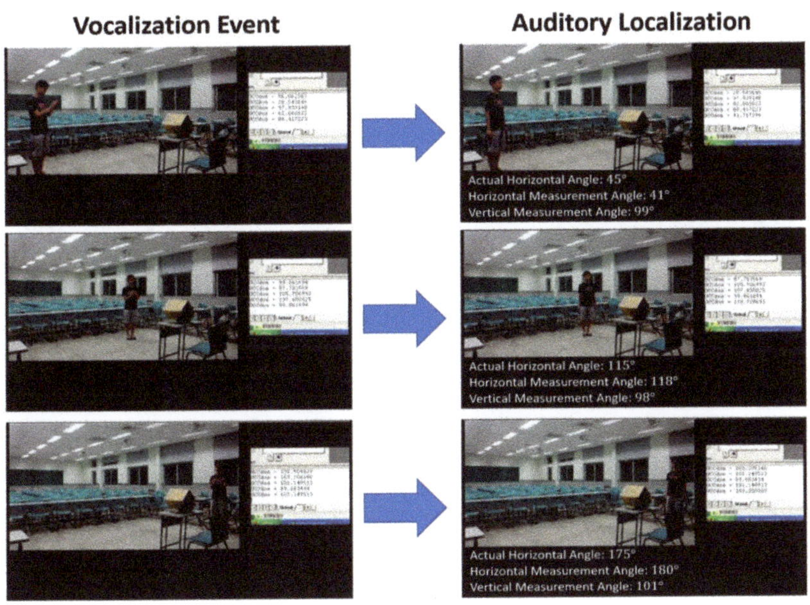

Figure 10. Experimental scenario of direction detection.

Based on the current execution results, most of the sound events that surpassed the threshold value during angle measurement achieved an error within 10 degrees. However, there were occasional instances where either no angle measurement was obtained for detected sound events or the measured result was close to 90 degrees. We speculate that the former is attributable to the development board being engaged in other tasks at the time of emitting sound. This resulted in no audio data being captured, thus classifying the signal as silent when determining threshold value passage. As for the latter, we infer that the emitted sound surpassed the threshold value, but it was either too soft and was overshadowed by noise, or the captured sound was too limited, leading to noise being mistakenly identified as a measurable sound event.

4.2. Embedded System for Mixed Sound Source Separation

4.2.1. Experimental Environment Setup

Experimental Setup 1: We utilized a classroom with dimensions of 5.5 m × 4.8 m × 3 m for the experiment. The microphone setup included the two built-in, omnidirectional MEMS microphones from the Cirrus Logic Audio Card, with a distance of 0.058 m between them. The sound sources, denoted as S1 and S2, were positioned at distances of 1 m and 2 m, respectively, from the center of the microphones. S1 was a male speaker, and S2 was a female speaker. Figure 11 illustrates the environment for the mixed sound source separation experiment, and Table 3 provides the setup details for the mixed sound source separation environment.

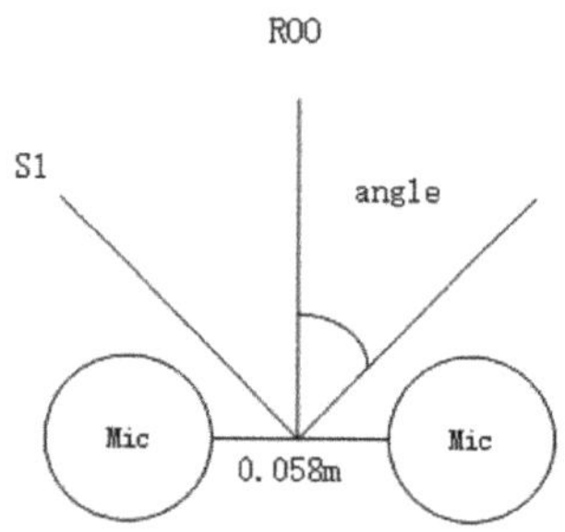

Figure 11. Experimental Environment 1 for mixed sound source separation.

Table 3. Details for mixed sound source separation in Environment 1.

Parameters	Value
angle	30°, 60°, 90°
microphone distance	58 mm
distance between source and microphone center	1 m, 2 m
sampling frequency	8 kHz

Experimental Setup 2: We utilized a classroom with dimensions of 5.5 m × 4.8 m × 3 m for the experiment. The microphone setup included the two built-in, omnidirectional MEMS microphones from the Cirrus Logic Audio Card, with a distance of 0.058 m between them. The speaker was positioned at a distance of 0.2 m from the center of the microphones, while the interfering sound source (a news broadcast) was positioned at a distance of 1 m from the center of the microphones. Figure 12 illustrates the environment for the mixed sound source separation experiment, and Table 4 provides the setup details for the mixed sound source separation environment.

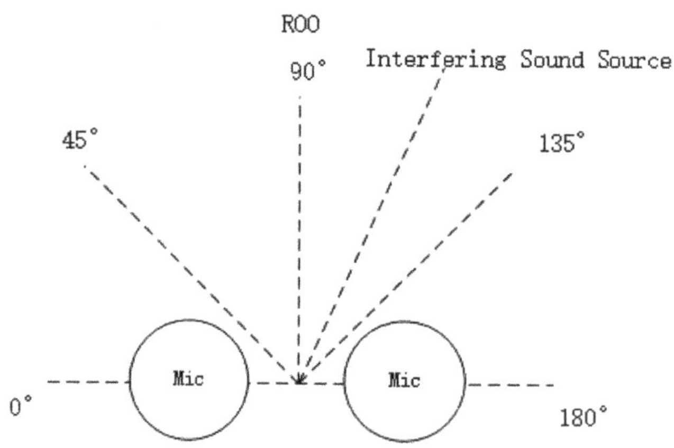

Figure 12. Experimental Environment 2 for mixed sound source separation.

Table 4. Setup Details for mixed sound source separation in Environment 2.

Parameters	Value
speaker angle	0°, 45°, 90°, 135°, 180°
microphone distance	58 mm
distance between speaker and microphone center	0.2 m
distance between interference source and microphone center	1 m
sampling frequency	8 kHz

4.2.2. Experimental Environment Equipment

We utilized a Raspberry Pi 7-inch touch screen (Figure 13) for display purposes, which could be connected to the Raspberry Pi 2 using DSI as the output. Additionally, we utilized the USB-N10 wireless network card for internet access and Google speech recognition.

Figure 13. Raspberry Pi 2, Cirrus Logic Audio Card, and 7-inch touch screen.

4.2.3. Experimental Results

For Experimental Setup 1, the signal-to-interference ratio (SIR) served as the performance evaluation metric. The formula for the SIR is as follows:

$$SIR = 10log_{10}\frac{\left\|y_{qtarget}\right\|}{\left\|e_{qinterf}\right\|^2} \quad (19)$$

where $y_{qtarget}$ represents the components of the source signal in the separated signal, and e_{qinterf} refers to the remaining interference components in the separated signal. Tables 5 and 6 present the SIR obtained at different distances.

Table 5. SIR for sound source at 1 m.

Parameters	Value							Average
sound source angle	30°, 30°	30°, 60°	30°, 90°	60°, 30°	60°, 60°	60°, 90°	90°, 90°	
SIR	17.35	17.13	15.64	17.87	17.21	15.76	16.04	16.72

Table 6. SIR for sound source at 2 m.

Parameters	Value						Average
sound source angle	30°, 30°	30°, 60°	60°, 30°	60°, 60°	60°, 90°	90°, 90°	
SIR	16.68	16.01	18.76	14.74	15.61	15.74	15.76

The sound source angles (30°, 30°) correspond to 90°, with S1 shifted to the left by 30° and S2 shifted to the right by 30°. Figure 14 displays the mixed signal for left and right channels at a distance of 1 m, while Figure 15 showcases the separated signal after mixed sound source separation at the same distance. Similarly, Figure 16 exhibits the mixed signal for left and right channels at a distance of 2 m, followed by Figure 17 demonstrating the separated signal after mixed sound source separation at the same distance.

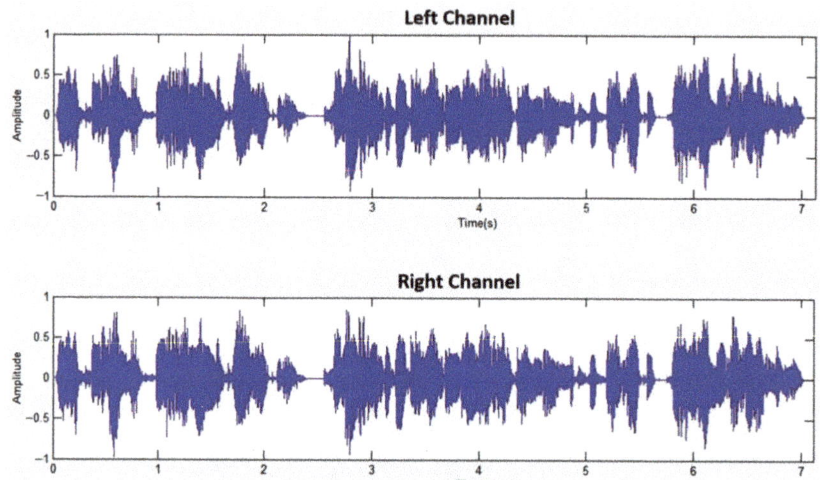

Figure 14. Mixed signal for left and right channels (1 m).

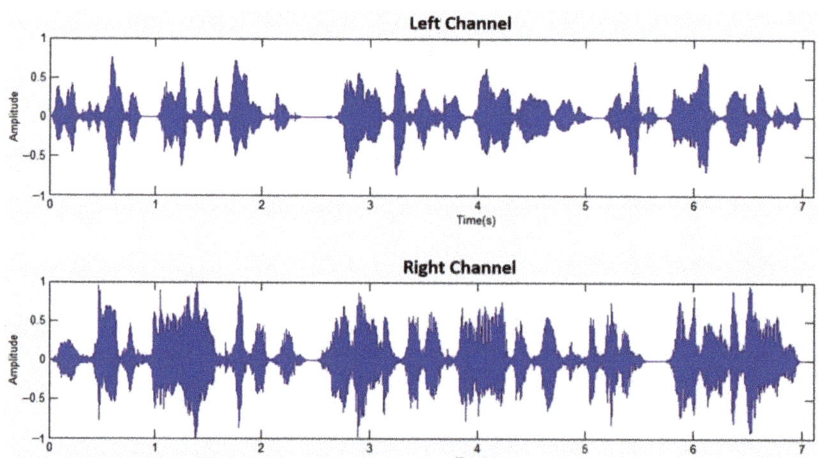

Figure 15. Separated signal after mixed sound source separation (1 m).

For Experimental Setup 2, we utilized the free Speech API provided by Google to evaluate the performance of the algorithm. We tested the algorithm using 20 common commands typically used in a general and simple smart home environment, such as "Open the window," "Weather forecast," "Turn off the lights," "Increase the volume," "Stock market status," and so on.

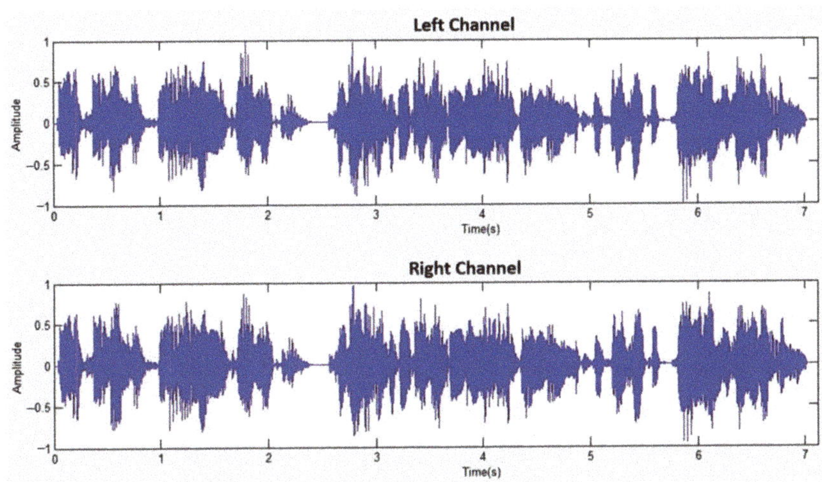

Figure 16. Mixed signal for left and right channels (2 m).

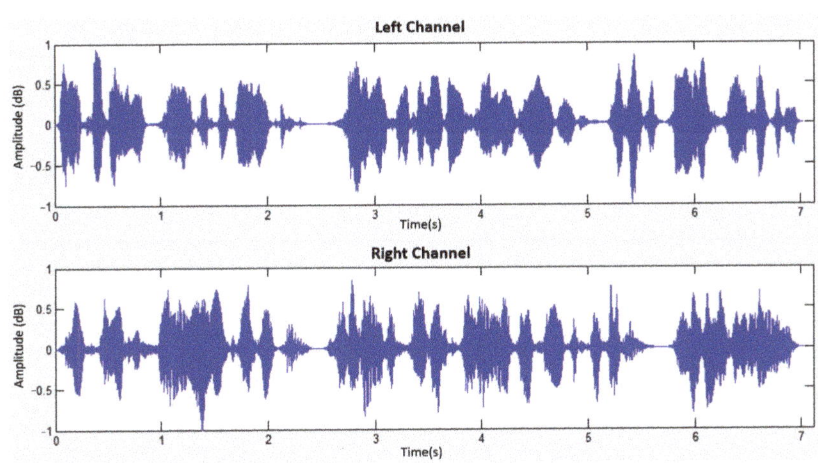

Figure 17. Separated signal after mixed sound source separation (2 m).

From Table 7, we can observe that the recognition accuracy of the separated signals is lower than that of the mixed signals. However, the recognition accuracy of the separated signals and the mixed signals does not completely overlap. Therefore, by combining the mixed signals and the separated signals (as shown in Figure 18) in a fusion set, we achieved a recognition accuracy of 95%.

Table 7. Speech recognition accuracy for mixed sound source separation.

Signal	Speech Recognition Accuracy
mixed signal	90%
separated signal	83%
mixed signal and separated signal	**95%**

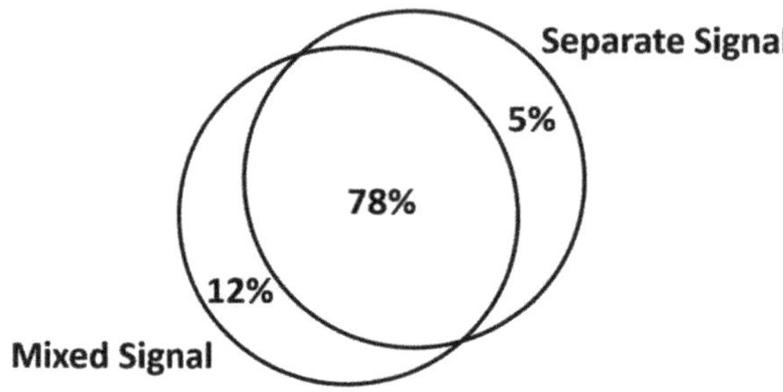

Figure 18. Speech recognition accuracy.

5. Conclusions and Future Research Directions

On the one hand, this study implemented the orientation detection system on the TI TMS320C6713 DSK development board, and on the other hand, it implemented the hybrid sound source separation system on the Raspberry Pi 2 development board. The experimental results show that the method proposed in this study is better than the hybrid signals, and separated mixed signals enhance the speech recognition capabilities of embedded systems in sensor networks and wearable devices suitable for people's activity and health monitoring. Firstly, we successfully implemented a direction detection system on the TI TMS320C6713 DSK development board. Secondly, a mixed sound source separation system was implemented on the Raspberry Pi 2 development board.

The received audio signal underwent preprocessing in the direction detection system using voice activity detection (VAD) to identify speech segments. Spectral subtraction was then applied to denoise the noisy audio. Finally, the denoised audio was passed to the DOA (Direction of Arrival) estimator for direction angle detection.

We designed a system with two microphones in the mixed sound source separation algorithm. The received mixed signal was initially transformed from the time domain to the frequency domain using the Fourier transform to exploit its sparsity. Then, feature extraction was performed and the signal was input into the K-Means algorithm for clustering. During the clustering process, a corresponding mask was generated. Here, we used binary masks for separation. The binary-masked mixed signal was then used to reconstruct the source signals in the frequency domain, which was subsequently transformed back to the time domain using the inverse Fourier transform to obtain the separated audio.

In future research, our goal is to further enhance and optimize the algorithm on the embedded board, reduce the computational load of the embedded system, and improve the embedded system's real-time performance. We also plan to enhance the separation signal quality of the mixed sound source separation algorithm to enhance speech recognition accuracy. In addition, we will try various situation settings and set sound sources at different distances to evaluate the system performance more comprehensively.

Author Contributions: Conceptualization, J.-C.W.; Methodology, W.-S.B.; writing—original draft preparation, W.-S.B.; writing—review and editing, J.-H.W., P.T.L., W.R.P., M.-H.S., K.-C.L., S.-L.C., J.-L.H., T.P. and Y.-H.L. All authors have read and agreed to the published version of the manuscript.

Funding: This research received no external funding.

Institutional Review Board Statement: Not applicable.

Informed Consent Statement: Not applicable.

Data Availability Statement: The data presented in this study are openly available.

Conflicts of Interest: The authors declare no conflicts of interest.

References

1. Wang, J.-C.; Lee, Y.-S.; Lin, C.-H.; Siahaan, E.; Yang, C.-H. Robust environmental sound recognition with fast noise suppression for home automation. *IEEE Trans. Autom. Sci. Eng.* **2015**, *12*, 1235–1242. [CrossRef]
2. Wang, J.-C.; Lee, H.-P.; Wang, J.-F.; Lin, C.-B. Robust environmental sound recognition for home automation. *IEEE Trans. Autom. Sci. Eng.* **2008**, *5*, 25–31. [CrossRef]
3. Lu, J.; Peng, Z.; Yang, S.; Ma, Y.; Wang, R.; Pang, Z.; Feng, X.; Chen, Y.; Cao, Y. A review of sensory interactions between autonomous vehicles and drivers. *J. Syst. Archit.* **2023**, *141*, 102932. [CrossRef]
4. Abdulhameed, Z.N. Design and Implement a Wireless Temperature Monitoring System using Noncontact IR Sensor Based on Arduino. In Proceedings of the 2023 IEEE International Conference on ICT in Business Industry & Government (ICTBIG), Indore, India, 8–9 December 2023; IEEE: New York, NY, USA, 2023; pp. 1–5.
5. Wang, J.-C.; Chin, Y.-H.; Hsieh, W.-C.; Lin, C.-H.; Chen, Y.-R.; Siahaan, E. Speaker identification with whispered speech for the access control system. *IEEE Trans. Autom. Sci. Eng.* **2015**, *12*, 1191–1199. [CrossRef]
6. Karunaratna, S.; Maduranga, P. Artificial Intelligence on Single Board Computers: An Experiment on Sound Event Classification. In Proceedings of the 2021 5th SLAAI International Conference on Artificial Intelligence (SLAAI-ICAI), Colombo, Sri Lanka, 6–7 December 2021; IEEE: New York, NY, USA, 2021; pp. 1–5.
7. Ahmad, M.; Amin, M.B.; Hussain, S.; Kang, B.H.; Cheong, T.; Lee, S. Health fog: A novel framework for health and wellness applications. *J. Supercomput.* **2016**, *72*, 3677–3695. [CrossRef]
8. Liu, X.; Zhang, M.; Xiong, T.; Richardson, A.G.; Lucas, T.H.; Chin, P.S.; Etienne-Cummings, R.; Tran, T.D.; Van der Spiegel, J. A fully integrated wireless compressed sensing neural signal acquisition system for chronic recording and brain machine interface. *IEEE Trans. Biomed. Circ. Syst.* **2016**, *10*, 874–883. [CrossRef] [PubMed]
9. Guo, Z.; Liu, Y.; Lu, F. Embedded remote monitoring system based on NBIOT. *J. Phys. Conf. Ser.* **2022**, *2384*, 012038. [CrossRef]
10. Liu, W.; Xu, W.; Mu, X.; Cheng, J. Design of Smart Home Control System Based on Embedded System. In Proceedings of the 2020 IEEE 9th Joint International Information Technology and Artificial Intelligence Conference (ITAIC), Shanghai, China, 6–8 November 2020; IEEE: New York, NY, USA, 2020; pp. 2209–2213.
11. Sehatbakhsh, N.; Alam, M.; Nazari, A.; Zajic, A.; Prvulovic, M. Syndrome: Spectral Analysis for Anomaly Detection on Medical IOT and Embedded Devices. In Proceedings of the 2018 IEEE International Symposium on Hardware Oriented Security and Trust (HOST), Washington, DC, USA, 30 April–4 May 2018; IEEE: New York, NY, USA, 2018; pp. 1–8.
12. Zhou, W.; Liao, J.; Li, B.; Li, J. A Family Medical Monitoring System Based on Embedded uC/OS-II and GPRS. In Proceedings of the 2012 IEEE International Conference on Information and Automation, Shenyang, China, 6–8 June 2012; IEEE: New York, NY, USA, 2012; pp. 663–667.
13. Ilmi, M.; Huda, M.; Rahardhita, W. Automatic Control Music Amplifier Using Speech Signal Utilizing by TMS320C6713. In Proceedings of the 2015 International Electronics Symposium (IES), Surabaya, Indonesia, 29–30 September 2015; IEEE: New York, NY, USA, 2015; pp. 163–166.
14. Manikandan, J.; Venkataramani, B.; Girish, K.; Karthic, H.; Siddharth, V. Hardware Implementation of Real-Time Speech Recognition System Using TMS320C6713 DSP. In Proceedings of the 2011 24th International Conference on VLSI Design, Chennai, India, 2–7 January 2011; IEEE: New York, NY, USA, 2011; pp. 250–255.
15. Mustafa, N.B.A.; Gandi, S.; Sharrif, Z.A.M.; Ahmed, S.K. Real-Time Implementation of a Fuzzy Inference System for Banana Grading Using DSP TMS320C6713 Platform. In Proceedings of the 2010 IEEE Student Conference on Research and Development (SCOReD), Kuala Lumpur, Malaysia, 13–14 December 2010; IEEE: New York, NY, USA, 2010; pp. 324–328.
16. Singh, J.; Singh, H.P.; Singh, S. Implementation of FIR Interpolation Filter on TMS320C6713 for VOIP Analysis. In Proceedings of the 2010 2nd International Conference on Computational Intelligence, Communication Systems and Networks, Liverpool, UK, 28–30 July 2010; IEEE: New York, NY, USA, 2010; pp. 289–294.
17. Texas Instruments. SPRS186L-TMS320C6713 Floating-Point Digital Signal Processing; December 2001. Revised November 2005; Texas Instruments, Texas, USA. 2005. Available online: https://media.digikey.com/pdf/Data%20Sheets/Texas%20Instruments%20PDFs/TMS320C6713.pdf (accessed on 20 June 2024).
18. Spectrum Digital. TMS320C6713 DSK Technical Reference; November 2003. Spectrum Digital, Texas, USA. 2003. Available online: https://electrical.engineering.unt.edu/sites/default/files/Spectrum_Digital_TMS320C6713_User_Guide.pdf (accessed on 20 June 2024).
19. Texas Instruments. Code Composer Studio Development Tools v3. 3 Getting Started Guide; SPRU509H; May 2008. Texas Instruments, Texas, USA. 2008. Available online: https://spinlab.wpi.edu/courses/ece4703_2010/spru509h.pdf (accessed on 20 June 2024).
20. Texas Instruments. Code Composer Studio User's Guide; Texas Instruments Literature Number SPRU328B. Texas Instruments, Texas, USA. 2000. Available online: https://software-dl.ti.com/ccs/esd/documents/users_guide/index.html (accessed on 20 June 2024).
21. Gupta, M.S.D.; Patchava, V.; Menezes, V. Healthcare Based on IOT Using Raspberry Pi. In Proceedings of the 2015 International Conference on Green Computing and Internet of Things (ICGCIoT), Greater Noida, India, 8–10 October 2015; IEEE: New York, NY, USA, 2015; pp. 796–799.

22. Hossain, N.; Kabir, M.T.; Rahman, T.R.; Hossen, M.S.; Salauddin, F. A Real-Time Surveillance Mini-Rover Based on OpenCV-Python-JAVA Using Raspberry Pi 2. In Proceedings of the 2015 IEEE International Conference on Control System, Computing and Engineering (ICCSCE), Penang, Malaysia, 27–29 November 2015; IEEE: New York, NY, USA, 2015; pp. 476–481.
23. Sandeep, V.; Gopal, K.L.; Naveen, S.; Amudhan, A.; Kumar, L.S. Globally Accessible Machine Automation Using Raspberry Pi based on Internet of Things. In Proceedings of the 2015 International Conference on Advances in Computing, Communications and Informatics (ICACCI), Kochi, India, 10–13 August 2015; IEEE: New York, NY, USA, 2015; pp. 1144–1147.
24. Paul, S.; Antony, A.; Aswathy, B. Android based home automation using raspberry pi. *Int. J. Comput. Technol.* **2014**, *1*, 143–147.
25. Upton, E.; Halfacree, G. *Raspberry Pi User Guide*; John Wiley & Sons: Hoboken, NJ, USA, 2016.
26. Knapp, C.; Carter, G. The generalized correlation method for estimation of time delay. *IEEE Trans. Acoust. Speech Signal Process.* **1976**, *24*, 320–327. [CrossRef]
27. Schmidt, R. Multiple emitter location and signal parameter estimation. *IEEE Trans. Antennas Propag.* **1986**, *34*, 276–280. [CrossRef]
28. Kwon, B.; Kim, G.; Park, Y. Sound Source Localization Methods with Considering of Microphone Placement in Robot Platform. In Proceedings of the RO-MAN 2007-The 16th IEEE International Symposium on Robot and Human Interactive Communication, Jeju, Republic of Korea, 26–29 August 2007; IEEE: New York, NY, USA, 2007; pp. 127–130.
29. Lv, X.; Zhang, M. Sound Source Localization Based on Robot Hearing and Vision. In Proceedings of the 2008 International Conference on Computer Science and Information Technology, Singapore, 29 August–2 September 2008; IEEE: New York, NY, USA, 2008; pp. 942–946.
30. Carter, G.C.; Nuttall, A.H.; Cable, P.G. The smoothed coherence transform. *Proc. IEEE* **1973**, *61*, 1497–1498. [CrossRef]
31. Hyvärinen, A.; Oja, E. Independent component analysis: Algorithms and applications. *Neural Netw.* **2000**, *13*, 411–430. [CrossRef] [PubMed]
32. Roberts, S.; Everson, R. *Independent Component Analysis: Principles and Practice*; Cambridge University Press: Cambridge, UK, 2001.
33. Wang, J.-C.; Wang, C.-Y.; Tai, T.-C.; Shih, M.; Huang, S.-C.; Chen, Y.-C.; Lin, Y.-Y.; Lian, L.-X. VLSI design for convolutive blind source separation. *IEEE Trans. Circ. Syst. II Express Briefs* **2015**, *63*, 196–200. [CrossRef]
34. Belouchrani, A.; Amin, M.G. Blind source separation based on time-frequency signal representations. *IEEE Trans. Signal Process.* **1998**, *46*, 2888–2897. [CrossRef] [PubMed]
35. Winter, S.; Kellermann, W.; Sawada, H.; Makino, S. MAP-based underdetermined blind source separation of convolutive mixtures by hierarchical clustering and-norm minimization. *EURASIP J. Adv. Signal Process.* **2006**, *2007*, 24717. [CrossRef]
36. Bofill, P. Underdetermined blind separation of delayed sound sources in the frequency domain. *Neurocomputing* **2003**, *55*, 627–641. [CrossRef]
37. Ying-Chuan, C. VLSI Architecture Design for Blind Source Separation Based on Infomax and Time-Frequency Masking. Master's Thesis, National Central University, Taoyuan, Taiwan, 2012.
38. Kinnunen, T.; Rajan, P. A Practical, Self-Adaptive Voice Activity Detector for Speaker Verification with Noisy Telephone and Microphone Data. In Proceedings of the 2013 IEEE International Conference on Acoustics, Speech and Signal Processing, Vancouver, BC, Canada, 26–31 May 2013; IEEE: New York, NY, USA, 2013; pp. 7229–7233.
39. Boll, S. Suppression of acoustic noise in speech using spectral subtraction. *IEEE Trans. Acoust. Speech Signal Process.* **1979**, *27*, 113–120. [CrossRef]
40. Kim, W.; Kang, S.; Ko, H. Spectral subtraction based on phonetic dependency and masking effects. *IEE Proc.-Vis. Image Signal Process.* **2000**, *147*, 423–427. [CrossRef]
41. Wang, J.-C.; Lee, Y.-S.; Lin, C.-H.; Wang, S.-F.; Shih, C.-H.; Wu, C.-H. Compressive sensing-based speech enhancement. *IEEE/ACM Trans. Audio Speech Lang. Process.* **2016**, *24*, 2122–2131. [CrossRef]
42. Tan, H.M.; Liang, K.-W.; Wang, J.-C. Discriminative Vector Learning with Application to Single Channel Speech Separation. In Proceedings of the ICASSP 2023-2023 IEEE International Conference on Acoustics, Speech and Signal Processing (ICASSP), Rhodes Island, Greece, 4–10 June 2023; IEEE: New York, NY, USA, 2023; pp. 1–5.
43. Tan, H.M.; Vu, D.-Q.; Wang, J.-C. Selinet: A Lightweight Model for Single Channel Speech Separation. In Proceedings of the ICASSP 2023-2023 IEEE International Conference on Acoustics, Speech and Signal Processing (ICASSP), Rhodes Island, Greece, 4–10 June 2023; IEEE: New York, NY, USA, 2023; pp. 1–5.
44. Varma, K.; Ikuma, T.; Beex, A.A. Robust TDE-Based DOA Estimation for Compact Audio Arrays. In Proceedings of the Sensor Array and Multichannel Signal Processing Workshop Proceedings, Rosslyn, VA, USA, 6 August 2002; IEEE: New York, NY, USA, 2002; pp. 214–218.
45. Cirrus Logic. *Cirrus Logic Audio Card User Documentation*; Cirrus Logic: Austin, TX, USA, 2015.
46. Cirrus Logic. *Cirrus Logic Audio Card for B+ and A+ Onwards Schematics*; Cirrus Logic: Austin, TX, USA, 2014.

Disclaimer/Publisher's Note: The statements, opinions and data contained in all publications are solely those of the individual author(s) and contributor(s) and not of MDPI and/or the editor(s). MDPI and/or the editor(s) disclaim responsibility for any injury to people or property resulting from any ideas, methods, instructions or products referred to in the content.

sensors

MDPI

Article

Proximity Sensor for Measuring Social Interaction in a School Environment

Tania Karina Hernández-Heredia [1], Cesar Fabián Reyes-Manzano [2], Diego Alonso Flores-Hernández [1], Gabriel Ramos-Fernández [3,4] and Lev Guzmán-Vargas [1,*]

1 Unidad Profesional Interdisciplinaria en Ingeniería y Tecnologías Avanzadas, Instituto Politécnico Nacional, Mexico City 07340, Mexico; tkhernandezh@ipn.mx (T.K.H.-H.); dfloreshe@ipn.mx (D.A.F.-H.)
2 Tecnológico Nacional de México, Tecnológico de Estudios Superiores de Ixtapaluca, Km. 7 Carretera Ixtapaluca-Coatepec S/N San Juan, Ixtapaluca 56580, Mexico; cesar.rm@ixtapaluca.tecnm.mx
3 Instituto de Investigaciones en Matemáticas Aplicadas y en Sistemas, Universidad Nacional Autonoma de Mexico, Mexico City 04510, Mexico; gabriel@aries.iimas.unam.mx
4 Centro de Ciencias de la Complejidad, Universidad Nacional Autonoma de Mexico, Mexico City 04510, Mexico
* Correspondence: lguzmanv@ipn.mx

Abstract: Social interactions are characterized by being very diverse and changing over time. Understanding this diversity and dynamics, as well as their emerging patterns, is of great interest from social, health, and educational perspectives. The development of new devices has been made possible in recent years by advances in applied technology. This paper presents the design and development of a novel device composed of several sensors. Specifically, we propose a proximity sensor integrated by three devices: a Bluetooth sensor, a global positioning system (GPS) unit and an accelerometer. By means of this sensor it is possible to detect the presence of neighboring sensors in various configurations and operating conditions. Profiles based on the Received Signal Strength Indicator (RSSI) exhibit behavior consistent with that reported by empirical relationships. The present sensor is functional in detecting the proximity of other sensors and is thus useful for the identification of interactions between people in relevant contexts such as schools.

Keywords: interaction; sensors; spatial configuration; received signal strength indicator

1. Introduction

Proximity sensors are devices designed to detect the presence or absence of an object within a certain range without physical contact. In general, the operation of this type of sensor is based on the principle of detecting changes in electromagnetic fields [1–3], i.e., changes in the signals emitted and/or received to determine the proximity of an object. These types of sensors have been used in various fields such as automotive, electronics, public safety and robotics, among others [4–7]. To date, there are several types of proximity sensors ranging from photoelectric to magnetic sensors. In the context of the study of social interactions, the use of sensors to detect interactions has been very useful in several studies [8–11]. In [12], a wearable sensor based on semi-supervised learning is presented to identify the interaction between university students to analyze the risk of exposure to SARS-CoV-2 transmission. The sensor is attached to the participant through a headband, including an image detector, an IR projector, and an RGB sensor. In addition, the components such as a power source and a microcomputer are mounted in the pocket of a specific jacket, reducing the portability and affecting the participant's feeling in the tests. Or, they are used to record data to perform various analyses in future studies, as shown in [13]. These sensors provide valuable data on social contact patterns and interactions within communities. Studies have shown that proximity sensors can capture high-resolution temporal data, allowing for detailed analysis of social dynamics and contact patterns [8]. For instance, radio frequency identification (RFID), which uses radio waves to transmit data wirelessly between an RFID transponder and an RFID reader, has been used to identify face-to-face interactions in many recent studies [10,14–16]. For example, Génois et al. [17] present the data collected at four science conferences for social interaction studies where the participants use cards and antennas with RFID technology, detecting other sensors in neighborhoods of a 1.5 m radius in face-to-face contact. This tool has the limitation of needing to be in certain areas with no obstacles in its way to be detected in order to obtain accurate measurements. One of the options that has often been used to identify proximity is Bluetooth, especially in cases where it is necessary to identify social interactions in controlled conditions. On the other hand, in [18], a face-to-face interaction study is presented using an IoT-based framework employing Bluetooth communication to represent social closeness and analyze the dynamics of social relationships through various models. Bluetooth proximity sensors provide some advantages such as low power consumption, cost-effectiveness and compatibility with a wide range of devices [19–23]. However, the accuracy of proximity detection can vary depending on factors such as signal strength, interference and the environment in which the sensors are deployed. The use of Bluetooth to identify social interactions has certain limitations because proximity can be misinterpreted as an interaction. However, this information can be extended by adding other sensors that can capture additional information such as angles of orientation, tilt and acceleration, so that if a representative displacement or rotation occurs, this information is useful for the detection of an interaction.

Since there exist limited options to obtain commercially available, low-cost proximity sensors that are adequate to detect social interactions, we present a viable option that fulfills these criteria [24–28]. Of particular interest is identifying social interactions in settings such as schools, where interactions may eventually impact the learning process and school performance. We propose a proximity sensor that consists of the integration of three devices: a Bluetooth sensor, a global positioning system (GPS) unit and an accelerometer. The components chosen in this work are from different companies and of commercial use.

In our tests, using the information obtained by the integrated devices, we have been able to detect the presence or proximity of neighboring sensors in various configurations and operating conditions. Specifically, we were able to reproduce the typical behavior of profiles based on the Received Signal Strength Indicator (RSSI), which is an indicator of the power level received by an antenna or receiver, for certain wireless sensor networks. We tested the sensor's performance by means of two experiments. The first one consisted in registering the RSSI strength between two sensors, increasing the distance among them, while the second involved measurements from eight sensors working simultaneously, placed in a specific star-type topology. From our experiments, which are based on the established configuration (master-slave) on each sensor, we were able to obtain the following results:

- High reading quality between the sensors.
- Minimum amount of loss of data read.
- Fast detection between sensors.
- Stability in the sensing range.

The remaining structure of this article is as follows. In Section 2 we present the main characteristics of the elements that constitute the sensor, as well as a brief explanation of their operation. In Section 3, we describe the basic aspects of the sensor operation, followed by the results (Section 4), where we describe the two experiments performed. Finally, we include our conclusions in Section 5.

2. Materials and Methods

The development of the proximity sensor was based on the methodology proposed in [29], which uses a systemic and concurrent approach, allowing performance improvement through constant validation and verification processes. Based on the definition of the problem, we built the functional architecture of the sensor, specifying the functions that the sensor should perform. Subsequently, we developed the physical architecture (see Figure 1), which is a description of the resources that the system comprises, grouped by modules. The sensor is made up of four modules:

1. Processing module: It is composed of an ESP32-DevKit V1 board, which performs the functions of information processing, decision making, signal conditioning, and internal and external connectivity of the proximity sensor. The ESP32 microcontroller of the board integrates Bluetooth, Bluetooth LE, and Wi-Fi, ensuring a wide range of applications with greater adaptability. The Wi-Fi channel allows a direct connection to the internet, while the Bluetooth module allows connecting and transmitting low-energy beacons for detection. The quiescent current is less than 5 μA, which makes it suitable for portable electronic applications as it is battery-powered. The module supports a data rate of up to 150 Mbps and an adjustable transmission power of 20 dBm at the antenna. The combination of features of the dual-core processor, Wifi and Bluetooth connectivity, as well as its general purpose pins (GPIO), provide the possibility of developing various applications, including IoT applications. The ESP32 can be programmed in several languages. In our case, we used the C language within the Arduino Integrated Development Environment (IDE), which provides a large number of libraries and tools for using the microcontroller in various contexts. The programming of the sensor includes particular functions that were programmed using different libraries allowed by the Arduino environment, so the sensor can work autonomously as long as it has the minimum energy needed to power all the devices that compose it. It is important to recall that the power supply of each of the sensors is different and therefore requires a specific current ($\approx$ 850 mW) to be able to work and function optimally. The tests performed with the sensor showed that it is necessary to maintain the mentioned current in the battery as a minimum. In case this requirement is not met, the sensor cannot complete the requested functions and stops communicating even though it may physically show that it is on.
2. Energetic module: This module performs the functions of storing electrical energy, conditioning, and distributing the energy to the other components of the proximity sensor. It is made up of the following components: a 3.7 V Lithium-Polymer (LiPo) battery with a capacity of 1200 mAh [30], a Standalone Linear Li-ion Battery Charger that is powered by a mini-USB port at 5 V [31], and a DC-DC switching boost converter that assists in controlling the battery output to a voltage of 3.3 V [32].
3. GPS module: This module determines the real time position of the sensor and monitors the movement, providing coordinates of the sensor in open spaces or near windows. This information can be useful to reconstruct the path of the sensor, which can help in the identification of a social interaction and accurate time measurements. The GPS module based on the Ublox NEO-6M chip, includes a GPS antenna with UFL connector, a small battery and an EEPROM memory that allows saving of the last positioning data [33]. The supply voltage is 5 V, it has a search consumption of 67 mA and a tolerance of $\pm$2.5 m [34]. This sensor measures longitude and latitude with a sampling rate of 0.2 Hz and sends the data through a UART port configured at a rate baud of 9600 bps.
4. Inertial Measurement Unit: It has the function of integrating motion information based on acceleration and angular velocity data acquired from the accelerometer and gyroscope, respectively. This enables calculation of inertial navigation as well as vibration, drop detector, distance and velocity measurement. The MPU6050 sensor is an IMU with 6 Degrees of Freedom (DoF) and is composed of a 3-axis accelerometer and a 3-axis gyroscope, which measure acceleration [m/s^2] and angular velocity [deg/s] in the x, y, and z axes, respectively. This sensor is powered by a 5 V power supply and was configured for a measurement range of 8 g in the accelerometer and 500 deg/s in the gyroscope. It also has a digital motion processor (DMP) that allows the calculation of the position of the sensor in pitch angle, roll angle, and yaw angle [35].

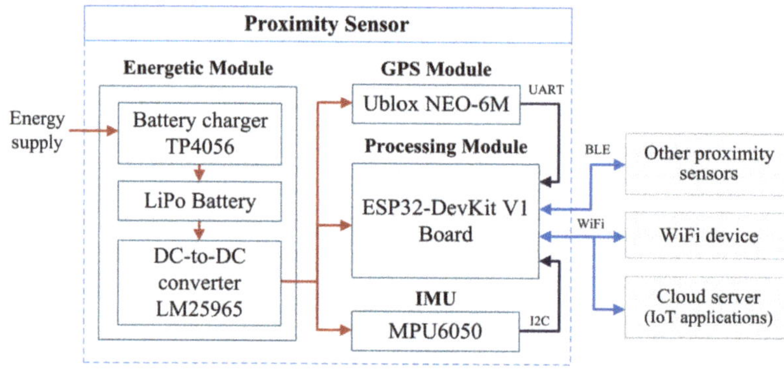

Figure 1. Physical architecture of the proximity sensor, including electrical power connections (red arrows) and data transfer (blue arrows).

Finally, once we analyzed the sensors that will integrate our proximity sensor, we carried out a unification process, which ensured that the operation of each module operated in an integrated manner. This is relevant because each of the modules operate with particular characteristics designed by the manufacturer, so that when unifying them, it must be ensured that they work within the range of the specifications indicated by the manufacturer. Figure 2 shows the design of the proximity sensor and the integrated modules.

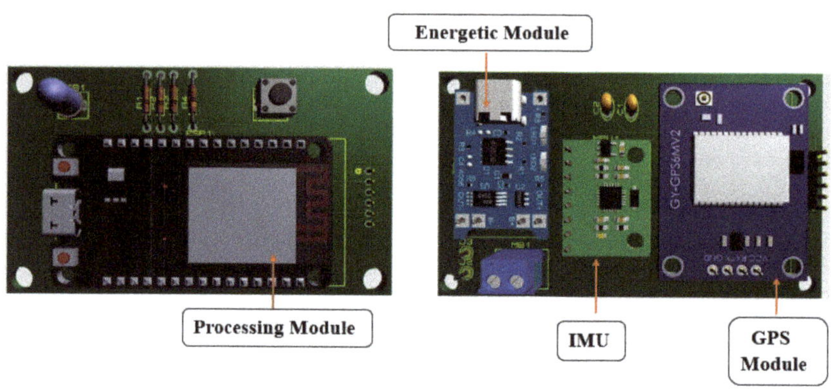

Figure 2. Design of the electronic hardware for the modules integration, top view (**left**) and bottom view (**right**).

3. Proximity Sensor Operation

3.1. Sensor Behavior

There are different electronic modules (sensors) on the market that can be integrated for the purposes of our proximity sensor. Therefore, when developing the present sensor, we have considered not only addressing a current need, but also ensuring that it can eventually be used in different conditions. Having a complete sensor that satisfies most of the needs is a complex task, so in our case we have considered the integration of several modules that are efficient in acquiring the key information for the identification of a social interaction.

The grouping of these electronic modules results in our proximity sensor based on parameters such as location, acceleration-gyroscope, and communication; as well as the possibility of bilateral BLE communication and RSSI measurement with other sensors. This information is achieved through Arduino programming and the use of libraries, where the ESP32 performs a scan every 5 s to detect nearby devices with its BLE module in Beacon configuration. This identification is conducted by assigning names in the format set in the programming (e.g., E[0 − 9][0 − 9][0 − 9]). Thus, once the contact is established, the receiving sensor retrieves the name of the identified device, the RSSI of the signal and its MAC address. In addition, it obtains the information registered by the GPS and the IMU, identifying the presence of the other sensors in its vicinity.

The sensor communicates through Bluetooth Low Energy (BLE) and Universal Asynchronous Receiver/Transmitter (UART), which allow asynchronous transmission and reception of data. This enables the connection to the Global Positioning System, allowing it to read signals transmitted by the satellite. On the other hand, the I2C Inter-Integrated Circuit connects to the IMU, which provides motion information (acceleration and angular velocity). All these modules are powered by an energy module containing a Lipo battery connected to a converter and its power module.

Figure 3 shows the flow diagram of the operation of the proximity sensor. In this diagram we can see the operating mode of the sensor and its configuration, which is necessary for an efficient detection of nearby devices. In order to explain the flow diagram in a general way, we begin by mentioning that the programming of the sensor is in series, that is, the programming is indicated in such a way that it must comply with the indications line by line. The way to proceed is as follows: When starting the sensor functions you have the possibility of accessing two modes: (i) Configuration mode or (ii) Data detection. To enter mode (i), it is necessary to detect a physical input (push button), giving access to the sensor and providing the necessary configuration parameters for the sensors. The parameters used are as follows: Name of the network to be connected; Name of the sensor and Port number.

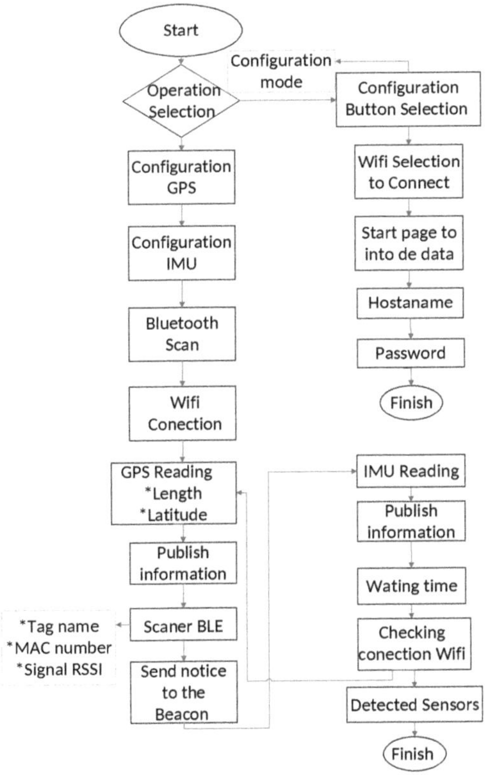

Figure 3. Flow diagram showing the configuration and working modes of the sensor.

Regarding mode (ii), this consists of configuring the sensors that make up the proximity sensor, that is, configuring the GPS, IMU, and Bluetooth of the ESP32. At the end of this process, the sensor connects to the configured network to send the readings of its next activity. Immediately after, it starts reading and publishing the GPS data (latitude and longitude). It also scans the Bluetooth BLE (name, MAC address, and RSSI(dB)). After this action, the IMU starts to operate and the readings obtained by the IMU are published. To obtain a separation between the readings, a timeout of 2 s is assigned. Then, the Wi-Fi connection is checked again and the cycle is repeated.

As can be seen in Figure 3, the process consists of a serial cycle, which limits the execution of any indication due to interruptions or possible interferences. For this reason, an indicator LED (RGB) was placed at specific points in the configuration, allowing the user to know the current state of the system. In addition, this configuration allows for obtaining power measurements (RSSI), indicating the power that can be registered by the sensors of the receiving system.

3.2. Sensor Calibration and Error Reduction

Due to the limited resolution of RSSI measurements, granularity errors can arise, which refer to inaccuracy or lack of precision in locating a device [36]. This type of error manifests mainly at distances greater than 0.5 m, where measurements tend to blend. To reduce these errors, an averaging filter is applied before calculating the distance to minimize fluctuations and obtain better resolution. In addition, each distance calculation is performed using two sets of measurements, one for each sensor. Another challenge is the multipath effect, where signals collide with obstacles present in the environment and are reflected, merging with the original signal, known as non-line-of-sight (NLOS) [37]. Since the sensors have been designed for people in motion, mapping the environment, as proposed by some authors, is not very functional [38,39]. One solution to overcome this problem is to rely on inertial sensors to obtain an approximation of the person's direction (based on a natural forward walking motion) and to give more weight to the sensors located just in front. On the other hand, packets are filtered based on the sensor name to avoid analyzing packets sent by the same sensor. Other relevant parameters to consider are weather conditions such as temperature and relative humidity, which influence RSSI measurements [40]. It has been shown that temperature has a negative correlation with RSSI at temperatures above 0 °C, with a loss of 1.3 to 2 dBm for every 10 °C increase [41–43]. In contrast, relative humidity (RH) has a positive correlation with RSSI, increasing approximately 0.3 dBm for every 10 °C increase in RH. It has also been shown that at temperatures below 0 °C, these values can be three times higher [42,44].

Finally, to avoid device heterogeneity errors, which occur when there are differences in device hardware, we use modules and components with the highest compatibility based on the manufacturer's specifications. Additionally, each sensor is calibrated to minimize errors due to device heterogeneity and environmental conditions. The calibration consists of using two sensors (S1 and S2) placed on a table in a classroom (an environment similar to where the future experiments will be performed). S1 is placed at the starting point, and S2 is placed at 1 m with respect to S1. RSSI measurements are registered in the position for 8 minutes. In this way, the reference value is obtained and later used to calibrate the other sensors (see details of the experiment (i) in Section 4).

4. Experimental Results

4.1. Experimental Setup

The RSSI behavior can be modelled by the so-called Friis transmission equation [45,46], which establishes the behaviour of the power of the signal recorded by a receiver as a function of distance. Formally, the Friis equation is given by $RSSI(D) = P_{r1} - \kappa \log_{10}(D)$, where P_{r1} is the received power at one meter, κ is a "path loss" parameter related with the transmission media and D is the distance. To empirically characterize the behavior of signal strength as a function of distance, we have performed two controlled experiments:

(i) when a pair of sensors are placed on the ground and controlling the distance between them; and (ii) when a group of eight sensors are placed on the ground in a predetermined network-like configuration.

The physical and time conditions of the scenario where the experiments were carried out can be seen in Table 1:

Table 1. Stage characteristic table.

Parameter	Experiment (i)	Experiment (ii)
Country	Mexico	Mexico
City	CDMX	CDMX
Location	UPIITA-IPN	UPIITA-IPN
Start time (GMT-6)	17:04 h	22:20 h
End time (GMT-6)	20:04 h	20:20 h
Day	17 February 2024	21 March 2024
Wind Speed	50–59 km/h	40–60 km/h
Maximum Environmental Temperature	24.7 °C	28.9 °C
Minimum Environmental Temperature	10.5 °C	14.1 °C
Wifi name	WI-Fi IPN	Wi-Fi IPN
Internet Speed	53 Mbps	58 Mbps
Presence of humans	Yes	Yes

4.2. Experiment (i)

Two devices are placed on the ground (Sensor 1 and Sensor 2), where we can control the relative distance between them, as shown in Figure 4. Our strategy is based on recording the interactions between the sensors during a certain time interval (8 mins.). Initially, they were placed 10 cm apart and this distance is increased by 10 cm, up to a maximum distance of 2 m. The resulting measurements are presented in Figures 5 and 6. In particular, Figure 5 shows the signal strength of the signal detected by sensor 1 coming from sensor 2 while the opposite is plotted in Figure 6. We observe that the measured signal intensity fluctuates with a representative mean value. For small distance values ($D \approx 10$ cm), the signal strength is relatively high -40 dBm, and with increasing distance, the signal strength decreases to values close to -80 dBm. We observe that the RSSI response provided values ranging from -40 to -90 dBm. According to RSSI quality tables [47], values greater than -65 dBm represent high quality, -65 dBm to -75 dBm represent good quality, -75 dBm to -85 dBm represent the limit of useful quality, and lower values represent poor data. Another important parameter for calibration purposes is the RSSI value when the separation between the sensors is 1 m. As shown in Figure 6, the average value $\langle RSSI \rangle_{1\text{m}} = -68.07 \pm 2.27$ represents the reference value for the calibration of the other sensors.

For a better identification of the signal intensity decay, we have calculated the mean value and standard deviation of each interval shown in Figure 7. It can be seen that both measurements present a similar profile as the distance increases. Then, using the mean values of the signal intensity and considering the different values of the distance, we performed a linear regression to identify the constants that characterise the straight line in a semi-logarithmic plane ($RSSI = b + a\log(D)$). These results are shown in Figure 8. The calculations lead to the following values for the parameters: $a_1 = -1.16$ and $b_1 = -7.25$ ($R^2 = 0.879$) for the data detected by S1, while the data sensed by S2 result in $a_2 = -1.10$ and $b_2 = -7.26$ ($R^2 = 0.851$); see Figure 8b. These responses agree with the values predicted by the Friis equation [46]. In particular, $a_1 = -1.16$ and $a_2 = -1.10$ correspond approximately to the values of the path loss parameter given in refs. [46–48].

It is important to mention that the behavior of the signal intensity measured by the device allowed us to determine an effective threshold value (in dBm) to identify sensors that, based on the received values, can be distinguishable and represent a social interaction. In our case, we choose RSSI= −75 dBm as the threshold value, which is approximately 1.5 m. The interval 0–1.5 m is in general agreement with previous studies suggesting that most social interactions occur within this interval [49,50].

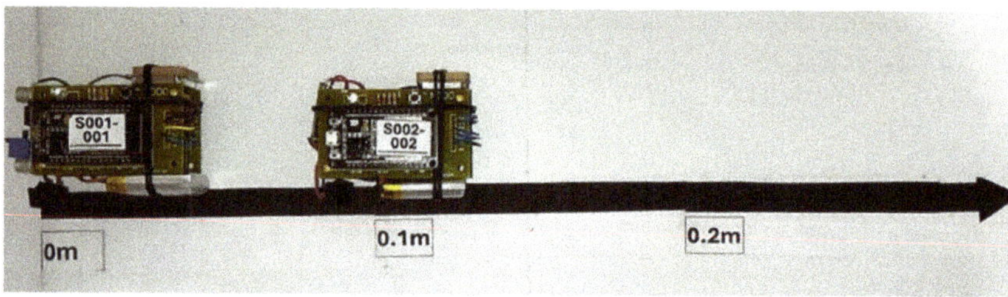

Figure 4. Image of the physical location of the sensors S1 and S2 during the measurement of the mutual signals for two distance values.

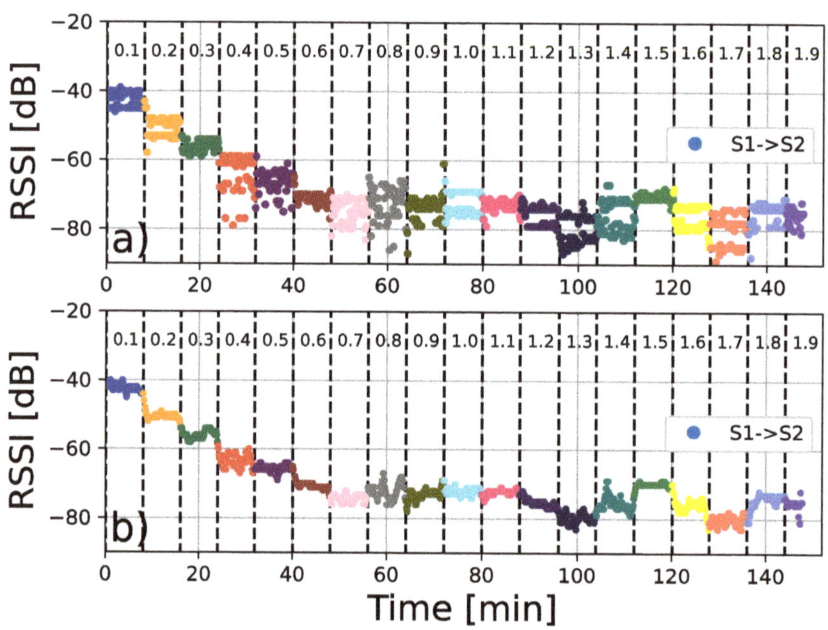

Figure 5. RSSI signal strength for several distance (meters) values and time intervals. (**a**) Scans of sensor 1 using the data emitted by sensor 2. The values shown are measurements per distance performed every five seconds during eighth-minute time intervals. (**b**) Similar to (**a**), but showing the mean values obtained from a moving average with one-minute windows.

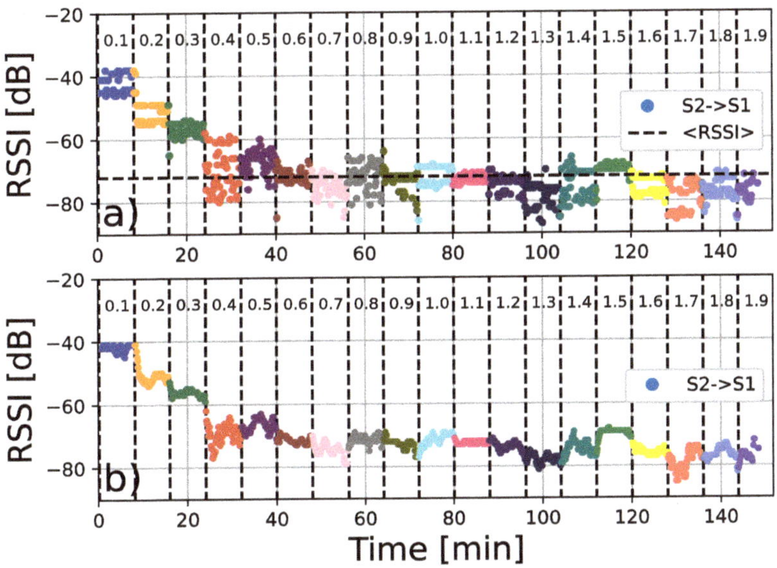

Figure 6. RSSI signal strength for several distance (meters) values and time intervals. (**a**) Scans of sensor 2 using the data emitted by sensor 1. The values shown are measurements per distance performed every five seconds during eight-minute time intervals. The reference value $\langle RSSI \rangle_{1\mathrm{m}} \approx -68$ is also shown, which corresponds to the average value at a distance of 1 m. (**b**) Similar to (**a**), but showing the mean values obtained from a moving average with one-minute windows.

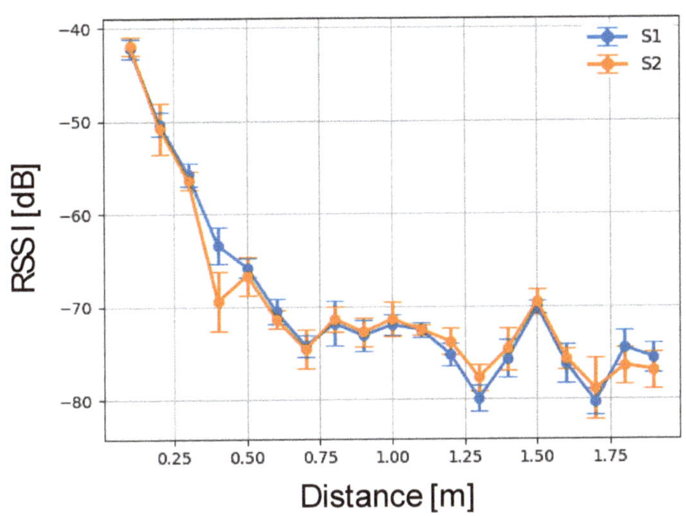

Figure 7. Signal strength vs. distance for the measurements from sensors S1 and S2. We observe that the signal intensity decreases as the distance increases. The vertical bars represent the standard deviation over the set of measurements.

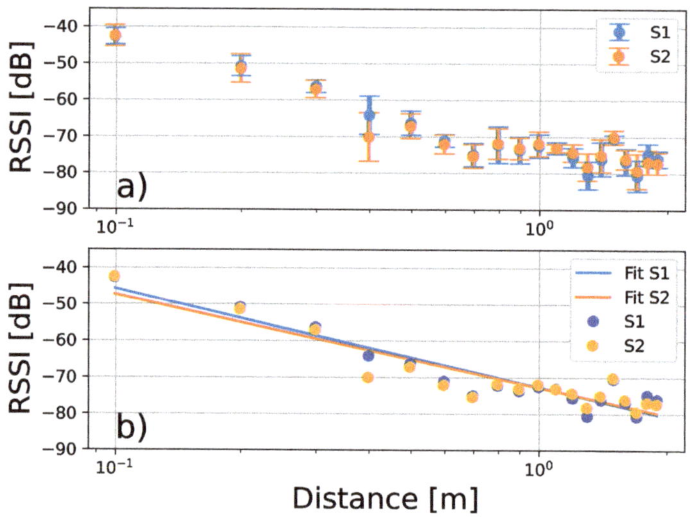

Figure 8. (**a**) Logarithmic-linear representation of the signal strength (RSSI) vs. distance for data obtained from sensors S1 and S2. This plot shows that the decay is logarithmic with an approximate linear behavior in this semi-log plane. (**b**) The calculations of the linear regression yield the values for the slopes: $a_1 = -1.16$ ($R^2 = 0.879$) for S1 and $a_2 = -1.10$ ($R^2 = 0.851$) for S2.

4.3. Experiment (ii)

To test the effectiveness of our device in a more realistic context, we have considered a non-simple configuration of a group of sensors, useful for statistical processing and elementary data verification. The configuration consists of eight sensors deployed in a star-like configuration (see Figure 9a). This configuration allows for observing the behavior of these sensors as the distance between sensors increases (as proposed below) and is a representation of sensor arrays in a real situation of social interaction. The distance between sensors of immediate neighbors is the same; initially, they are placed at 0 m between first neighbors, and then this distance increases by 0.20 m until it reaches 1.20 m (see Figure 9a), for a duration of 10 min per distance. This results in five different distance values.

The results of the number of measurements from three representative sensors for different values of distance are depicted in Figure 9b–d. Then, the distance between the neighboring sensors is increased while keeping the initial topology. As expected, when the distance between the sensors is small ($D < 0.20$), most of the sensors detect each other (see Figure 9a), including those located at opposite ends of the array (S7 and S12). In contrast, for larger distances between first neighboring sensors ($D > 0.30$) the closest neighboring sensors still register the highest power intensity, and some sensors no longer identify the presence of distant sensors.

For the case when the neighboring sensors are very close, most of the sensor pairs detect each other, as the distance increases, distant sensors are no longer detected and the number of measurements decreases significantly. To test the overall validity of the RSSI behavior in our sensor array, we considered all readings from all pairs of sensors for a given distance and calculate the mean values as a function of the distance. The results are shown in Figure 9c. As it can be seen in this semi-log plot, the intensity decays as the distance increases with a rate modulated by constants that can be determined with a linear regression (see figure caption for details). We also note that, as in experiment (i), there is a good agreement with the empirical Friis equation.

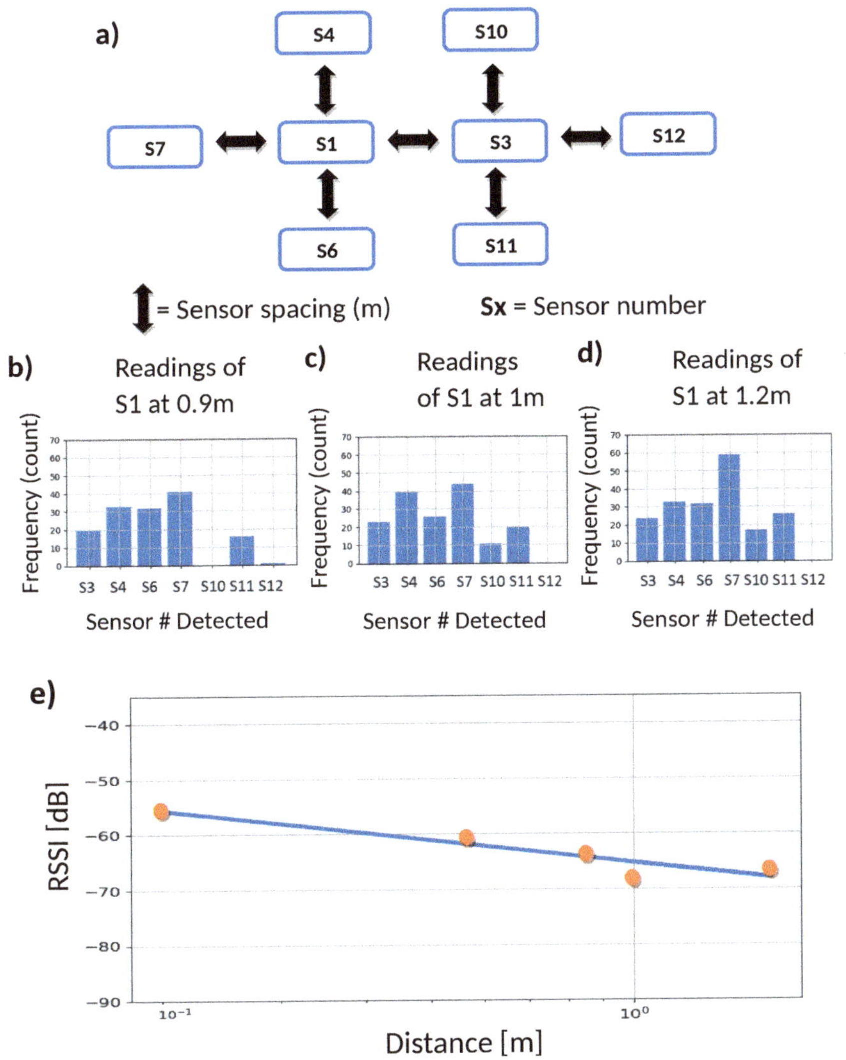

Figure 9. (**a**) Array of sensors in a pre-established star-like configuration. Initially, the distance between the nearest neighboring sensors is set at 0.2 m, then increased in 0.2 m intervals, until 1.2 m is reached. (**b–d**) Histograms of the number of measurements by S1 for three distance values. (**e**) Behavior of the average signal strength (RSSI) of all sensors versus distance in a semi-logarithmic plane. The solid line represents a linear regression that leads to the slope $a_r = -1.3$ ($R^2 = 0.90$).

It is important to note that for this experiment, the information recorded by the GPS sensor and the accelerometer is complementary since the sensors were kept static and at a predetermined distance. In this sense, the information provided by the GPS and the accelerometer will be useful when considering dynamic situations.

4.4. Comparison with Other Sensors

The present sensor design allows the user to freely place the sensors at different distances. We have analyzed different configurations in which the range of distance between sensors is varied, which allowed us to determine sensing ranges that are useful for the identification of a proximity in a real context of social interaction. The data recorded by the sensors are consistent with the behavior reported for wireless sensor networks. Specifically, Table 2 presents a comparison of capabilities and elements between similar sensors reported in the literature and the current proposal. The proposed sensor with these features offers improved performance, range and connectivity, as well as acceptable power consumption. As shown in Table 2, several similar sensor developments were designed for use in different contexts but share similarities in design and operation.

Table 2. List of sensors and their main characteristics reported in the literature. The first three rows correspond to sensors designed to identify social proximity or social interaction. The characteristics of the sensors include the space (indoor or outdoor) where the sensor was developed; the type of positioning system used by the sensor; as well as the type of internal measurement system for the systems; another parameter observed is the intensity indicator of the signal received. In this case, the question was whether it is used or not?. Another relevant data point is the number of sensors detected or the number of cards that maintain the communication. In addition to the range of coverage that the sensor has, we consider the following parameters: from 0 m to 1 m (short) or >1 m (high), and the cost.

Systems	Space	Positioning Systems	Inertial Measurement Unit	Received Signal Strength Indicator	Target BLE	Capacity Number Sensors	Range	Cost	Communication
[12]	I	IR	No	No	No	56	Short	High	Wifi
[17]	I	RFID	No	No	No	40	Short	High	Wifi
[18]	I	BLE	No	Yes	Yes	14	High	High	No
[51]	I	IPS	No	No	Yes	1	Short	Low	MQTT
[52]	E	No	No	No	Yes	1	High	Low	Wifi
[53]	I	UWB	No	No	Yes	1	High	High	Wifi
[54]	E	No	No	Yes	Yes	1	High	Low	MQTT
[55]	I	IPS	No	Yes	Yes	1	Short	Low	Wifi
[56]	I	UWB	Yes	No	No	1	Short	High	IoT
[57]	I	No	No	Yes	Yes	6	High	Low	Wifi
[58]	I and E	WMN and BLE	No	No	Yes	6	High	Low	IoT
Proposal	**E**	**GPS**	**Yes**	**Yes**	**Yes**	8	**Short**	**Low**	**Wifi**

5. Conclusions

In this paper we have presented a proximity sensing system, based on the integration of wireless devices, position sensors, and motion communication devices. The proposed embedded system is robust and can record sensor proximity in different configurations. Also, it can work in a stand-alone manner, remotely and with low cost. We have performed controlled tests with predetermined configurations, where we have been able to verify that neighboring sensors detect each other up to a certain distance threshold, relevant for normal social interactions between people. Additionally, we have verified that the intensity of the activity of the sensors, in terms of the distance between them, approximately follows the behavior predicted by well-accepted theoretical models. The system is functional and could potentially have a variety of uses. Although the proposed sensor represents an important step in collecting valuable information about social interactions in a school context, several aspects deserve further exploration. One possibility for future research could be to deploy the sensors in a wearable form (badge) during school hours on a group of undergraduate students in at least three stages of their studies. From this information, we propose to construct interaction networks, which would potentially allow a more complete understanding of how interaction patterns impact school performance, taking into account the dynamic nature of social interaction networks as the vehicle for the flow of information or knowledge among students. In addition, incorporating qualitative research methods, such as interviews and questionnaires, could provide a deeper understanding of relationships among students. By incorporating the nuances of their interactions and behaviors in a qualitative manner, it is possible to gain valuable insights into the underlying mechanisms that impact possible correlations between interaction patterns and school performance outcomes. In conclusion, our sensor represents a foundation for future research to contribute to a better understanding of the complex interactions between students and their possible relationship to school learning outcomes.

Author Contributions: Conceptualization, T.K.H.-H., C.F.R.-M., G.R.-F. and L.G.-V.; methodology, T.K.H.-H., C.F.R.-M., D.A.F.-H., G.R.-F. and L.G.-V.; software, T.K.H.-H. and C.F.R.-M.; validation, T.K.H.-H., C.F.R.-M., D.A.F.-H., G.R.-F. and L.G.-V.; formal analysis, T.K.H.-H., C.F.R.-M., D.A.F.-H., G.R.-F. and L.G.-V.; investigation, T.K.H.-H., C.F.R.-M., D.A.F.-H., G.R.-F. and L.G.-V.; writing—original draft preparation, T.K.H.-H. and C.F.R.-M.; writing—review and editing, T.K.H.-H., C.F.R.-M., D.A.F.-H., G.R.-F. and L.G.-V.; visualization, T.K.H.-H., C.F.R.-M. and D.A.F.-H.; supervision, C.F.R.-M., D.A.F.-H., G.R.-F. and L.G.-V.; All authors have read and agreed to the published version of the manuscript.

Funding: This research was funded by Consejo Nacional de Humanidades, Ciencias y Tecnologías (CONAHCYT, Mexico), grant CF-2019-263958 and Secretaría de Investigación y Progrado, Instituto Politécnico Nacional, Mexico.

Informed Consent Statement: Not applicable.

Data Availability Statement: The data generated and analyzed in this study are included in this published article or are available from the corresponding author upon request.

Acknowledgments: We thank COFAA-IPN, EDI-IPN and CONAHCYT-Mexico for the partial support received during the development of this work.

Conflicts of Interest: The authors declare no conflicts of interest.

References

1. Moheimani, R.; Hosseini, P.; Mohammadi, S.; Dalir, H. Recent advances on capacitive proximity sensors: From design and materials to creative applications. *J. Carbon Res.* **2022**, *8*, 26. [CrossRef]
2. Moreno, M.V.R.; Muñoz, D.G.R. Sensor de Acoplamiento Inductivo para la Medida de Pulsos de Corriente de alta Frecuencia. Aplicación para la Medida y Detección de Descargas Parciales. Ph.D. Thesis, Departamento de Ingeniería Eléctrica, Universidadcarlosi i Idemadrid, Getafe, Spain, 2011.
3. Ramírez, L.G.C.; Jiménez, G.S.A.; Carreño, J.M. *Sensores y Actuadores*; Grupo Editorial Patria: Mexico City, Mexico, 2014.
4. Elsisi, M.; Rusidi, A.L.; Tran, M.Q.; Su, C.L.; Ali, M.N. Robust Indoor Positioning of Automated Guided Vehicles in Internet of Things Networks with Deep Convolution Neural Network Considering Adversarial Attacks. *IEEE Trans. Veh. Technol.* **2024**, *73*, 7748–7757. [CrossRef]
5. Sharma, A.; Chauhan, R.C.S.; Kaur, J. Performance Evaluation of SC-FDMA in Fading Channels with PAPR, QAM, and RSSI Analysis. *Res. Sq.* **2024**. [CrossRef]
6. Zhou, H.; Yang, J.; Deng, S.; Zhang, W. VTIL: A multi-layer indoor location algorithm for RSSI images based on vision transformer. *Eng. Res. Express* **2024**, *6*, 015069. [CrossRef]
7. Chen, F.; Nguyen, H.V.; Taggart, D.A.; Falkner, K.; Rezatofighi, S.H.; Ranasinghe, D.C. ConservationBots: Autonomous aerial robot for fast robust wildlife tracking in complex terrains. *J. Field Robot.* **2024**, *41*, 443–469. [CrossRef]
8. Ozella, L.; Paolotti, D.; Lichand, G.; Rodríguez, J.P.; Haenni, S.; Phuka, J.; Leal-Neto, O.B.; Cattuto, C. Using wearable proximity sensors to characterize social contact patterns in a village of rural Malawi. *EPJ Data Sci.* **2021**, *10*, 46. [CrossRef]
9. Ma, J.H.; Cha, S.H. A human data-driven interaction estimation using IoT sensors for workplace design. *Autom. Constr.* **2020**, *119*, 103352. [CrossRef]
10. Cattuto, C.; Van den Broeck, W.; Barrat, A.; Colizza, V.; Pinton, J.F.; Vespignani, A. Dynamics of person-to-person interactions from distributed RFID sensor networks. *PLoS ONE* **2010**, *5*, e11596. [CrossRef] [PubMed]
11. Zhang, H.; Parsia, B.; Poliakoff, E.; Harper, S. Tracking social behaviour with smartphones in people with Parkinson's: A longitudinal study. *Behav. Inf. Technol.* **2023**, 1–20. [CrossRef]
12. Zhang, N.; Liu, L.; Dou, Z.; Liu, X.; Yang, X.; Miao, D.; Guo, Y.; Gu, S.; Li, Y.; Qian, H.; et al. Close contact behaviors of university and school students in 10 indoor environments. *J. Hazard. Mater.* **2023**, *458*, 132069. [CrossRef]
13. Diallo, D.; Schönfeld, J.; Blanken, T.F.; Hecking, T. Dynamic Contact Networks in Confined Spaces: Synthesizing Micro-Level Encounter Patterns Through Human Mobility Models from Real-World Data. *Preprints* **2024**, 2024041998. [CrossRef]
14. Elmer, T.; Chaitanya, K.; Purwar, P.; Stadtfeld, C. The validity of RFID badges measuring face-to-face interactions. *Behav. Res. Methods* **2019**, *51*, 2120–2138. [CrossRef] [PubMed]

15. Bianco, G.M.; Occhiuzzi, C.; Panunzio, N.; Marrocco, G. A survey on radio frequency identification as a scalable technology to face pandemics. *IEEE J. Radio Freq. Identif.* **2021**, *6*, 77–96. [CrossRef]
16. Elmer, T.; Stadtfeld, C. Depressive symptoms are associated with social isolation in face-to-face interaction networks. *Sci. Rep.* **2020**, *10*, 1444. [CrossRef] [PubMed]
17. Génois, M.; Zens, M.; Oliveira, M.; Lechner, C.M.; Schaible, J.; Strohmaier, M. Combining sensors and surveys to study social interactions: A case of four science conferences. *Personal. Sci.* **2023**, *4*, e9957. [CrossRef]
18. Elhamer, Z.; Suzuki, R.; Arita, T. An IoT-based experimental framework for studying continuous social dynamics in a game-theoretical and face-to-face situation with human participants. *Psychologia* **2023**, *65*, 211–232. [CrossRef]
19. Zhou, S.; Pollard, J.K. Position measurement using Bluetooth. *IEEE Trans. Consum. Electron.* **2006**, *52*, 555–558. [CrossRef]
20. Montanari, A.; Nawaz, S.; Mascolo, C.; Sailer, K. A study of bluetooth low energy performance for human proximity detection in the workplace. In Proceedings of the 2017 IEEE International Conference on Pervasive Computing and Communications (PerCom), Kona, HI, USA, 13–17 March 2017; pp. 90–99.
21. Yu, W.; Zhang, J.; Cai, J.; Xu, J. A Novel iBeacon Deployment Scheme for Indoor Pedestrian Positioning. In Proceedings of the 2021 IEEE 27th International Conference on Parallel and Distributed Systems (ICPADS), Beijing, China, 14–16 December 2021; pp. 8–17.
22. Lubis, A.F.; Basari. Proximity-based COVID-19 contact tracing system devices for locally problems solution. In Proceedings of the 2020 3rd International Seminar on Research of Information Technology and Intelligent Systems (ISRITI), Yogyakarta, Indonesia, 10–11 December 2020; pp. 365–370.
23. Tiwari, K.R.; Singhal, I.; Mittal, A. Smart social distancing solution using bluetooth® low energy. In Proceedings of the 2020 5th International Conference on Computing, Communication and Security (ICCCS), Patna, India, 14–16 October 2020; pp. 1–5.
24. Raento, M.; Oulasvirta, A.; Eagle, N. Smartphones: An emerging tool for social scientists. *Sociol. Methods Res.* **2009**, *37*, 426–454. [CrossRef]
25. Shimizu, K.; Iwai, M.; Sezaki, K. Social link analysis using wireless beaconing and accelerometer. In Proceedings of the 2013 27th International Conference on Advanced Information Networking and Applications Workshops, Barcelona, Spain, 25–28 March 2013; pp. 33–38.
26. Girolami, M.; Mavilia, F.; Delmastro, F. A bluetooth low energy dataset for the analysis of social interactions with commercial devices. *Data Brief* **2020**, *32*, 106102. [CrossRef]
27. Girolami, M.; Mavilia, F.; Delmastro, F. Sensing social interactions through BLE beacons and commercial mobile devices. *Pervasive Mob. Comput.* **2020**, *67*, 101198. [CrossRef]
28. Otsason, V.; Varshavsky, A.; LaMarca, A.; De Lara, E. Accurate GSM indoor localization. In *International Conference on Ubiquitous Computing*; Springer: Berlin/Heidelberg, Germany, 2005; pp. 141–158.
29. Buede, D.M.; Miller, W.D. *The Engineering Design of Systems: Models and Methods*; John Wiley & Sons: Hoboken, NJ, USA, 2024.
30. Honcell. Rechargeable Lithium ion Polymer Battery. 2020. Available online: https://www.lithium-polymer-battery.net/wp-content/uploads/2021/03/LP503759-1200mAh-datasheet.pdf (accessed on 1 February 2024).
31. ASIC. TP4056 1 A Standalone Linear Li-Lon Battery Charger with Thermal Regulation in SOP-8. 2018. Available online: https://dlnmh9ip6v2uc.cloudfront.net/datasheets/Prototyping/TP4056.pdf (accessed on 1 February 2024).
32. XLSEMI. 400 KHz 42 V 5 A Switching Current Boost DC/DC Converter. 2020. Available online: https://www.ti.com/lit/ds/symlink/lm2596.pdf (accessed on 1 February 2024).
33. U-blox. NEO-6 6 GPS Modules Data Sheet. 2011. Available online: https://content.u-blox.com/sites/default/files/products/documents/NEO-6_DataSheet_%28GPS.G6-HW-09005%29.pdf (accessed on 1 February 2024).
34. Zeimpekis, V.; Giaglis, G.M.; Lekakos, G. A taxonomy of indoor and outdoor positioning techniques for mobile location services. *ACM SIGecom Exch.* **2002**, *3*, 19–27. [CrossRef]
35. InvenSense. MPU-6000 and MPU-6050 Product Specification. 2013. Available online: https://invensense.tdk.com/wp-content/uploads/2015/02/MPU-6000-Datasheet1.pdf (accessed on 1 February 2024).
36. Asaad, S.M.; Maghdid, H.S. A comprehensive review of indoor/outdoor localization solutions in IoT era: Research challenges and future perspectives. *Comput. Netw.* **2022**, *212*, 109041. [CrossRef]
37. Hayward, S.; van Lopik, K.; Hinde, C.; West, A.A. A survey of indoor location technologies, techniques and applications in industry. *Internet Things* **2022**, *20*, 100608. [CrossRef]
38. Siyang, L.; de Lacerda, R.; Fiorina, J. WKNN indoor Wi-Fi localization method using k-means clustering based radio mapping. In Proceedings of the 2021 IEEE 93rd Vehicular Technology Conference (VTC2021-Spring), Helsinki, Finland, 25–28 April 2021; pp. 1–5.
39. Kawecki, R.; Hausman, S.; Korbel, P. Performance of fingerprinting-based indoor positioning with measured and simulated RSSI reference maps. *Remote Sens.* **2022**, *14*, 1992. [CrossRef]
40. Szyc, K.; Nikodem, M.; Zdunek, M. Bluetooth low energy indoor localization for large industrial areas and limited infrastructure. *Ad Hoc Netw.* **2023**, *139*, 103024. [CrossRef]
41. Boano, C.A.; Tsiftes, N.; Voigt, T.; Brown, J.; Roedig, U. The impact of temperature on outdoor industrial sensornet applications. *IEEE Trans. Ind. Inform.* **2009**, *6*, 451–459. [CrossRef]

42. Luomala, J.; Hakala, I. Effects of temperature and humidity on radio signal strength in outdoor wireless sensor networks. In Proceedings of the 2015 Federated Conference on Computer Science and Information Systems (FedCSIS), Lodz, Poland, 13–16 September 2015; pp. 1247–1255.
43. Bannister, K.; Giorgetti, G.; Gupta, S. Wireless sensor networking for hot applications: Effects of temperature on signal strength, data collection and localization. In Proceedings of the 5th Workshop on Embedded Networked Sensors (HotEmNets'08), Charlottesville, VA, USA, 2–3 June 2008; pp. 1–5.
44. Guidara, A.; Fersi, G.; Derbel, F.; Jemaa, M.B. Impacts of Temperature and Humidity variations on RSSI in indoor Wireless Sensor Networks. *Procedia Comput. Sci.* **2018**, *126*, 1072–1081. [CrossRef]
45. Friis, H.T. A note on a simple transmission formula. *Proc. IRE* **1946**, *34*, 254–256. [CrossRef]
46. Heurtefeux, K.; Valois, F. Is RSSI a good choice for localization in wireless sensor network? In Proceedings of the 2012 IEEE 26th International Conference on Advanced Information Networking and Applications, Fukuoka, Japan, 26–29 March 2012; pp. 732–739.
47. Mohsin, H.; Abdulameer, K.; Khudhair, Z. Study and performance analysis of received signal strength indicator (rssi) in wireless communication systems. *Int. J. Eng. Technol.* **2017**, *6*, 195–200.
48. Xu, J.; Liu, W.; Lang, F.; Zhang, Y.; Wang, C. Distance measurement model based on RSSI in WSN. *Wirel. Sens. Netw.* **2010**, *2*, 606. [CrossRef]
49. Holl, B.; Audard, M.; Nienartowicz, K.; Jevardat de Fombelle, G.; Marchal, O.; Mowlavi, N.; Clementini, G.; De Ridder, J.; Evans, D.W.; Guy, L.P.; et al. Gaia Data Release 2. Summary of the variability processing and analysis results. *Astron. Astrophys.* **2018**, *618*, A30. [CrossRef]
50. Sekara, V.; Lehmann, S. The strength of friendship ties in proximity sensor data. *PLoS ONE* **2014**, *9*, e100915. [CrossRef] [PubMed]
51. Misal, S.R.; Prajwal, S.R.; Niveditha, H.M.; Vinayaka, H.M.; Veena, S. Indoor Positioning System (IPS) Using ESP32, MQTT and Bluetooth. In Proceedings of the 2020 Fourth International Conference on Computing Methodologies and Communication (ICCMC), Erode, India, 11–13 March 2020; pp. 79–82. [CrossRef]
52. Höchst, J.; Baumgärtner, L.; Kuntke, F.; Penning, A.; Sterz, A.; Sommer, M.; Freisleben, B. Mobile device-to-device communication for crisis scenarios using low-cost lora modems. In *Disaster Management and Information Technology: Professional Response and Recovery Management in the Age of Disasters*; Springer: Berlin/Heidelberg, Germany, 2023; pp. 235–268.
53. Sidiropoulos, A.; Bechtsis, D.; Vlachos, D. A Real-Time Locating System with Intelligent Position Correction for Harsh Environments. Available online: https://ssrn.com/abstract=4124531 (accessed on 1 February 2024).
54. Quiroga, D.; Diaz, S.; Pastrana, H.F. Integration of Wireless Sensor Networks and IoT for air quality monitoring. In Proceedings of the 2023 IEEE Colombian Caribbean Conference (C3), Barranquilla, Colombia, 22–25 November 2023; pp. 1–6.
55. Goh, B.S.; Mahamad, A.K.; Saon, S.; Isa, K.; Ameen, H.A.; Ahmadon, M.A.; Yamaguchi, S. IoT Based Indoor Locating System (ILS) using Bluetooth Low Energy (BLE). In Proceedings of the 2020 IEEE International Conference on Consumer Electronics (ICCE), Las Vegas, NV, USA, 4–6 January 2020; pp. 1–4. [CrossRef]
56. Naheem, K.; Kim, M.S. A Low-Cost Foot-Placed UWB and IMU Fusion-Based Indoor Pedestrian Tracking System for IoT Applications. *Sensors* **2022**, *22*, 8160. [CrossRef] [PubMed]
57. Fernández-Caramés, T.; Rodas, J.; Iglesia, D.; Escudero, C. Estabilización del RSSI en una red de sensores Bluetooth usando múltiples antenas. In Proceedings of the URSI Meeting, Ottawa, ON, Canada, 22–26 July 2007.
58. Fazio, M.; Celesti, A.; Villari, M. Improving Proximity Detection of Mesh Beacons at the Edge for Indoor and Outdoor Navigation. In Proceedings of the 2020 IEEE 25th International Workshop on Computer Aided Modeling and Design of Communication Links and Networks (CAMAD), Pisa, Italy, 14–16 September 2020; pp. 1–6. [CrossRef]

Disclaimer/Publisher's Note: The statements, opinions and data contained in all publications are solely those of the individual author(s) and contributor(s) and not of MDPI and/or the editor(s). MDPI and/or the editor(s) disclaim responsibility for any injury to people or property resulting from any ideas, methods, instructions or products referred to in the content.

sensors

MDPI

Communication

Physical Frailty Prediction Using Cane Usage Characteristics during Walking

Haruki Toda [1,]* and Takaaki Chin [2]

[1] Robot Rehabilitation Center, The Hyogo Institute of Assistive Technology, Kobe 651-2134, Japan
[2] Hyogo Prefectural Rehabilitation Center, Kobe 651-2134, Japan; chin@assistech.hwc.or.jp
* Correspondence: toda@assistech.hwc.or.jp; Tel.: +81-78-927-2727

Abstract: This study aimed to determine the characteristics of accelerations and angular velocities obtained by an inertial measurement unit (IMU) attached to a cane between older people with and without physical frailty. Community-dwelling older people walked at a comfortable speed using a cane with a built-in IMU. Physical frailty was assessed using exercise-related items extracted from the Kihon Check List. The efficacy of five machine learning models in distinguishing older people with physical frailty was investigated. This study included 48 older people, of which 24 were frail and 24 were not. Compared with the non-frail participants, the older people with physical frailty had a small root mean square value in the vertical and anteroposterior directions and angular velocity in the anteroposterior direction ($p < 0.001$, r = 0.36; $p < 0.001$, r = 0.29; $p < 0.001$, r = 0.30, respectively) and a large mean power frequency value in the vertical direction ($p = 0.042$, r = 0.18). The decision tree model could most effectively classify physical frailty, with an accuracy, F1 score, and area under the curve of 78.6%, 91.8%, and 0.81, respectively. The characteristics of IMU-attached cane usage by older adults with physical frailty can be utilized to effectively evaluate and determine physical frailty in their usual environments.

Keywords: cane; decision tree; frail; frequency analysis; gait; inertial measurement unit; machine learning; older people; root mean square

1. Introduction

In Japan, 29.0% of the population is 65 years old or more [1]. Age-related frailty is a factor that causes older adults to require nursing care. Frailty is a condition characterized by a reduction in various physiological reserves, thereby decreasing the capacity to effectively respond to stress [2]. Generally, people who experience frailty show a more rapid functional decline and lower resistance to illness and disability than healthy individuals [3,4]. Early detection of and intervention for frailty are essential for extending older people's healthy life expectancy.

The Kihon Check List (KCL) is a primary frailty screening tool used in Japan [5]. It is a 25-item, self-administered questionnaire that includes items on exercise, nutrition, oral health, confinement, cognition, and depression. However, questionnaire scoring may have been influenced by bias [6]. In addition, repeated assessments using a questionnaire over a short period may cause fatigue in respondents, thereby affecting the quality and reliability of the responses. Recently, wearable technologies have been used to evaluate the gait performance of older adults [7]. Wearable sensing technologies that can conduct a quantitative assessment daily may be useful for the early detection of frailty among older people. These technologies provide a cost-effective, efficient, and accessible method for standardizing data collection and analysis, both in clinical settings and within the living communities of frail older people [7].

A cane is an application of wearable technology. A cane is one of the most commonly used walking aids for older people. Consequently, previous studies have evaluated movements using an inertial measurement unit (IMU) built into a cane [8–11]. The IMU can be attached to the cane without any body contact, thus causing no pressure on the device and less strain on the user. In addition, evaluation using cane usage provides a less intrusive method that can continuously monitor changes in gait and balance without requiring daily active participation by the individual. This method reduces response fatigue to frailty assessments and provides a more sustainable approach in long-term monitoring with minimal impact on natural gait patterns. Thus, integrating sensing technology into canes provides a sustainable approach to early frailty detection. Characteristic waveforms observed during cane use [8] enable estimation of walking abilities and terrains using IMU data [12], and the frequency characteristics of acceleration provide further evaluative metrics. Since physical frailty in older adults is generally associated with decreased walking ability [13], changes in cane use may reflect this decline. However, to the best of our knowledge, physical frailty in older adults has not yet been evaluated using an IMU built into a cane.

If the differences in cane usage depending on physical frailty can be clarified and if physical frailty can be accurately classified, daily gait sensing using a cane can quickly detect a decline in the ability to walk due to physical frailty. This study aimed to compare the accuracy of some machine learning techniques in predicting physical frailty among older people based on the characteristics of the IMU sensor built into the cane. By analyzing these characteristics, we determine the most effective machine-learning method for classifying the presence and absence of physical frailty.

2. Materials and Methods

2.1. Participants

This study recruited older adults living in Harima-cho, Kako-gun, in Hyogo from September to October 2023. The inclusion criteria were as follows: ability to walk independently, a level of cognitive function sufficient to understand the research content, and no severe pain that significantly impacted walking. This study conformed to the principles of the Declaration of Helsinki [14] and obtained approval from the local ethics committee (R2303). All participants signed informed consent forms before data acquisition.

2.2. Data Collection

A cane, which is frequently used by older adults daily, was used for the measurements. An IMU (MTw; Xsens Technologies Inc., Enschede, The Netherlands) was fixed at the distal end of the cane using a jig fabricated using a 3D printer (Figure 1). Accelerations and angular velocities were sampled at 120 Hz. Before the measurement, the length of the cane was adjusted according to each participant's height. The participants identified the side of the cane that was comfortable for them to hold. If they experienced pain or disability, they held the cane on the opposite side. They were allowed to sufficiently practice walking to become accustomed to their environment. Furthermore, they were instructed to walk 1–2 round trips on a 7 m straight path at a comfortable speed with a two-movement gait in which the cane and the opposite foot were simultaneously grounded. The number of measurements depended on the participant's availability and the nature of the study conditions. For instance, if a participant was available for two round trips, we took measurements across these trials. Conversely, if a participant could attend only one round trip, we took a single measurement. Two trials were extracted from the round trip.

2.3. Data Analysis

The IMU data were filtered using a fourth-order Butterworth low-pass filter with a cutoff frequency of 10 Hz [15]. The ground contacts were detected using the jerk norm according to a previous study [8]. The waveform of the jerk norm shows that a large spike occurs when the cane touches the ground (Figure 2). Subsequently, 512 frames were extracted from the time of the first cane ground contact. The gait cycle was defined as the period between the cane contact. Subsequently, we counted the gait cycle time. In each gait cycle, the root mean square (RMS) of the acceleration was calculated in the vertical (VT), anteroposterior (AP), and mediolateral (ML) directions. In addition, the peak angular velocity per gait cycle was extracted for each direction. The calculated values were averaged.

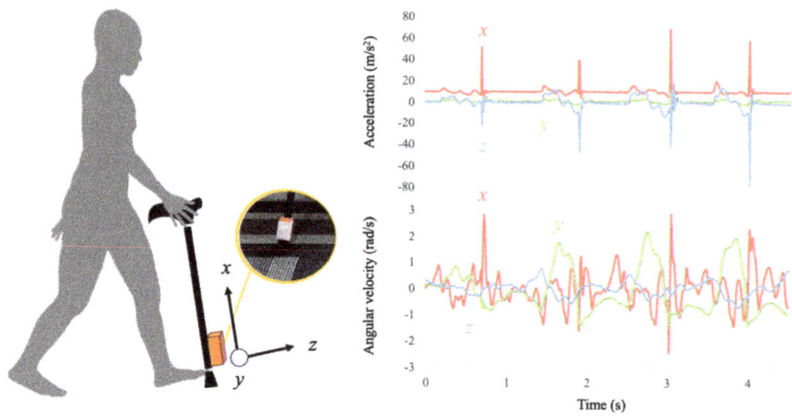

Figure 1. The proposed system using an inertial measurement unit built into a cane to evaluate physical frailty among older adults. The presented data are typical of measurement data.

To calculate the power spectrum, we applied a 512-point Fast Fourier Transform to the filtered acceleration through the periodic Hamming windows. The mean power frequency (MPF) was calculated within the frequency band of 0.5–10 Hz.

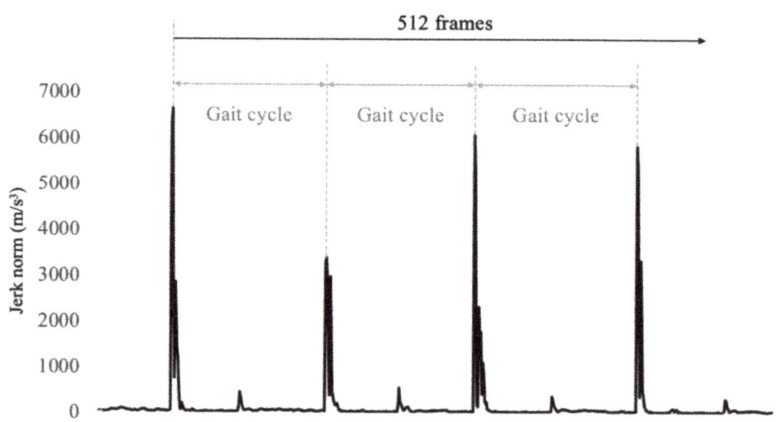

Figure 2. A typical example of time series data of jerk norm into measurements. The peak values indicate the ground contact of the cane. Peak value intervals are shown in walking cycles.

2.4. Frailty Assessment

We used the KCL to assess physical frailty and extracted its exercise-related section [16], considering that gait patterns were the most associated with physical frailty. This section was composed of five items: "Do you normally climb stairs without using handrail or wall for support?", "Do you normally stand up from a chair without any aids?", "Do you normally walk continuously for 15 min?", "Have you experienced a fall in the past year?", and "Do you have a fear of falling while walking?". Those who answered three or more negative answers to these five items were judged to have physical frailty.

2.5. Statistical Analysis

Before the dependent variables were compared, the normality assumptions of each data point were confirmed using the Shapiro–Wilk test [17]. Subsequently, the differences in the variables were compared between older adults with and without physical frailty, using the Mann–Whitney U test [18]. The effect size (r-value) was also calculated as the amplitude of the differences, which were classified as small ($0.1 \leq r < 0.3$), moderate ($0.3 \leq r < 0.5$), and large ($0.5 \leq r$) [19]. Statistical significance was set at $p < 0.05$. All statistical data were analyzed using SPSS statistical software (version 29.0; SPSS Japan Inc., Chicago, IL, USA).

In this study, we created five machine learning models to classify physical frailty in older adults, using MATLAB R2023b (MathWorks, Inc., Natick, MA, USA), decision tree (DT), linear discriminant analysis (LDA), k-nearest neighbors (k-NNs), support vector machine (SVM) with Gaussian kernel, and random forest (RF) [20]. These models were developed on a Mac (macOS version 14.6) equipped with an Apple M1 Max processor, 64 GB of RAM, and 2 TB of storage.

- DT: A simple and highly interpretable model that allows for a visual understanding of feature importance. In this study, the number of splits was set to 5.
- LDA: Used to evaluate whether the data are linearly separable. LDA is computationally efficient and effective when strong linear relationships are present.
- k-NN: A non-parametric method that is well suited to capturing local patterns in the data. It directly leverages similarities between different physical frailty states. The datasets were normalized to z-scores for this model.
- SVM: Capable of handling non-linear data and effective in classifying high-dimensional data. Using a Gaussian kernel, it can effectively classify data with complex boundaries. The datasets were normalized to z-scores for this model as well.
- RF: An ensemble learning method that helps prevent overfitting and offers high classification performance. It also enables the evaluation of feature importance. In this study, the ensemble method was performed using the following parameters: LogitBoost as the method, 243 learning cycles, and a learning rate of 0.59256. This ensemble model combines multiple weak learners to improve classification performance.

Hyperparameters that minimized the tenfold cross-validation loss were determined through automatic hyperparameter optimization. The dataset was randomly divided into 10 equal subsets (folds). Each fold was used once as test data, while the remaining folds were used to train the model. Because the split was random, data from the same individual could have appeared in both the training and test sets. Each participant contributed data from multiple trials, and these trials were treated as independent instances in the analysis. The classification accuracy was calculated as 1 − misclassification rate, and area under the curve (AUC), precision, recall, and F1 score (weighted average of precision and recall) were averaged across the ten models obtained through cross-validation.

3. Results

Table 1 presents the characteristics of the participants. Out of the 48 older people enrolled, 3 were excluded because their data could not be measured. Ultimately, we included 37 trials from 14 older adults with physical frailty and 94 trials from 31 older adults without physical frailty. Demographic data showed no significant differences between the older adults with and without physical frailty.

Figure 3 and Table S1 show a comparison of the parameters obtained using the IMU built into the cane in older adults with and without physical frailty. The RMS in the VT and AP directions and the angular velocity in the AP direction were significantly smaller in older adults with physical frailty than in those without physical frailty ($p < 0.001$, r = 0.36; $p < 0.001$, r = 0.29; $p < 0.001$, r = 0.30, respectively). The MPF in the VT direction was significantly greater in older adults with physical frailty than in those without physical frailty ($p = 0.042$, r = 0.18). Moreover, the RMS in the ML direction and MPF in the AP and ML directions, the angular velocity in the VT and ML directions, and stride time were not significantly different between older adults with and without physical frailty ($p = 0.50$, r = 0.06; $p = 0.543$, r = 0.05; $p = 0.111$, r = 0.14; $p = 0.177$, r = 0.12; $p = 0.971$, r = 0.00; $p = 0.427$, r = 0.07).

Table 1. Demographic characteristics of participants.

	Total	Physical Frail	Robust	*p*-Value
Age (years)	81.0 (7.5)	85.0 (9.0)	80.0 (7.5)	0.063
Sex, *n* (%)				
Male	13 (28.2)	4 (26.6)	9 (30.0)	
Female	32 (71.8)	11 (73.4)	21 (70.0)	
Height (m)	1.53 (0.17)	1.52(0.07)	1.53 (0.12)	0.238
Weight (kg)	50.5 (11.6)	48.4 (20.6)	51.0 (10.6)	0.887
BMI (kg/m^2)	22.3 (3.8)	22.4 (3.7)	22.5 (3.4)	0.198

Values: Median (interquartile).

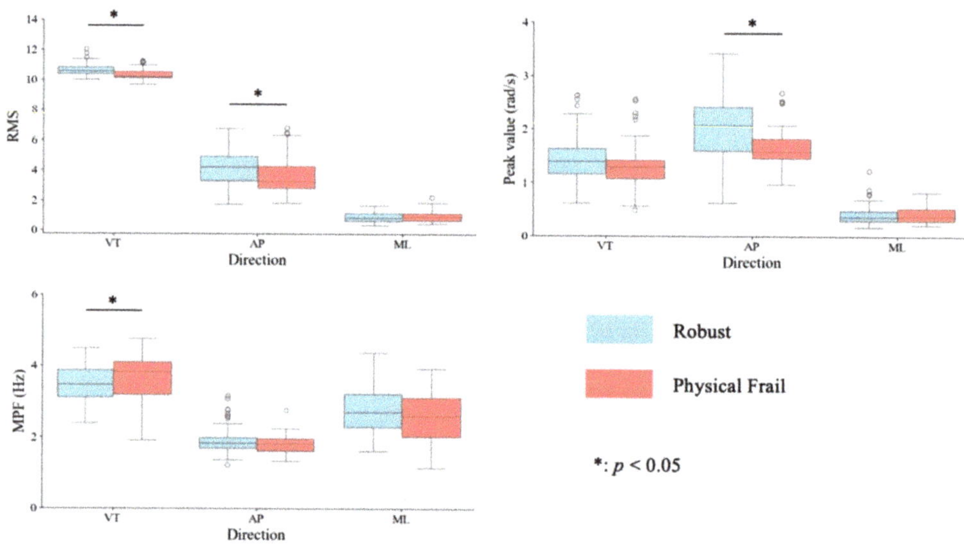

Figure 3. The box plots represent the root mean square (RMS) and mean power frequency (MPF) of acceleration and peak values of angular velocity in the vertical (VT), anteroposterior (AP), and mediolateral (ML) directions in older people with and without physical frailty. The box indicates the interquartile range. The horizontal line signifies the median, whereas the vertical lines represent the maximum and minimum values. The small circles indicate outliers. * indicates a significant difference between older people with and without physical frailty at $p < 0.05$.

Table 2 summarizes the performance, Table 3 indicates the confusion matrix, and Figure 4 illustrates the receiver operating characteristic (ROC) curve of models in classifying physical frailty among older people. The classification accuracies of the models ranged from 71.8 for the SVM to 78.6 for the DT model. The F1 score ranged from 83.6 for the SVM and RF to 91.4 for the DT. The AUC ranged from 0.50 for SVM to 0.96 for RF. Among the five machine learning algorithms, the DT model was the most suitable for classifying physical frailty.

Table 2. The model performance of the five machine learning algorithms.

	Decision Tree	Linear Discriminant Analysis	k-Nearest Neighbors	Support Vector Machine	Random Forest
Accuracy (%)	78.6	73.3	73.3	71.8	76.3
Precision (%)	96.7	92.1	90.6	100	99
Recall (%)	87.5	82.8	88.9	71.8	97.8
F1 score	91.8	87.0	89.7	83.6	98.3
AUC [1]	0.81	0.72	0.81	0.5	0.96

[1] AUC: area under curve.

Table 3. Confusion matrix of the five machine learning algorithms.

		Decision Tree		Linear Discriminant Analysis		k-Nearest Neighbors		Support Vector Machine		Random Forest	
		Actual		Actual		Actual		Actual		Actual	
		True	Negative	True	Negative	True	Negative	True	Negative	True	Negative
Predicted	True	90.9	13.0	86.6	18.0	85.2	10.6	94.0	37.0	93.1	2.2
	Negative	3.1	24.0	7.4	19.0	8.8	26.4	0.0	0.0	0.9	34.8

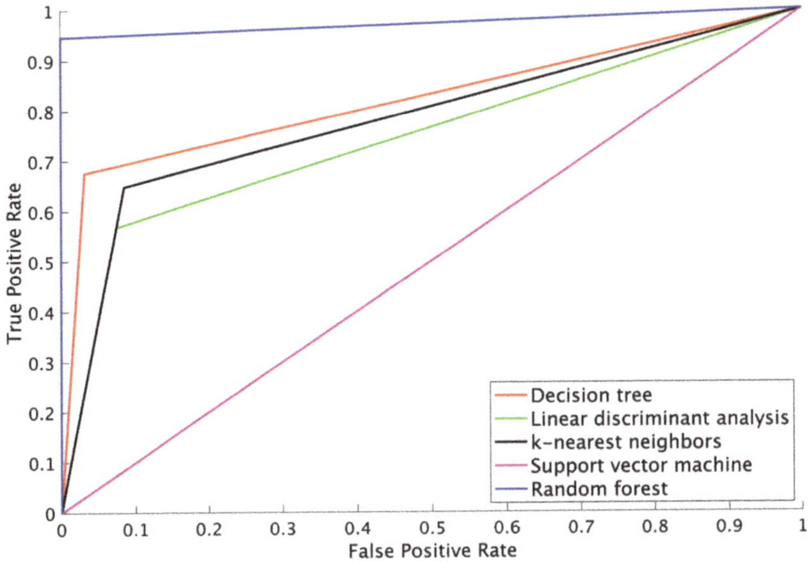

Figure 4. Receiver operating characteristic (ROC) curve of the five machine learning algorithms.

4. Discussion

This study examined the difference in cane use between older adults with and without physical frailty and created a classification model for physical frailty using the IMU built into the cane. The results showed that the characteristics of acceleration and angular velocity were significantly different and that the DT algorithm can classify older adults with and without physical frailty. Therefore, a cane with a built-in IMU can be utilized for the daily assessment of physical frailty in older people.

Older adults with physical frailty had significantly lower RMS and angular velocities in the AP direction than those without physical frailty. The RMS had approximately moderate effect sizes. Generally, stride length is shorter among older adults with physical frailty than among healthy older adults [21]. Hence, older adults with physical frailty have a decreased propulsive force during walking. Additionally, when the cane is grounded because of a small stride length, the posture angle is smaller [11]. Therefore, the forward momentum and swing of the cane decreased in older adults with physical frailty. These results were consistent with those of a previous study. Thus, in older adults with physical frailty, the ability to walk can be evaluated using the acceleration and angular velocity of the cane in the AP direction.

The small RMS of the cane in the VT direction reflects a small collision of the cane with the ground. At the initial contact, vertical ground reaction force is greater in a faster walking speed [22]. Therefore, collision intensifies when using a cane [11]. However, stride length is expected to be shorter in older adults with physical frailty than in those without physical frailty [23]. Thus, in our study, the RMS value in the VT direction became smaller among older adults with physical frailty than among the healthy ones.

In the frequency analysis, only the MPF in the VT direction of older adults with physical frailty was significantly smaller than that of older adults without physical frailty. According to Espy et al., compensating for the unstable walking patterns of older adults with physical frailty by reducing their stride length, which results in a shorter stride, seems logical [24]. For them, the cane becomes a stronghold to support their balance. In posture adjustment, the high-frequency component of time-series data reflects a reactive and adaptive balance strategy [25]. In this study, the MPF in the VT direction may have increased among older adults with physical frailty to support their unstable balance with a cane.

Among the five machine-learning models developed, the DT model demonstrated the highest efficacy in classifying physical frailty. This model could classify older people with physical frailty, with an accuracy rate of 78.6%, a precision of 96.8%, a recall of 88.3%, an F1 score of 92.4%, and an AUC of 0.82. Hence, the DT model had a very low rate of false positives, suggesting that it can reliably detect and correctly identify instances of physical frailty. Our metrics are comparable to the results of previous studies that classified physical frailty using an insole sensor [26,27] and sit-to-stand motion patterns [28]. In addition, a model with an AUC of 0.801–0.841 could be judged excellent [29]. The DT model can visually represent hierarchical interconnections among the parameters [20]. Considering the benefits of using an IMU-fitted cane to assess gait, the physical frailty of people using this cane can be evaluated daily.

On the other hand, although the RF also demonstrated large metrics, these results were obtained from the model using tenfold partitions, and its averaged accuracy was lower than that of the DT model. This indicates that the RF model may have a lower generalization performance and could overfit the data. Therefore, this study concluded that the DT model was the most suitable for classifying physical frailty from both performance and interpretability perspectives. Improving the performance of the RF model may require increasing the number of subjects, which should be addressed in future research.

This study had some limitations. First, this study did not collect objective biomechanical data, such as gait speed, kinematics, or the kinetics of walking. Thus, the differences in the RMS, angular velocity, and frequency characteristics measured with the cane were attributable to the differences in the gait of older adults with physical frailty. This limitation may have introduced a potential source of bias as more detailed biomechanical assessments could offer deeper insights into the underlying mechanisms of gait in this population. Future studies should aim to include detailed kinematic and kinetic measures, as well as comfortable walking speed and other spatiotemporal measures, to better understand the underlying mechanisms of gait differences between frail and non-frail older adults. Second, this study had a small sample size; thus, the results could not be generalized. Nevertheless, the differences in some parameters obtained through the IMU built into the cane had moderate effect sizes. Thus, if gait can be clearly assessed using the IMU built into the cane, developing this into a large cohort or longitudinal data measurement may be possible. Third, the participants in this study comprised both habitual cane users and non-users. Future studies should isolate the effects of physical frailty on gait characteristics without confounding factors of varying levels of experience with cane use. Fourth, only the exercise-related section of the KCL was used to determine the presence of physical frailty. This method does not provide a complete picture of physical frailty, which is a multi-dimensional condition [5]. Finally, this study used only the IMU to minimize the computational cost and weight impact. Previous studies have used a strain gauze and pressure sensor [8,9,30]. We were able to consider the type of sensor attached to the cane to improve the evaluation accuracy.

5. Conclusions

The characteristics of using a cane in older people with physical frailty include acceleration, angular velocity, and the acceleration's frequency response in the VT and AP directions. These characteristics were used in creating a machine learning model to determine physical frailty. We ultimately found that the DT model demonstrated the most suitable performance based on evaluation metrics such as accuracy, precision, recall, and AUC. The DT model's simplicity and interpretability make it highly practical for clinical settings, where understanding the decision-making process is crucial. This study suggests that physical frailty can be effectively assessed daily using an IMU attached to the cane in older adults.

However, this study has limitations, including limited sample size and data collected from a single facility, which may affect the generalizability of the results. Future work should involve validating the model with larger and more diverse datasets and exploring other machine learning methods and evaluation metrics to enhance model performance and clinical applicability.

6. Patents

The results of this study validate the recently published patent, "Frail state evaluation assist device, frail state evaluation assist system, frail state evaluation assist method, and frail state evaluation assist program". JP Patent No. 7455263 was filed on 10 August 2023 and issued on 25 March 2024.

Supplementary Materials: The following supporting information can be downloaded at: https://www.mdpi.com/article/10.3390/s24216910/s1, Table S1: Values obtained by inertial measurement unit attached to a cane in older people with and without physical frailty.

Author Contributions: Conceptualization was performed by H.T. and T.C. Funding acquisition, validation, data curation, investigation, formal analysis, and methodology, resources, and software were performed by H.T. Supervision was performed by T.C. The original draft was written by H.T., and the final version was reviewed by all authors. All authors have read and agreed to the published version of the manuscript.

Funding: This study was supported by the JSPS KAKENHI Grant Number JP 23K16508, Japan.

Institutional Review Board Statement: This study was conducted in accordance with the Declaration of Helsinki and was approved by the Institutional Review Board of The Hyogo Institute of Assistive Technology (R2303).

Informed Consent Statement: Informed consent was obtained from all subjects involved in this study.

Data Availability Statement: Data available on request due to restrictions (ethical reasons).

Conflicts of Interest: The authors declare no conflicts of interest.

References

1. Annual Report on the Ageing Society. Available online: https://www8.cao.go.jp/kourei/whitepaper/index-w.html (accessed on 20 May 2024).
2. Watanabe, D.; Yoshida, T.; Watanabe, Y.; Yamada, Y.; Miyachi, M.; Kimura, M. Validation of the Kihon Checklist and the frailty screening index for frailty defined by the phenotype model in older Japanese adults. *BMC Geriatr.* **2022**, *22*, 478. [CrossRef] [PubMed]
3. Satake, S.; Shimokata, H.; Senda, K.; Kondo, I.; Toba, K. Validity of total Kihon Checklist score for predicting the incidence of 3-year dependency and mortality in a community-dwelling older population. *J. Am. Med. Dir. Assoc.* **2017**, *18*, 552.e1–552.e6. [CrossRef] [PubMed]

4. Hirashiki, A.; Shimizu, A.; Suzuki, N.; Nomoto, K.; Kokubo, M.; Hashimoto, K.; Sato, K.; Kondo, I.; Murohara, T.; Arai, H. Composite biomarkers for assessing frailty status in stable older adults with cardiovascular disease. *Circ. Rep.* **2022**, *4*, 123–130. [CrossRef]
5. Arai, H.; Satake, S. English translation of the Kihon Checklist. *Geriatr. Gerontol. Int.* **2015**, *15*, 518–519. [CrossRef] [PubMed]
6. Satake, S.; Senda, K.; Hong, Y.J.; Miura, H.; Endo, H.; Sakurai, T.; Kondo, I.; Toba, K. Validity of the Kihon Checklist for assessing frailty status. *Geriatr. Gerontol. Int.* **2016**, *16*, 709–715. [CrossRef]
7. Fan, S.; Ye, J.; Xu, Q.; Peng, R.; Hu, B.; Pei, Z.; Yang, Z.; Xu, F. Digital health technology combining wearable gait sensors and machine learning improve the accuracy in prediction of frailty. *Front. Public Health* **2023**, *11*, 1169083. [CrossRef]
8. Dang, D.C.; Suh, Y.S. Walking distance estimation using walking canes with inertial sensors. *Sensors* **2018**, *18*, 230. [CrossRef]
9. Gill, S.; Seth, N.; Scheme, E. A multi-sensor matched filter approach to robust segmentation of assisted gait. *Sensors* **2018**, *18*, 2970. [CrossRef]
10. Nohara, R.; Toda, H.; Tada, M.; Fujita, K. Relationship between walking ability and cane movement detectable by inertial sensors. *Trans. Jpn. Soc. Med. Biol. Eng.* **2021**, *59*, 14–23.
11. Ribeiro, N.F.; Mouta, P.; Santos, C.P. Two kinematic data-based approaches for cane event detection. *J. Ambient. Intell. Humaniz. Comput.* **2023**, *14*, 567–579. [CrossRef]
12. Gill, S.; Seth, N.; Scheme, E. A multi-sensor cane can detect changes in gait caused by simulated gait abnormalities and walking terrains. *Sensors* **2020**, *20*, 631. [CrossRef]
13. Studenski, S.; Perera, S.; Patel, K.; Rosano, C.; Faulkner, K.; Inzitari, M.; Brach, J.; Chandler, J.; Cawthon, P.; Connor, E.B.; et al. Gait speed and survival in older adults. *JAMA* **2011**, *305*, 50–58. [CrossRef] [PubMed]
14. World Medical Association. World Medical Association Declaration of Helsinki. Ethical principles for medical research involving human subjects. *Bull. World Health Organ.* **2001**, *79*, 373.
15. Antonsson, E.K.; Mann, R.W. The frequency content of gait. *J. Biomech.* **1985**, *18*, 39–47. [CrossRef]
16. Sewo Sampaio, P.Y.; Sampaio, R.A.C.; Yamada, M.; Arai, H. Systematic review of the Kihon Checklist: Is it a reliable assessment of frailty? *Geriatr. Gerontol. Int.* **2016**, *16*, 893–902. [CrossRef] [PubMed]
17. Shapiro, S.S.; Wilk, M.B. An analysis of variance test for normality (complete samples). *Biometrika* **1965**, *52*, 591–611. [CrossRef]
18. Mann, H.B.; Whitney, D.R. On a test of whether one of two random variables is stochastically larger than the other. *Ann. Math. Stat.* **1947**, *18*, 50–60. [CrossRef]
19. Cohen, J. *Statistical Power Analysis for the Behavioral Sciences*; Routledge: London, UK, 2013.
20. Imura, T.; Toda, H.; Iwamoto, Y.; Inagawa, T.; Imada, N.; Tanaka, R.; Inoue, Y.; Araki, H.; Araki, O. Comparison of supervised machine learning algorithms for classifying home discharge possibility in convalescent stroke patients: A secondary analysis. *J. Stroke Cerebrovasc. Dis.* **2021**, *30*, 106011. [CrossRef]
21. Kawai, H.; Obuchi, S.; Ejiri, M.; Ito, K. Association between daily life walking speed and frailty measured by a smartphone application: A cross-sectional study. *BMJ Open* **2023**, *13*, e065098. [CrossRef]
22. Nilsson, J.; Thorstensson, A. Ground reaction forces at different speeds of human walking and running. *Acta Physiol. Scand.* **1989**, *136*, 217–227. [CrossRef]
23. Kressig, R.W.; Gregor, R.J.; Oliver, A.; Waddell, D.; Smith, W.; O'Grady, M.; Curns, A.T.; Kutner, M.; Wolf, S.L. Temporal and spatial features of gait in older adults transitioning to frailty. *Gait Posture* **2004**, *20*, 30–35. [CrossRef]
24. Espy, D.D.; Yang, F.; Bhatt, T.; Pai, Y.C. Independent influence of gait speed and step length on stability and fall risk. *Gait Posture* **2010**, *32*, 378–382. [CrossRef] [PubMed]
25. Stoffregen, T.A.; Pauelsen, M.; Jafari, H.; Strandkvist, V.; Nyberg, L.; Gustafsson, T.; Vikman, I.; Röijezon, U. Frequency domain shows: Fall-related concerns and sensorimotor decline explain inability to adjust postural control strategy in older adults. *PLoS ONE* **2020**, *15*, e0242608.
26. Anzai, E.; Ren, D.; Cazenille, L.; Aubert-Kato, N.; Tripette, J.; Ohta, Y. Random forest algorithms to classify frailty and falling history in seniors using plantar pressure measurement insoles: A large-scale feasibility study. *BMC Geriatr.* **2022**, *22*, 746.
27. Kraus, M.; Saller, M.M.; Baumbach, S.F.; Neuerburg, C.; Stumpf, U.C.; Böcker, W.; Keppler, A.M. Prediction of physical frailty in orthogeriatric patients using sensor insole-based gait analysis and machine learning algorithms: Cross-sectional study. *JMIR Med. Inform.* **2022**, *10*, e32724. [CrossRef]
28. Park, C.; Mishra, R.; Sharafkhaneh, A.; Bryant, M.S.; Nguyen, C.; Torres, I.; Naik, A.D.; Najafi, B. Digital biomarker representing frailty phenotypes: The use of machine learning and sensor-based sit-to-stand test. *Sensors* **2021**, *21*, 3258. [CrossRef]
29. Mandrekar, J.N. Receiver operating characteristic curve in diagnostic test assessment. *J. Thorac. Oncol.* **2010**, *5*, 1315–1316. [CrossRef]
30. Wade, J.W.; Boyles, R.; Flemming, P.; Sarkar, A.; De Riesthal, M.; Withrow, T.J.; Sarkar, N. Feasibility of automated mobility assessment of older adults via an instrumented cane. *IEEE J. Biomed. Health Inform.* **2019**, *23*, 1631–1638. [CrossRef]

Disclaimer/Publisher's Note: The statements, opinions and data contained in all publications are solely those of the individual author(s) and contributor(s) and not of MDPI and/or the editor(s). MDPI and/or the editor(s) disclaim responsibility for any injury to people or property resulting from any ideas, methods, instructions or products referred to in the content.

sensors

Review

A Review on Assisted Living Using Wearable Devices

Grazia Iadarola, Alessandro Mengarelli, Paolo Crippa, Sandro Fioretti and Susanna Spinsante *

Dipartimento di Ingegneria dell'Informazione, Università Politecnica delle Marche, 60131 Ancona, Italy; g.iadarola@staff.univpm.it (G.I.); a.mengarelli@staff.univpm.it (A.M.); p.crippa@staff.univpm.it (P.C.); s.fioretti@staff.univpm.it (S.F.)
* Correspondence: s.spinsante@staff.univpm.it

Abstract: Forecasts about the aging trend of the world population agree on identifying increased life expectancy as a serious risk factor for the financial sustainability of social healthcare systems if not properly supported by innovative care management policies. Such policies should include the integration within traditional healthcare services of assistive technologies as tools for prolonging healthy and independent living at home, but also for introducing innovations in clinical practice such as long-term and remote health monitoring. For their part, solutions for active and assisted living have now reached a high degree of technological maturity, thanks to the considerable amount of research work carried out in recent years to develop highly reliable and energy-efficient wearable sensors capable of enabling the development of systems to monitor activity and physiological parameters over time, and in a minimally invasive manner. This work reviews the role of wearable sensors in the design and development of assisted living solutions, focusing on human activity recognition by joint use of onboard electromyography sensors and inertial measurement units and on the acquisition of parameters related to overall physical and psychological conditions, such as heart activity and skin conductance.

Keywords: wearable devices; assisted living; human activity recognition; electromyography; skin conductance; biomedical measurement systems

1. Introduction

According to the "2024 Revision of World Population Prospects" published by the United Nations [1], the total global population aged 65 and older is projected to reach 2.2 billion, surpassing the number of children (under age 18) by the late 2070s, and a similar increasing trend will be observed for the population aged 80 and over, which is projected to reach 265 million as early as the mid-2030s. This aging of people has an ever-increasing impact on families, communities, and societies in most countries worldwide. While, on the one hand, a longer life expectancy is the result of social progress and impressive improvements in medicine and healthcare, on the other hand, the rate of older adults living longer but losing their autonomy because of poor health conditions has a relevant impact on the financial sustainability of modern public healthcare and welfare systems [2,3]. Therefore, ever-increasing efforts and new approaches are needed to assist the independent daily lives of the elderly, resulting in a growing interest in research and development of methods, technologies, and solutions for active and assisted living (AAL) [4].

Among the chronic conditions that deeply affect the quality of life of older adults, it is worth mentioning age-related neural diseases, such as Parkinson's and Alzheimer's diseases. They are often characterized by a progressive, and sometimes quick, decline in functional and cognitive abilities. Therefore, ensuring independent living for people suffering from these diseases is very difficult, especially due to the progressive behavioral and psychiatric symptoms that are often present in affected subjects. As a result, caring for these individuals is particularly stressful, and the quality of life experienced by family caregivers is strongly compromised. Since a lower quality of life often translates into increased absences from work and reduced work productivity, it is very important to identify devices and systems that can help and support caregivers in the daily activities of caring. This way, the quality of life of people in need of assistance and those who care for them will improve, and the social costs linked to diseases will be reduced [5,6]. Needs expressed by caregivers of people affected by dementia or related diseases pertain to the physical, psychological, and social demands of providing assistance [7]. Among them, it is of paramount importance to provide caregivers with assistive devices and systems that can help detect potentially harmful, unsafe, or dangerous activities in order to generate alerts and prevent accidents. Human activity recognition (HAR) is an engineering research area devoted to the implementation of devices, systems, and methods for the automatic recognition and classification of human activities based on sensor data. It has been demonstrated that HAR applied to simple household activities (e.g., sleeping or walking, going up or down the stairs, and opening or closing doors) as well as to complex activities (e.g., riding a bicycle or driving a car) can strongly support the surveillance, care, and treatment of the elderly or people suffering from age-related diseases [8]. In addition to the above considerations, HAR also plays a fundamental role in the design and assessment of assistive robotic devices, dealing with locomotion and other body movements [9], which are important components of modern AAL systems.

In a similar manner, AAL strongly relies on the capability of monitoring people's conditions, i.e., physical, mental, and emotional, by the continuous, long-term, and minimally invasive acquisition of health-related signals, from which quantitative parameters can be extracted. The latter may be of huge importance, for example, in checking the progression of diseases, evaluating the steadiness of physical and cognitive functionalities, and detecting acute episodes that could lead to a decrease in autonomy and quality of independent living. In this context, modern wearable devices can help break the paradigm of traditional medicine based on single visits, scattered over time, to clinics and health facilities for the execution of specific tests by healthcare operators, using clinical instruments in controlled environments instead to promote approaches based on long-term multidimensional observations in living environments and in everyday life conditions. The data collected from wearable devices include a variety of body signals that can aid in the implementation of personalized interventions, supporting preventive measures and lifestyle modifications against the development of chronic conditions limiting elderly well-being [10]. However, some aspects that can still limit the contribution that wearable devices are able to provide to the development of AAL systems must be considered, in particular with regard to the quality of the signals acquired in the presence of disturbances determined by the conditions of use, the environment, and the lack of industry standards [11].

In this work, the attention is focused on the possibilities opened up by wearable devices for the implementation of effective and user-acceptable AAL solutions by combining different embedded sensors made available by recent developments in electronic design. Starting from Section 2, we provide an overview of the electronic components and sensors found in wearable devices, such as accelerometers and sensors for the acquisition of surface electromyographic (sEMG) and other body signals, also looking at commercially available devices. In Section 3, the related processing approaches enabling the monitoring of both activity recognition and physiological parameters by the integrated use of sensors, with different scopes and applications, are presented. Section 4 provides an overall discussion of the role of wearable devices in AAL, and finally, Section 5 draws the main conclusions of the work.

2. Wearable Devices Enabling AAL Solutions

AAL solutions are based on information and communication technology (ICT) systems to support and enhance the quality of life and well-being of older or impaired people in both indoor and outdoor environments using a distributed network of sensors and actuators to maintain the people's independence and autonomy, while ensuring safety and providing assistance. HAR is considered a pivotal technology within the field of AAL in that it is responsible for the automatic detection and classification of the activities performed by individuals using sensor-based systems. Researchers and engineers have exploited cutting-edge advances in micro- and nano-electronics, flexible and integrated circuits, signal processing, and communication network protocols to create small, low-power, and low-cost sensors that can be easily worn or placed directly on the body skin to acquire large amounts of data, such as inertial body information, heart rate (HR), muscle fatigue, skin conductance (SC), and so on [12–14]. These tiny sensors and devices may even power themselves from human motion while sensing [15].

2.1. Acceleration Sensors

The simplest way to recognize human activities is to acquire body speed and orientation based on accelerometers placed on the body. Often, in embedded systems, accelerometers are augmented with gyroscopes that measure the angular velocity along three axes, which can be used to increase the accuracy of electronic compasses and improve the estimate of the orientation, otherwise performed only with the accelerometer.

Usually, for a given set of activities of interest in AAL applications, a single accelerometer sensor, positioned on the wrist, ankle, or alternatively embedded in a smartwatch, may be sufficient to monitor elderly living at home or in residential centers [16–20] in a very simple manner. In the last decade, the huge development of the global smartwatch market has made wearable and autonomous accelerometer sensors available everywhere and to everyone. Many HAR systems have been developed by exploiting data acquired by these personal devices, and the placement of multiple accelerometer sensors on the body has been considered as well [21].

More details about the experimental setup used in the above-mentioned studies, together with some details about the sensor characteristics, like full scale and sampling frequency, are given in Table 1. In the case of elderly people, these systems could be used to detect alarming conditions generated by the unusual behavior of the person (inability to get out of bed, absence of activity for a defined period of time) or changes in routine activities related to psychomotor or neurodegenerative diseases. In applications dealing with rehabilitation, these systems could be used to monitor training routines, e.g., to count repetitions of exercises and determine the fatigue or energy expenditure associated with individual movements.

Table 1. Experimental setup information of works using wearable accelerometers for human activity recognition (SR stands for sampling rate).

Reference	Number, Type, How It Is Used	Sensor Characteristics	Sensor location	Activities Considered for Recognition
Rodriguez-Martin et al. [16]	single, triaxial, used alone	model LIS3LV02DQ, ±6 g full scale, SR: 200 Hz	wearable, on waist	walking, stand, sit, lying, sit to stand, stand to sit, bending up/down, lying from sit and sit from lying transitions
Vakacherla et al. [17]	single, triaxial, used alone	-2000 to 2000 cm s^{-2} full scale, SR: 200 Hz	wearable, on user chest	standing, walking on level ground, walking on an incline, running, squatting
Tian et al. [18]	single, triaxial, used alone	±6 g full scale, SR: 150 Hz	wearable, on waist	sitting, jumping, walking, standing, lying, running, ascending stairs, descending stairs

Table 1. *Cont.*

Reference	Number, Type, How it is Used	Sensor Characteristics	Sensor Location	Activities Considered for Recognition
Meng et al. [19]	single, triaxial, used alone or joint with gyroscope and sEMG	SR: 148 Hz for accelerometer and gyroscope, SR: 1260 Hz for sEMG sensor	wearable, on waist	walking, tooth brushing, face washing, drinking
Mannini et al. [20]	two, triaxial, used independently	±4 g full scale, SR: 90 Hz	wearable, on dorsal aspect of dominant wrist and on outside of the ankle	many (23 to 26) laboratory-based physical activities and common daily activities, grouped into four classes: sedentary, cycling, ambulation, other activities
Mannini et al. [21]	five, biaxial, on-body network	SR: 76.25 Hz	wearable, on: hip, wrist, arm, ankle, thigh	sitting, lying, standing, walking, stair climbing, running, cycling

2.2. EMG Sensors

The surface electromyography (sEMG) signal, generated by the electrical potential of muscle contractions, is one of the easiest biological signals to acquire using non-invasive sensor systems; therefore, it has proven to be very useful in monitoring muscle strength, health status, fitness levels, and physical performance [22–26].

The sensors used to acquire sEMG signals can easily be derived from those used for acquiring ECG signals [27–29], even if they are quite different from those used to detect cardiac activity [30] due mainly to the larger bandwidth of the EMG signals that have a frequency content that extends up to 500 Hz [31] and is more likely to be acquired during movements, being therefore affected by artifacts induced by body and cable movements at frequencies usually below 5 Hz. This must be taken into account when designing sensors capable of acquiring both EMG and ECG signals [32]. In particular, low-frequency motion artifacts (MAs) must be rejected by the high-input-impedance low-noise amplifier, which should be carefully designed to capture the relatively low amplitude of the sEMG signal without incurring saturation [28,33,34]. Furthermore, data derived from integrated triaxial accelerometers can be used to reduce the MAs [35] and improve the performance of the sEMG acquisition node when used for activity detection and classification.

Innovative and diverse health monitoring systems and interfaces [36] have been made possible thanks to advances in several design aspects, such as micro- and flexible electronic components [37], new nanotechnologies, tiny wireless communication interfaces, as well as powerful but lightweight machine learning algorithms [38–40]. Body data acquisition has become easier, more affordable, and reliable thanks to sensors that, being small, ultra-low-power, and wireless-enabled, can be embedded into small devices (smartwatches, smartphones, wristbands, rings, and glasses), clothing, or directly placed on the skin [27,41,42]. For these reasons, healthcare and AAL domains have started to increasingly use and rely on wearable sensors [43]. As an example, in [44], a wearable, lightweight, and low-cost sEMG sensor is implemented in combination with a machine learning classifier to identify neck postural risks caused by prolonged work or human behavior involving upper extremities. Furthermore, in [45], a wearable system using an e-textile substrate combines signals from sEMG sensors with inertial data from IMUs acquired from the cervical region to monitor in real-time the occurrence of forward head posture that is a habitual bad posture of the neck, consisting of the forward translation of the cervical vertebrae and hyperextension of the upper ones.

Table 2 provides a summary of studies involving the development and use of miniaturized multimodal sensors, often implemented as systems-on-chip (SoCs), for different acquisition and monitoring purposes.

Table 2. Summary of studies describing miniaturized sensors, and sensors able to perform multimodal signal acquisition for different monitoring purposes.

Reference	Acquisition Capability	Acquired Signals	Notes
Doheny et al. [40]	multimodal, compressed sensing and wireless powering capabilities	EMG, ECG	battery-free, standard 0.18-μm CMOS technology occupying a silicon area of 4.25 mm^2
Dow et al. [36]	multimodal	EMG, ECG, vibration, temperature	analog front-end in 180 nm CMOS technology, applied in elderly care and precision sports study
Orguc et al. [34]	modular, low-voltage, ultra-low-power (3.8 nW)	EMG	65 nm CMOS technology, on 0.22 mm^2 silicon area
Jani et al. [46]	multimodal, low-power	EMG, ECG	8.85 cm^2 sensor readout PCB area, sensor weight $\leq$ 10 g
Biagetti et al. [47]	multimodal, three channels, low-power	EMG, ECG, IMU	13.77 cm^2 sensor readout PCB area, sensor weight: 40 g
Said et al. [48]	multimodal, real-time data acquisition	EMG, ECG, triaxial acceleration, electrodermal activity (EDA), temperature	used to implement real-time monitoring systems
Pinto et al. [49]	multimodal	sEMG, ECG, triaxial acceleration	used in aquatic environments
Zhang et al. [50]	multimodal	five channel EMG, triaxial acceleration	used in hand gesture recognition
Wu et al. [51]	multimodal, real-time data transmission or local storage	four channel EMG, nine-axis motion sensor	100 Hz sampling rate for inertial signals, IEEE 802.15.4 wireless module
Tanweer et al. [52]	multimodal	EMG, ECG, IMU	to be applied on user chest, inertial sensor signals used for motion artifact removal
Das et al. [53]	multimodal	EMG, ECG, electrooculography (EOG), EEG	custom-designed ultra-low-noise instrumental amplifier used as an analog front-end with both programmable gain and bandwidth
Ding et al. [54]	hybrid (electrophysiological, acoustic, optical)	sEMG, mechanomyography (MMG), near-infrared spectroscopy (NIRS)	used to study interaction among grip strength, blood oxygen metabolism, and acquired signals
Ke et al. [55]	modular and multimodal, onboard signal conditioning and amplification	EMG, forcemyography (FMG)	floating electrodes used, A/D signal conversion at the base station

Typically, wearable sEMG sensing nodes for AAL applications must efficiently transmit the acquired data to base stations wirelessly using low-power and possibly low-cost solutions. Therefore, several wireless EMG signal acquisition systems use 2.4 GHz radio links based on the IEEE 802.15.4 standard protocol [56] to transmit their data [57,58], exploiting the well-known ZigBee technology. In [28], a low-cost wearable system based on several ultralight wireless sensor nodes has been proposed to acquire and process EMG, ECG, and acceleration signals and transmit data toward base stations through a custom communication protocol designed on the IEEE 802.15.4 physical layer to exploit current technology and improve the throughput and synchronization allowed by the standard.

In [47,59], a three-channel wireless EMG/ECG/IMU sensor using a Bluetooth Low Energy (BLE) radio for real-time data transmission has been proposed in a basic hardware configuration to minimize power consumption. Real-time operation is ensured by an adaptation layer on top of BLE that enables tight time synchronization and reliability of data transmission, while energy efficiency is ensured by motion-triggered wake-up capability, thanks to the onboard inertial measurement unit (IMU) that switches on the system only when worn.

2.3. Skin Conductance Sensors

The human skin exhibits time-varying electrical characteristics, as captured by a continuous-time signal known as skin conductance (SC), or, alternatively, the galvanic skin response (GSR), i.e., the reciprocal of skin conductance, namely skin resistance [60], or electrodermal activity (EDA). Those changes in the skin's electrical properties are due to body sweating [61], which indicates psychological or physiological arousal reflecting people's emotions. Physiology underneath classical approaches to SC analysis relates SC variations to the status of sweat glands spread over the skin and regulated by the autonomic nervous system (ANS): high arousal of the ANS sympathetic branch increases the activity of the sweat glands, thus enhancing SC. Conversely, elevated sweat gland activity indicates both physiological and psychological arousal, which may be captured by changes in the SC. On the other hand, the activity of sweat glands is autonomously influenced by the sympathetic nervous system (SNS) branch driving subconscious human behavior so that the SC signal may be exploited by devices aimed at supporting meditation, monitoring stress, and the overall physical and psychological condition of a subject [62].

Given these characteristics of human physiology, it can be observed that slow changes of the SC, known as the tonic level or skin conductance level (SCL), mostly depend on skin dryness, hydration, and automatic body temperature regulation, while rapid variations of the SC, i.e., the so-called phasic level or skin conductance response (SCR), represent dynamic changes associated with the reaction of a subject to different stimuli, either physical, emotional, or cognitive [61], thus reflecting their psychological [63] and emotional status. SC has been discovered to have a positive correlation with various affective and cognitive encounters, such as stress, anxiety, fear, physical exertion, attention, and high cognitive load. As a result, SC serves as a readily available and sensitive indicator of sympathetic drive, making it an ideal measure of psychophysiological arousal in a conscious and healthy individual [64].

While it is possible to obtain direct recordings of sympathetic nerve bundles by inserting a thin microelectrode under the skin, more comfortable and non-invasive methods are preferred for longitudinal recordings. One such method involves passing a small external current through two electrodes (dry or wet) placed on the skin's surface and calculating electrical conductance, which is the reciprocal of resistance. Electrodes are typically located on the second and third fingertips of the same hand (usually the dominant one). The variations of the voltage generated by a small applied current circulating through the skin between the electrodes provide a measure of the SC. Such a measuring approach is implemented in laboratory equipment, where a proper DAQ ensures quite a high sampling frequency (typically, 1000 Hz [65]). With the recent advent of comfortable-to-wear commercial healthcare devices, usually designed as bracelets, watches, or rings, the SC measure has also become available in uncontrolled settings to support research in the neuroscience domain. In this method, referred to as exosomatic, an external energy source, such as a constant voltage or current, is employed. The signals detected primarily arise from alterations in resistance and allow for an estimation of increases in sympathetic nerve activity, even in the absence of observable sweating. It is worth noting that both the skin and sweat glands possess resistive and capacitive properties that can be effectively modeled [66]. It is important to note that there are various other techniques available for measuring electrodermal activity. However, the most extensively studied and wearable approach involves the method discussed above. Typically, researchers measure and report the SCL, which represents a slow-moving average of the electrodermal activity signal over time.

Additionally, they also analyze SCRs, which capture more rapid fluctuations and phasic changes in the activity [64,67].

Serving as a crucial link between the brain and behavior, peripheral physiological signals offer valuable insights into the workings of the body and mind. Nevertheless, the current range of peripheral physiology sensing devices faces several obstacles when it comes to effectively utilizing their data. For example, not all the available devices provide access to raw SC data, which are needed to design custom processing algorithms in research studies. Some commercial devices provide so-called stress indexes, which are derived from raw data by algorithms that are typically not disclosed by manufacturers. Being truly wearable means being able to wear something comfortably, and smartwatches provide a convenient and socially acceptable option. They have easily readable displays and can be customized through various apps and interconnected devices [64]. For scientists, these small wearable sensors are minimally intrusive and can be easily incorporated into a wide range of experimental designs. Additionally, they can be used over a long period of time by multiple individuals simultaneously and by vulnerable populations that are typically not well-represented in traditional neuroscience research. Furthermore, they are significantly more affordable compared with most neuroimaging equipment. Consumer-grade devices are remarkable engineering achievements that are priced affordably, making them accessible to a wide range of users. However, they have limited access to derivative data, which are calculated using undisclosed algorithms. This limitation restricts their usefulness for innovative scientific investigations.

When exploring the market of commercial portable and wearable devices capable of acquiring SC, which could be used in human monitoring applications, not so many options emerge; those available can be categorized as research devices and consumer-grade devices. One of the wearable devices most commonly used in research is the Empatica E4 (Empatica Inc., Cambridge, MA, USA), which is now going to be replaced by the newest Embrace one from the same manufacturer. Both are clinically validated devices (Class IIa Medical Device according to 93/42/EEC Directive), which can be used to measure the impact of different stimuli [68,69] or to classify elicited emotions [70]. Similar to the E4 device, Embrace hosts four sensors (a triaxial accelerometer, a skin thermometer, a triaxial gyroscope, and an SC sensor) and also provides several software tools to process the collected data on the Cloud platform where they are stored. Shimmer3 GSR+ (Shimmer Research Ltd., Dublin, Ireland) is another well-adopted device in research studies, thanks to the pre-amplified channel available for SC acquisition, which makes it possible to measure the electrical characteristics of the skin as well as an optical pulse/PPG signal on a different but synchronized channel for heart rate estimation. As research-grade devices, Empatica, Shimmer, and BioPac support extensive access to high-quality data.

However, these devices are priced at unaffordable levels. In fact, they offer access to raw data, which is different from consumer-grade devices, allowing in-depth analysis but with expensive licensing. Conversely, consumer-grade devices, conveniently worn on the wrist, only offer limited access to consumer-grade data derived through signal processing algorithms that are often not openly disclosed. Consequently, this can pose challenges when attempting to interpret the data within research contexts. The GoBe3 smart band (Healbe Corporation, Redwood City, CA, USA), for example, acquires a so-called emotional tension related to feelings and mood, somehow derived from the SC by analyzing the changes in cutaneous sweating. Additionally, three other embedded sensors are also provided, namely a bioimpedance sensor, an accelerometer, and a piezo sensor. Other consumer-grade devices that allow to acquire SC data are Fitbit Charge and Fitbit Sense (FitBit Inc., San Francisco, CA, USA), Amazfit Helio Ring (Amazfit, Hefei, China), Nuanic (Nuanic Oy, Tampere, Finland) and EmotiBit (Connected Future Labs, LLC, New York, NY, USA). Fitbit Charge and Fitbit Sense are devices equipped with an SC sensor that can measure the user stress level, and apps to provide insights on how to manage it. Amazfit Helio Ring and Nuanic are rings that can measure SC. Finally, EmotiBit is a device designed to provide an affordable solution that grants users access to raw data. Importantly, the data obtained from EmotiBit are fully owned by the user, ensuring privacy and control. Additionally, EmotiBit operates within an open-source ecosystem, allowing for flexibility and customization to explore and ask new types of questions about the physiological signals generated by bodies and minds. EmotiBit bridges the gap between research and consumer-grade devices by offering affordable access to raw data. This empowers users to delve deeper into their physiological signals and encourages the exploration of novel scientific inquiries [71]. Table 3 lists the commercially available wearable and portable devices for SC monitoring. It is interesting to notice they come in different options, from wrist-worn devices to portable acquisition systems to minimally invasive rings, usually intended for elite athletes, such as the Amazfit Helio Ring.

Table 3. Comparison of commercial wearable and portable devices for EDA monitoring.

Device	Measured Quantity	Sampling Frequency	Availability of Raw Signal Samples	Type of Device
Amazfit Helio Ring	skin conductance	not provided	no	wearable
BioPac MP36R (BIOPAC Systems, Inc., Goleta, CA, USA)	skin conductance	500 Hz	yes	portable
EmotiBit	skin conductance	not provided	yes	wearable
Empatica Embrace	skin conductance	4 Hz	yes	wearable
Fitbit	skin conductance	not provided	no	wearable
GoBe3	skin resistance	10 Hz	no	wearable
Nuanic	skin conductance	not provided	no	wearable
ProComp Infiniti	skin conductance	256 Hz	yes	portable
Shimmer3 GSR+	skin resistance	32 Hz	yes	wearable

Klimek et al., in [72], analyzed several studies involving the acquisition of SC, identifying sixteen different wearable devices, among which the Empatica wristband was the most common (40.1% of the screened articles), followed by the Shimmer3 and Affectiva Q (Affectiva Inc., Boston, MA, USA) sensors. Overall, the vast majority of studies (almost 70%) used a wristband device, while a glove with integrated sensors was used in very few devices. Regarding the position of the electrodes, most of the studies referred to fingertips (Shimmer3 and BioPac systems, Neurobit Optima 4 (Neurobit Systems, Gdynia, Poland)), and a few experiments used palmar electrodes (Mindplace Thoughstream (Mindplace, Eastsound, WA, USA) or collected SC signals from the chest (using the RespiBAN Professional (PLUX wireless biosignals S.A., Lisbon, Portugal)), or the torso, by custom manufactured garment. Betancourt et al. [73] measured SC on the ankle using the Affectiva Q sensor.

2.4. PPG Sensors

Remote monitoring of subjects in AAL applications, where it is important to guarantee the personal safety of people who, perhaps due to age or cognitive disorders, could carry out risky activities, is certainly a research topic of crucial importance. Joint with activity monitoring and recognition, it is important to keep track of vital signs, such as blood oxygenation (SpO2), HR, or respiratory rate [74], while still allowing the person as much freedom of movement as possible [75]. Thus, it is important to collect and process a combination of data coming from heterogeneous physical domains, from which it is possible to extract much more significant and precise information than using just one type at a time. This can be done by exploiting photoplethysmography, an optical non-invasive measurement technique of peripheral blood flow commonly used to acquire parameters such as SpO2, HR, or heart rate variability (HRV).

A photoplethysmograph works by shining light, possibly at different wavelengths, through the skin of the person wearing the device and detecting the amount of light reflected or passed through, depending on the mechanical arrangement of the light source and detector. A green light source is commonly used for HR estimation from changes in blood volume during heartbeats, as this usually provides the highest signal-to-noise ratio. Both red and infrared light sources are instead used to measure blood oxygenation (oxygenated hemoglobin absorbs infrared light more and red light less, unlike de-oxygenated hemoglobin). Therefore, for greater flexibility, integrating all three types of light sources into the same sensor is the best choice for commercial and research devices that use the PPG signal and its second derivative, the acceleration plethysmography (APG) [76], for estimating HR and HR variations, as well as oxygen saturation and blood pressure [77–80].

When PPG signals are acquired in controlled clinical settings where sensors are often applied to the fingers, head, or earlobe, MAs are inherently weak. In this case, signal processing techniques based on time-frequency algorithms, such as empirical mode decomposition [81] and its modifications [82], or wavelets [83] and ad hoc filtering techniques, such as adaptive step-size least mean squares filter [84], can be successfully used.

In AAL applications where, instead, comfort and low invasiveness are important, PPG sensors are embedded in wearable devices that can be worn on the wrist, arm, or fingers, such as smartwatches, smart bracelets, or smart rings. Onboard these devices, miniaturized, low-power semiconductor technologies and multimodal signal acquisition and processing techniques have made it possible to obtain clean PPG signals by removing strong MAs derived from motion activities carried out by the people wearing them [85]. For example, a 2.76 mm^2 chip that implements a light-to-digital converter for long-term continuous PPG monitoring was fabricated in TSMC 180 nm technology [86]. The widespread diffusion of these commercial wearable devices has also suggested the use of PPG signals alone [87,88] or, better, in combination with inertial signals [89] for the recognition of human activities, as it will be better explained in Section 3.5.

3. Integration and Processing of Wearable Sensors Data for AAL

As highlighted in previous sections, HAR is one of the most important functions of AAL systems and applications. HAR of the upper limb pursued by inertial data found widespread diffusion, but many other studies focused on the fusion of such kind of data with physiological signals able to provide further information regarding the upper limbs, hands, and finger movements since many activities performed in the daily living involve fine gestures, e.g., deployed for achieving grasping of different objects. Indeed, physiological biosignals, such as EMG and SC, are strongly related to HAR and within an AAL context, can provide additional, valuable information for the physiological state functional assessment of the user in terms of motor activity, mental stress, and, in general, human well-being. In particular, muscular activity recorded by surface EMG represents one of the major ways for HAR implementation. The following sections present different sensor integration modalities and related signal-processing approaches.

3.1. Integration of Inertial and EMG Sensors for HAR

The effectiveness of EMG signals, recorded with a sparse or high-density setup, for performing hand gesture identification using pattern recognition modalities is well-acknowledged [90]. Myoelectric activity represents one of the principal solutions for measuring and assessing physical activity in healthy and pathological individuals [91], but the widespread diffusion and availability of portable devices for EMG recording, such as armbands for forearm muscle recording [92], made this kind of biosignal a suitable solution for HAR in AAL applications. Indeed, EMG-based architectures have been extensively employed for developing human–machine interfaces for robotic exoskeletons and power prostheses [90]. However, they found extensive usage also for AAL-related applications [93], such as activity recognition, often in conjunction with inertial sensing data [94], affective computing, e.g., for micro-expression recognition and emotional response assessment [95], and eating behavior during the daily living [96–98]. Moreover, recent advancements in terms of probes design for biosignal recording [99] have allowed their embedding within garments or clothing, enlarging the field of applicability of the EMG signals toward home-based healthcare applications [42].

Integration of inertial sensors and EMG sensors was largely leveraged for hand gesture recognition, aimed at prosthetic and assistive device control. Shahzad et al. [100] used IMU information to make the myoelectric-based pattern recognition system aware of the position of the arm in order to enhance classification performances. Two magnetic, angular rate, and gravity (MARG) sensors were placed on the wrist and on the biceps brachii muscle and raw data were processed to obtain the orientation of the arm segments. Four EMG probes were placed instead on the flexor-extensor muscles on the forearm, and six hand motions were performed by volunteers during static and dynamic movements of the arm along defined trajectories. Thus, in this case, inertial information was leveraged to reconstruct the upper limb pose and not for feature extraction and pattern recognition. EMG and inertial data fusion, due to augmented information available, enhance the pattern recognition outcomes for gesture-related tasks, and thus, it was employed when a particularly challenging recognition objective must be achieved.

For instance, this configuration was explored for allowing the correct identification of sign language to enable communication and bridge the gap between deaf people and hearing people [51]. In the latter work, instead of using an instrumented glove, the forearm was chosen as the sensor location, and it was instrumented with four EMG probes and a single IMU sensor placed on the wrist, resembling a smartwatch location. From the latter, accelerometer and gyroscope data were considered. An adaptive auto-segmentation method was also proposed in order to potentially allow the system to work in real-time, based on the EMG signal energy computed on windows of 128 ms. Two different sets of features were extracted from myoelectric signals and inertial signals, outlining that the different nature of these kinds of data require dedicated feature selection to be computed. Overall, 9 features were computed on EMG and 13 from the tri-axial acceleration and angular velocity and from their magnitude, providing the dimensionality of the final feature space equal to 268. Classical machine learning models (DT, SVM, kNN, and NB) were used for classifying 80 sign language words repeated by four volunteers 25 times. Features were ranked separately for EMG and IMU channels in order to check which source of information was the most useful for the purposes of the study. Outcomes showed that acceleration was the most important modality in terms of feature ranking, followed by gyroscope data. EMG resulted in the least important, as confirmed also by the classification results, where the inclusion of EMG features only slightly improved the performance (about 4%), which with inertial data were only just above 90% for the 80 signs. However, EMG features improved the recognition of some specific signs that likely share some characteristics in terms of acceleration or angular rate, and thus, additional information was needed for correct identification. This indicates once more the value of inertial data for HAR, but also highlights that the selection of source information to be used within the classification pipeline is strongly dependent on the gestures or activities that are taken into account for a certain application.

3.2. Integrated HAR for Rehabilitation

Rehabilitation is an important field of research where multi-sensor fusion was explored. Song et al. [101] focused their efforts on the development of the wearable multimodal serious game for hand rehabilitation in stroke patients by proposing a multi-sensor fusion for hand movement classification involving IMU, force myography, and EMG. The system encompasses six EMG probes (Trigno wireless system (Delsys Inc., Natick, MA, USA)) placed on the forearm, eight barometric sensors, and one IMU mounted on the wrist. Hand activities considered for rehabilitative purposes involved twelve grasping tasks and forearm movements related to activities of daily living. Force myography and IMU were recorded at 36 Hz, whereas EMG data were collected at 1926 Hz. Features were computed on 200 ms windows with an increment of 50 ms, which is also suitable for real-time purposes. The mean absolute value was used as the only feature from IMU and force myography, together with eight additional features from EMG channels. LDA was employed as the preferred model for the classification of hand movements. In this case, IMU data alone showed the lowest performance, well below 60%, while the addition of EMG features raised the accuracy to about 75%, only slightly below the best configuration given by all three sources of information (80% accuracy). These results indicate that the selection of the appropriate source of information is crucial for HAR and is strongly dependent on the final objective of recognition. In this case, movements to be recognized were mainly hand gestures, which are more suitable for recognition by the myoelectric activity of the forearm. Indeed, EMG alone provided an accuracy comparable to that observed when both EMG and IMU data were used.

The same authors [102] investigated another setup, where two IMUs (MTw Awinda and Xsens (Xsens Technologies B.V., Enschede, The Netherlands)) were placed on the upper arm and the forearm in order to detect gross movements of the arm, whereas finer gestures of the wrist and hand were identified through six EMG probes (Trigno wireless EMG system) placed around the forearm. Also, in this case, eight barometric sensors (MPL115A2 and the Freescale semiconductor) were attached around the wrist. The system was composed of a serious interactive game for shoulder and elbow rehabilitation, and this is why gross movements needed to be identified. A total of six shoulder and elbow movements were considered, together with seven hand gestures. Features were only extracted from EMG and barometric sensors, whereas IMU data were used to measure the upper limb pose using quaternion estimation. This work highlights the wide potential of inertial data since they allow us to not only extract meaningful features from raw acceleration, gyroscope, and even magnetometer readings but also to obtain information related to the pose in the 3D space of the human segment where the sensor is mounted, thus significantly extending the type of information that can be retrieved from this kind of portable devices.

Eventually, it deserves to be mentioned that integration between inertial and EMG data was also leveraged by choosing the wrist as the location for IMU sensors, as in [103], where a hierarchical method for hand motion recognition was proposed, with a probe setup made by two EMG probes attached on the forearm, and a single IMU sensor mounted on the wrist, from which acceleration and gyroscope data were recorded for estimating device orientation. The proposed method showed high recognition accuracy, but it should be noted that only six common hand gestures were considered, mainly involving forearm and wrist motions, which can be relatively easily recognized when 3D pose data are available from an inertial device, outlining once more the potential value of a wrist-mounted IMU device for HAR related to the upper limb.

3.3. Integrated HAR for Industrial Applications

Due to their complexity, EMG and inertial data were also explored in the context of industry-related applications. For instance, in [104], an extensive dataset of working activities was collected using 17 IMU sensors (Xsens Technologies B.V., Enschede, The Netherlands) and 16 EMG channels prototyped in laboratory, located on the upper limb while performing seven activities that involve object manipulations, resembling industrial tasks. Human activities that involve full-body movements were also investigated for producing reliable recognition by integrating different sources of biosignals. In [105], a ResNetXt DL model, with three kernel sizes in each multi-kernel module, was proposed for the recognition of a series of activities related to walking, such as walking upstairs and downstairs, curving, spinning, running, and some others. A publicly available dataset was used in [106], where sensing technology encompasses IMU sensors embedding an accelerometer and gyroscope located on the thigh and shank, together with four EMG probes placed on thigh muscles and gastrocnemius. Although good results were achieved with the proposed learning scheme, the proposed setup can hardly be employed in a real usage scenario due to the high number of probes and multiple locations that could be uncomfortable to use on a daily basis.

Locomotion-related activities were also considered by Zhou et al. [107], whose objective was the recognition of four ambulation modalities, namely jumping, walking, ascending stairs, and descending stairs. For this purpose, eighteen volunteers were instrumented with three IMU sensors (sampling rate 100 Hz) and five EMG probes (sampling rate 1 kHz). IMUs were placed on the chest, the lateral side of the thigh, and the lateral side of the shank, whereas EMG electrodes were located on three thigh muscles and two calf muscles. A total of 15 features in multiple domains were extracted from EMG and acceleration and gyroscope signals, with 200 ms windows and an increment of 40 ms. Feature selection was performed by a Markov random field-based Fisher–Markov selector that minimizes the within-class distance of features and maximizes the between-class distance. Then, a series of ML models were compared (kNN, SVM, ensemble learning, and the back propagation neural network) with 10-fold cross-validation. Also, in this case, IMU-based features showed high discriminative capabilities, and the EMG feature only provided recognition accuracy less than 80% for the locomotion modalities, whereas IMU features showed about 90% accuracy for each classifier, confirming their value for this kind of pattern recognition problems, where similar human movement have to be correctly identified. However, sensor placement targeted the proposed pipeline for specifically recognizing walking-related movements, being hardly generalizable towards a different kind of movement, e.g., of the upper limb.

A focused analysis of inertial and myoelectric information was proposed in [108] in order to unveil if the compensatory balance responses needed for avoiding falls in free-living conditions are better detected with IMU, EMG, or a combination of them. Three Shimmer ExG wireless units with 3D IMU and two EMG probes were used in this study, and four additional EMG probes were used in this study, mounted on the shank, thigh, and sternum, whereas the myoelectric activity of two thighs and two shank muscles was recorded. Volunteers were asked to walk along a 10 m path, and they were randomly perturbed by lateral pushes to the right shoulder during the left and right swing phases of gait in order to deliver a side-step and a cross-over reaction, respectively. Then, features were extracted separately from IMU data and EMG data, and a set of hybrid features was also computed based on spatial and temporal distance between EMG and inertial signals. Three different kinds of classification tasks were tested involving the distinction between normal gait and perturbed gait and the type of perturbation, but also, in this case, outcomes showed that inertial information has the potential to be leveraged alone for recognizing such kinds of highly dynamic activities, outperforming EMG information.

3.4. HAR by Commercially Available Devices

In the past few years, devices such as armbands became commercially available for a variety of possible applications, from gaming to the interaction between humans and machines. In this field, the Myo armband (Thalmic Labs Inc., Kitchener, ON, Canada) was one of the first examples of this kind of technology, followed in more recent years by other devices such as the MindRove. In general, such wearable devices include several EMG probes and one IMU or MARG sensor embedded in a circular structure that should be worn by the subject on their forearm. Thus, a significant amount of information can be recorded since the disposition of the EMG probes allows us to record the forearm's myoelectric activity, even if sensor location cannot be accurate, as this happens when gold-standard EMG probes are directly applied to the muscles of interest. Due to the novelty of the Myo armband, many studies tried to leverage this device, mainly for gesture recognition, due to the constraint of wearing the armband on the forearm. In [92,109], the Myo armband was leveraged for gesture recognition using only eight EMG probes, whereas in [110], the same device was used as a human–machine interface for the control and trajectory planning of a robotic arm; in this case, using EMG data and IMU as well but considering only a limited set of hand and wrist poses. A control system for the robotic arm was developed in the Simulink environment, and IMU data were used for obtaining the position and orientation of the robot. With an inverse kinematic analysis, the angular position of the robot joint was estimated to satisfy the desired end-effector trajectory. The effectiveness of the Myo armband for gesture recognition was also investigated in [111], where the addition of the IMU data was considered to address the problem of the multi-user usability of pattern recognition architectures since the latter represents a well-acknowledged problem in the field of myoelectric-based gesture and activity recognition. Further, the influence of different upper limb positions on the recognition of seven hand gestures was also explored. In this case, the improvement given by IMU data was strongly dependent on the specific ML model used for classification, highlighting, in this case, the non-negligible role of the pattern recognition architecture that has to be implemented.

The use of the Myo armband for recognizing human activities not limited to hand and wrist gestures was investigated in [112], where a feature selection method was employed for identifying the most meaningful features to recognize ten human activities, namely handshaking, hugging, bouncing, running, sitting, bending forward, standing, walking, sprawling, and applauding. However, in this case, only EMG data were processed, without relying on IMU data provided by the armband, and no classification was carried out. However, the investigation was limited to feature ranking. On the other hand, HAR by the Myo armband was exploited in [94], where a series of activities strongly related to the daily living scenario were considered. It is worth noting that the choice of the above-mentioned activities was based on their clinical relevance, on the basis of the functional arm activity behavioral observation system (FAABOS), a taxonomy to quantify the relevance of upper limb activities on various levels of significance [113]. Thus, a total of 17 activities were selected and grouped into four categories, i.e., non-functional activities, non-task-related activities, task-related activities, and high-exertion activities. EMG data were recorded at 200 Hz, whereas inertial data were recorded at 50 Hz. A set of simple time domain features were extracted from tri-axial acceleration, angular velocity, i.e., mean value, and magnitude area, whereas, for the eight-channel EMG, only the root-mean-square value was considered as a feature. Ten unimpaired volunteers were recruited for the study, who each performed the activity six times. As a pattern recognition algorithm, the kNN was used, where the number of neighbors was set as three. In this case, inertial data alone provided a promising 78.3% accuracy, outlining the value of this kind of data for HAR. As expected, the inclusion of myoelectric information boosted the performance to about 85%, but surprisingly, the best outcomes were observed when EMG was used in conjunction with gyroscope data, with 89.2% accuracy in recognizing all activities. Thus, the angular rate of change can be itself a valuable source of information for activities that involve upper limb motions, but this indicates once more that an a priori selection of data sources for HAR application is hardly feasible and can be accomplished only after preliminary evaluations, related to the specific kind of activities, are made. Finally, selecting the set of human activity based on recognized criteria, i.e., using the FAABOS taxonomy in this case, provides advantages in terms of study replicability and would allow the direct comparison of results obtained from different works, thus strengthening the field of HAR based on wearable devices. The FAABOS taxonomy is becoming an acknowledged basis for the choice of activities to be included in experimental setups when dealing with activity recognition in a scenario close to real contexts [114] because such kinds of movements are also meaningful from a clinical viewpoint.

Finally, it is interesting to note that integration between EMG and inertial data has also been employed for estimating the physiological state of individuals, such as fatigue. In [115], the Myo armband was employed to monitor fatigue in construction workers. In particular, the aim was the prediction of the aerobic fatigue threshold using EMG and inertial data of the forearm by estimating the average oxygen consumption that is related to the fatigue threshold by the maximum aerobic capacity. The fatigue threshold was estimated on windows of five minutes. In addition to the Myo armband, a metabolic analyzer and a heart rate monitor were also used, the former as ground-truth and the latter for comparative purposes. For classification, a bidirectional RNN-LSTM network was exploited. Fourteen scaffold-building activities were selected for eliciting fatigue in the ten recruited volunteers, with five different fatigue levels, and each activity was performed for no more than 10 min. The proposed methodology provided good results in terms of both the classification of the five levels of fatigue and the estimation of the threshold over time, suggesting the suitability of myoelectric and IMU data integration for monitoring fatigue levels over time.

The recognition of fatigue using inertial devices was also investigated in [116], where the focus was given to the gait task. Volunteers were asked to perform three different modalities: the first one was a common walking task alongside a straight path. The second one was a fatigued walking task where fatigue was elicited previously by a tiptoeing task and performed until the subject was no longer able to perform the task. Then, the same previously described walking task was performed. The cadence of the fatigued gait was recorded, and, on the second day, the subjects were asked to walk with a cadence equal to that recorded during the fatigue task, imposed by a metronome. Volunteers were instrumented with six IMU sensors (Trigno Avanti, Delsys Inc.) placed on the head, the second sacral vertebra, the left and right toes, and the heels. Additionally, six EMG probes were used for recording shank muscle activity and confirming the fatigued state of the subject by observing the median frequency over time and its descending rate. An LSTM network was used to classify the three gait modalities. The best performances were given by the combination of IMUs mounted on the toe and sacrum; IMU on the toes alone provided the best recognition of the three gait modalities (about 95%), whereas head-mounted devices showed the worst identification performances (around 60%). Although very promising results were shown in this study, the sensor setup required for the proposed application appears cumbersome for practical purposes, where the objective should be the reduction of the sensor's number in order to avoid obtrusiveness and enhance user comfort. A summary of the works where inertial and myoelectric signals were fused for HAR is reported in Table 4.

3.5. SC and PPG Processing for AAL

As previously mentioned, the SC is measured to assess the sympathetic branch activity of the ANS that relates to attention and reaction to stress conditions. While the SCL is made up of the signal gradual tidal variation level and spontaneous signal changes, i.e., SCRs, also known as electrodermal responses (EDRs) or SC peaks, which are expressions of the subject's reaction to specific stimuli.

A de facto standard method to process the SC signal is called trough-to-peak (TTP). The method searches for events in which the SC signal goes from a local minimum to a local maximum within a fixed response time window (typically 4 s). The SCR amplitude is defined as the difference between the SC values at the peak and at the preceding trough. A minimum amplitude criterion (e.g., 0.05 or 0.01 μS) is often applied to identify a peak. The TTP method is only used for analyzing acquired signals, and it does not rely on deconvolution, i.e., it does not require making an assumption about the shape of the SC peak signal generated by the underlying physiological reaction to a stimulus. A different approach decomposes SC data into SCL and SCR parts, the latter of which includes SC peaks that are of interest to analyze. This way, unsuperposed response components may be extracted to quantify SC peak characteristics in terms of amplitude and time distribution. SCR extraction may be performed according to so-called continuous decomposition analysis (CDA) or discrete decomposition analysis (DDA). The former aims at retrieving traces of the underlying ANS activity from the signal characteristics: SC data are deconvolved by a general response shape and then decomposed into continuous phasic and tonic components. The latter decomposes SC data into distinct phasic components and a tonic component by means of non-negative deconvolution, which is mainly aimed at studying the physiological models of the SC peaks. Both methods may also find application in the synthetic generation of SC signals by numerical simulation. Several free and open-source software (FOSS) are available for processing measured SC signals. Some of the most popular ones are OpenSignals (ver. 2.2.5), BioPatRec (2009 release), and OpenBCI GUI (ver. 6.0.0-beta.1). All of them can be used to record and analyze physiological signals such as SC, ECG, EMG, and more. They are also compatible with various sensors and devices and can be used for research and educational purposes.

Table 4. Summary of experimental setup information from works in which the fusion of inertial and myoelectric data was applied for human activity recognition.

Reference	Type and Number of Sensors	Probes Location	Raw Inertial Signals	Activities	Sensors Characteristics
Totty and Wade [94]	Myo Armband (8 EMG probes and 1 IMU)	Forearm	Triaxial ACC and GYRO. Feature extraction	4 categories of FAABOS activities	SR: 50 Hz (IMU), 200 Hz (EMG)
Shazad et al. [100]	4 EMG probes and 2 MARG	Forearm and upper arm	Triaxial ACC, GYRO, and MAG. Pose estimation	Six hand gestures for six forearm positions	SR: 1 kHz (EMG)
Wu et al. [51]	1 MARG and 4 EMG probes	Forearm (EMG) and wrist (MARG)	Triaxial ACC and GYRO. Feature extraction	80 ASL	SR: 100 Hz (MARG), 1 kHz (EMG), RES: 0.4 μV (EMG)
Bassani et al. [104]	17 MARG and 4 EMG probes	Full body (MARG), upper arm and forearm (EMG)	Triaxial ACC, GYO, and MAG	8 working activities	Full scale ± 160 m s^{-2} (ACC), $\pm 2000\ ^\circ$ s^{-1} (GYRO), SR: 240 Hz (MARG), 500 Hz (EMG)
Liu et al. [106]	4 EMG probes and 2 IMU	Thigh, shank	Triaxial ACC, GYRO	22 walking-related activities	SR: 1 kHz (EMG), 100 Hz (IMU), RES: 16 bit
Zhou et al. [107]	5 EMG probes and 3 IMU	Chest (IMU), thigh (EMG and IMU), shank (EMG and IMU)	Triaxial ACC and GYRO. Feature extraction	4 walking-related activities	SR: 1 kHz (EMG), 100 Hz (IMU)
Nouredanesh and Tung [108]	3 MARG and 4 EMG probes	Chest (IMU), thigh and shank (EMG and IMU)	Triaxial ACC and GYRO. Feature extraction	Detection of compensatory balance responses	SR: 512 Hz (EMG, IMU)
Song et al. [101]	6 EMG probes, 1 MARG, 8 barometric pressure sensors	Forearm (EMG), wrist (MARG, barometric sensors)	Triaxial ACC, GYRO, and MAG. Pose estimation	11 hand gestures for rehabilitation with serious game	SR: 1926 Hz (EMG), 36 Hz (MARG, barometer)
Song et al. [102]	2 MARG, 6 EMG probes, 8 barometric pressure sensors	Forearm (EMG, MARG), upper arm (MARG), wrist (barometric sensors)	ACC, quaternions. Pose estimation and feature extraction	6 gross arm movement and 7 hand gestures	SR: 2 kHz (EMG), 90 Hz (barometer), 40 Hz (MARG)
Chang et al. [103]	2 EMG probes, 1 IMU	Forearm (EMG), wrist (IMU)	Triaxial ACC and GYRO. Segmentation and feature extraction	6 hand gestures	SR: 1.5 kHz (EMG), 50 Hz (IMU)

Table 4. *Cont.*

Reference	Type and Number of Sensors	Probes Location	Raw Inertial Signals	Activities	Sensors Characteristics
Biagetti et al. [117]	3 EMG and acceleration probes	Upper arm	Triaxial ACC. Feature extraction	4 strength exercises	SR: 2 kHz (EMG, ACC), full scale ±12 g (ACC)
Schabron et al. [110]	Myo Armband (8 EMG probes and 1 IMU), 1 IMU	Forearm (EMG), robot end-effector (IMU)	Triaxial ACC and GYRO. Pose estimation	5 hand gestures for robotic control	SR: 50 Hz (IMU), 200 Hz (EMG)
Colli Alfaro and Trejos [111]	Myo Armband (8 EMG probes and 1 IMU)	Forearm	Triaxial ACC and GYRO. Feature extraction	7 hand gestures in 4 arm positions	SR: 50 Hz (IMU), 200 Hz (EMG)
Bangaru et al. [115]	Myo Armband (8 EMG probes and 1 IMU)	Forearm	Triaxial ACC and GYRO. Feature extraction	Estimation of 5 levels of fatigue	SR: 50 Hz (IMU), 200 Hz (EMG)
Lee et al. [116]	6 IMU, 6 EMG probes	Shank (EMG), Head, lower back, feet (IMU)	Triaxial ACC and GYRO. Feature extraction	Fatigue identification	Not reported

ACC: accelerometer; ASL: American sign language; FAABOS: functional arm activity behavioral observation system [113]; GYRO: gyroscope; MAG: magnetometer; SR: sampling rate.

While the responses of individuals in terms of SC and their physiological reactions may vary, wearable devices present an opportunity to customize treatment plans by providing real-time information on activities that either enhance or reduce arousal levels. This allows for targeted interventions aimed at specific attention-related and cognitive outcomes. Wearable technology is increasingly being tested in studies addressing the identification of personalized biomarkers in populations of subjects affected by anxiety, depression, or autism spectrum disorder (ASD) [64]. Remarkably, the acceptance and adherence to the use of wearables strongly depend on the body position they require to work, which is usually minimally invasive (e.g., wrist, hand, and fingertips) but not necessarily best accepted. In this regard, the proper positioning of SC electrodes to maximize the quality of the collected signals but also the user acceptance was investigated in [118] by comparing foot and fingers and in [119], where 16 different body locations were compared to measure the SC following emotional stimulation. Positioning electrodes on fingers was the choice resulting in the best signal quality, but finger sensors, which can be doable in laboratory conditions, may limit the freedom of subjects in daily life settings. It is interesting to notice that sensors in the form of jewellery-like rings, recently introduced to the market (e.g., the Nuanic smart ring), have not yet been extensively investigated for use in studies dealing with stress monitoring. Other body locations that exhibited good SC responsiveness in experiments are the forehead, shoulders, neck, chest, and even shoulders [119]. Wearable technologies can be developed for these bodily locations, making them more suitable for usage. The accuracy of devices to measure SC can vary depending on the device and the adopted method [120]. It is important to note that the accuracy of the device can be affected by various factors, such as the quality of the signal, the calibration of the device, and the placement of the electrodes [121].

Regarding the processing of PPG signals acquired by wearable devices outside of clinical contexts, as in most of AAL applications, ambient light and MAs sum up the semiconductor noise of the sensing element and greatly affect the collected signal quality in terms of the signal-to-noise ratio (SNR) and dynamic range (DR). Specific circuit blocks are consequently added to the sensor design to attenuate these interferences, further increasing power consumption. To extend the PPG sensor's lifetime, ultra-low-power designs, circuit techniques, and sampling schemes have been proposed in the literature, as presented in [122]. The need for algorithms that meet real-time and low-power constraints for MA removal directly onboard the small and lightweight wearable devices commonly used for AAL applications has yielded excellent results in PPG acquisition while performing normal daily human activities [123]. A list of established real-time and low-power state-of-the-art methods used to remove MAs from PPG signals, which may or may not require the support of an additional IMU or a triaxial accelerometer embedded in the same device, is reported in Table 5.

Table 5. Summary of real-time and low-power processing approaches to remove artifacts from PPG signals with and without the use of an additional accelerometer.

Reference	Artifact Removal Technique or Method	Target Artifact	Additional Accelerometer Needed?
Kim et al. [124]	independent component analysis combined with block interleaving and low-pass filtering	MA	Yes
Foo [125], Gibbs et al. [126]	adaptive filtering	MA	Yes
Tanweer et al. [127]	reference generation using singular value decomposition and multistage application of filtered X-LMS algorithm	MA during jogging	Yes

Table 5. *Cont.*

Reference	Artifact Removal Technique or Method	Target Artifact	Additional Accelerometer Needed?
Chowdhury et al. [128]	multiple reference adaptive noise cancellation algorithm	MA, by four versions of cleaned PPG signal	Yes
Galli et al. [129], Lee et al. [130]	Kalman filtering	MA, improves HR tracking from PPG signal	Yes
Zhang et al. [131]	general framework tailored to hardware design	extremely strong MAs, improves HR estimation	Yes
Zhang et al. [132]	JOint Sparse Spectrum reconstruction (JOSS)	extremely strong MAs, improves HR estimation with no extra signal processing needed	Yes
Islam et al. [133]	cascade of adaptive, recursive-least-squares filters, aided by singular spectrum analysis	MA, improves HR tracking	Yes
Zhang et al. [134]	spectrum subtraction and empirical mode decomposition	MA, improves HR changes tracking	Yes
Islam et al. [135]	spectrum subtraction and empirical mode decomposition combined with a higher-accuracy wavelet-Fourier frequency estimator	MA, improves HR changes tracking	Yes
Biagetti et al. [136]	geometric separation of signal subspaces complemented with an automatic activity intensity classifier	extremely strong MAs, improves HR estimation	Yes
Zhao et al. [137]	short-time Fourier transform and statistical-based tracking algorithm	MA	No
Brophy et al. [138]	convolutional neural network without inertial data	MA in HAR and HR estimation	No
Dubey et al. [139]	specialized approach	MA in HR monitor during running	No
Wójcikowski et al. [140]	time-domain approach	MA in HR monitoring	No
Temko et al. [141]	Viterbi-based tracker in offline processing	MA in HR monitoring	No
Cajas et al. [142]	model of PPG noise components from database of corrupted PPG signals	MA in HR monitoring	No

Approaches based on the geometric separation of signal subspaces have also been presented in [136]. These have been designed to have a lower computational complexity than [131] while still offering comparable accuracy. Indeed, the dominant computational burden of these approaches lies in the SVD, which can be implemented leveraging any of many specialized algorithms [143,144], some of which are optimized for frequency tracking and so especially efficient in this context.

Figure 1 provides a summary pictorial representation of the body sites where the above-mentioned and presented sensors are typically located according to the analyzed literature and commercial state of the art.

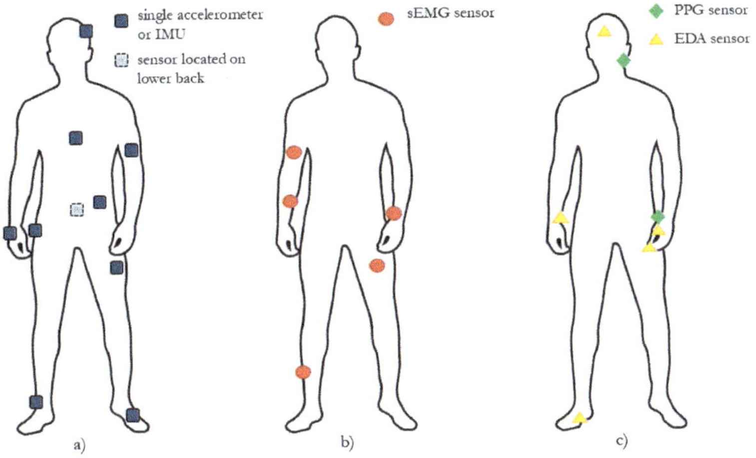

Figure 1. Typical body sites where sensors used in analyzed studies have been placed: (**a**) single accelerometer or IMU, (**b**) sEMG sensor, and (**c**) PPG and EDA sensors. Please note that the left or right side position of the marker representing the sensor is not relevant but only chosen for better readability of the figure.

4. Discussion

The technological advancement of miniaturized sensors has allowed their embedding within mobile and wearable devices that can be continuously worn by subjects during daily living, potentially enabling a continuous monitoring of physiological state and motor habits. Physical activity represents one of the first aspects whose monitoring and assessment were performed by wearable devices, and applications included a wide spectrum of physical tasks, such as walking, standing, fall detection, rehabilitation, and physical activity recognition [91,145]. Although many other human behaviors and activities have been investigated in order to produce reliable monitoring through smart sensing, such as sleep, feeding behavior, and even human emotional condition by affective computing [93,146], the outbreak of the COVID-19 pandemic contributed to stressing two different aspects related to HAR. First, the importance of correctly identifying a series of activities related to human well-being and health, such as those activities related to personal hygiene [147]. Secondly, the value of technical solutions capable of remotely monitoring user activities, habits, and behavior within an in-home scenario was highlighted. Two common approaches are leveraged for HAR within a closed environment, i.e., wearable sensors and external or ambient devices, such as presence sensors and cameras. However, the latter approach exhibits two main potential drawbacks that can hamper a widespread diffusion for HAR purposes, namely the need for instrumenting homes with sensors and video devices and necessary data transmission capabilities, which in principle should be installed within each room where the user is supposed to perform the activities of interest. This would lead to undesired additional costs for a private user that cannot always be faced. Secondly, video cameras record images not only of the internal space of the room but also of the subject and bystanders, implying several issues regarding privacy aspects, even about storing this kind of sensitive information [148]. On the other hand, wearable devices provide information related to the activities performed, which prevent any direct identification of a specific user, and can be processed with small computational delay, thus potentially allowing also an online identification of specific activities performed by the subject.

When testing a wearable device-based scheme for HAR, by leveraging portable devices that can include EMG and inertial sensors, the aim should be developing a solution that would be able to correctly identify activities of daily living that mainly involve upper limb and hand, with particular focus on some human activities that have a specific role for subject well-being and health. While conceiving such kinds of systems, many different experimental and technical points deserve attention since they can lead to different final framework configurations, each of them tailored for specific applications in the field of HAR.

The first point regards the human activities to be recognized by the system and those activities that should be included in the whole set. As outlined before, a wide spectrum of possibilities is available in this case, ranging from activities involving full body movements, such as walking, running, standing, lying, and many others [145], to those that mainly rely on upper limb motion, including interaction with objects, and several kinds of manipulations [149]. For the purposes of AAL, many activities to be considered involve upper limbs and can be performed commonly during daily living. The rationale for this choice is based on their importance for many aspects related to health and well-being: for instance, the correct identification of the drinking gesture can be useful for monitoring fluid intake and hydration levels [150], and the recognition of bringing the hand to the mouth with a certain grasping shape can be linked to pill intake, useful for medical adherence assessment [151], and also eating. Thus, specific attention shall be given to such kinds of gestures and activities, and within the selected set, other daily living activities that can have similar execution mechanics should be included in order to test the robustness of the recognition methodologies to potentially confounding activities. However, it should be noted that the inclusion of additional activities that involve the whole body movement can be considered as well in order to enlarge the spectrum of detectable activities through the specific sensor configuration chosen. The latter aspect is another point of primary importance since it strongly impacts the overall performance of HAR systems. Many possible locations for sensor placement on the human body are possible, but the focus on the upper limb naturally drives the choice towards three main possible locations, i.e., the forearm, the wrist, and the hand. Although the forearm represents perhaps the most preferred choice for gesture recognition of the upper limb when also EMG information is leveraged [90], from a practical viewpoint, this configuration is not optimal since wearing sensors or even a smart device like an armband continuously during the day could be hardly acceptable by a user, being an unnatural condition. The latter aspect is also valid for the hand, where the presence of sensors would likely modify the physiological way of making gestures. Therefore, the wrist appears to be a preferred location for sensor placement, also considering that some previous studies have already investigated HAR with wrist-mounted devices [151,152] and that smart devices embedding miniaturized inertial sensors, such as smartwatches, have widespread diffusion and are now well accepted by users who can wear this kind of devices for the entire day and sometimes even during the sleeping periods. However, attention will also be devoted to the forearm location in order to establish a benchmark for comparing the performances of HAR attained by wrist-mounted sensors. The focus on the wrist for sensor placement also leads to the choice of leveraging mainly inertial information for HAR. This is justified in part by the fact that myoelectric probes are not available on commercial devices, even if some attempts have been made to develop wrist-mounted EMG sensors [153]. However, more importantly, myoelectric information that can be retrieved from the wrist is of poorer quality with respect to that from the forearm, where muscles responsible for wrist and finger movements can be easily recorded with surface EMG probes.

It is worth noting that even if a single inertial unit is used, different information can be extracted and then processed from this kind of device since each embedded sensor provides tri-axial measurements of acceleration and angular velocity, in the case of an IMU, and also of the magnetic field, in the case of a MARG sensor. This means that different characteristics related to movements and activities are potentially available from nine channels using only a single sensor, thus providing significant information for feature extraction and training of learning models for automatic HAR. In addition, 3D orientation of an inertial sensor can be estimated, also in real-time, by different algorithms [154] in the form of Euler angles or quaternions, thus representing further information that can be used for HAR purposes. Additional points that could be investigated within the proposed solution are related to the artificial intelligence pipeline that has to be employed for classification. In this field, two main aspects should be taken into account, namely the feature extraction and the classification architectures. In the first case, points that can be investigated are related to the optimal windowing for feature extraction, which is strongly related to the temporal duration of the various activities and different schemes for feature extraction since recently, various techniques have been proposed that also account for temporal and spatial dependencies between signals [90], rather than a more straightforward feature extraction in time and frequency domain with fixed window length. Regarding classification methodologies, relatively simple machine learning architectures have found widespread usage in the field of HAR, having low computational expense that makes this kind of model particularly attractive to applications that must be run online. However, comparisons between different ML models have shown that their performances can vary depending on the considered application [152,155]; thus, finding the best classification model for the selected sensors setup and activities to be recognized represents one of the potential points to be investigated when dealing with the pattern recognition aspects of a HAR framework. On the other hand, unsupervised approaches, such as deep learning models, would be also worth investigating in this context, since they would allow us to avoid handcrafted feature extraction by relying directly on sensors' raw data. Further, this kind of methodology showed a generally good performance for HAR based on inertial data only, often outperforming machine learning architectures [151,156]. However, deep learning for pattern recognition has some drawbacks related to the amount of data required for model training and validation, higher computational costs with respect to machine learning models, and also the fact that deep learning models are, in general, black box-like models for feature extraction, which can limit the interpretability of the learned features for the final pattern recognition, and for the extraction of meaningful and relevant information for the purpose of AAL.

5. Conclusions

This paper focused on the importance of wearable devices in the design of AAL systems and applications, with a specific discussion of the advantages they provide in capturing information regarding human activities performed in uncontrolled scenarios and physiological parameters pertaining to motor capabilities, health status, and emotional well-being. In fact, the advances in electronic components' miniaturization in addition to increased signal processing efficiency have led to the widespread adoption of commercial wearable devices that well beyond the originally targeted applications of fitness or leisure activities.

Despite the potential for groundbreaking discoveries, wearable technology has its limitations. Many wearable devices currently available on the market utilize proprietary algorithms and do not provide access to raw data. Consequently, claims cannot be fully verified, research inquiries are restricted to the available aggregated data, and it becomes challenging to distinguish actual data from artifacts. Additionally, the performed analysis of the literature shows that relevant information about sensor selectivity and sensitivity may affect the quality of the acquired signals, which is, in turn, reflected in more accurate classification of activities performed or better detection of parameters of interest related to the subject being monitored, is typically missing in research studies, but also unavailable for many commercial devices as well. Implementing HAR techniques using wearable devices into AAL solutions should thus turn our attention to a set of specific activities commonly performed during daily living within a home scenario, focusing on a specific subset whose recognition would be valuable for the monitoring of various aspects related to the health and well-being of a user. The measurement setup should be minimized as much as possible in order to improve the comfort of the user, the portability of the proposed technology for continuous monitoring during the day, low power consumption, and the extended lifetime of the device.

Author Contributions: G.I., A.M., P.C., S.F. and S.S. contributed equally to the conceptualization, investigation, and writing of this manuscript. All authors have read and agreed to the published version of the manuscript.

Funding: This research received funding from the project Vitality—Project Code ECS00000041, CUP I33C22001330007—funded under the National Recovery and Resilience Plan (NRRP), Mission 4 Component 2 Investment 1.5—Creation and strengthening of "innovation ecosystems", construction of "territorial leaders in R&D"—Innovation Ecosystems—Project "Innovation, digitalization and sustainability for the diffused economy in Central Italy—VITALITY" Call for tender No. 3277 of 30/12/2021, and Concession Decree No. 0001057.23-06-2022 of Italian Ministry of University funded by the European Union—NextGenerationEU.

Institutional Review Board Statement: Not applicable.

Informed Consent Statement: Not applicable.

Data Availability Statement: No new data were created or analyzed in this study. Data sharing is not applicable to this article.

Conflicts of Interest: The authors declare no conflicts of interest.

References

1. United Nations, Department of Economic and Social Affairs. World Population Prospects 2024—Population Division. Available online: https://population.un.org/wpp/ (accessed on 22 July 2024).
2. Komp-Leukkunen, K.; Sarasma, J. Social Sustainability in Aging Populations: A Systematic Literature Review. *Gerontologist* **2023**, *64*, gnad097. [CrossRef] [PubMed]
3. Mostepaniuk, A.; Akalin, T.; Parish, M.R. Practices Pursuing the Sustainability of A Healthcare Organization: A Systematic Review. *Sustainability* **2023**, *15*, 2353. [CrossRef]
4. Cicirelli, G.; Marani, R.; Petitti, A.; Milella, A.; D'Orazio, T. Ambient Assisted Living: A Review of Technologies, Methodologies and Future Perspectives for Healthy Aging of Population. *Sensors* **2021**, *21*, 3549. [CrossRef] [PubMed]
5. McKhann, G.M.; Knopman, D.S.; Chertkow, H.; Hyman, B.T.; Jack, C.R.; Kawas, C.H.; Klunk, W.E.; Koroshetz, W.J.; Manly, J.J.; Mayeux, R.; et al. The diagnosis of dementia due to Alzheimer's disease: Recommendations from the National Institute on Aging-Alzheimer's Association workgroups on diagnostic guidelines for Alzheimer's disease. *Alzheimer'S Dement.* **2011**, *7*, 263–269. [CrossRef]
6. Hazzan, A.A.; Ploeg, J.; Shannon, H.; Raina, P.; Oremus, M. Association between caregiver quality of life and the care provided to persons with Alzheimer's disease: Protocol for a systematic review. *Syst. Rev.* **2013**, *2*, 17. [CrossRef]
7. Queluz, F.N.F.R.; Kervin, E.; Wozney, L.; Fancey, P.; McGrath, P.J.; Keefe, J. Understanding the needs of caregivers of persons with dementia: A scoping review. *Int. Psychogeriatr.* **2020**, *32*, 35–52. [CrossRef]
8. Vuong, N.K.; Chan, S.; Lau, C.T.; Chan, S.Y.W.; Yap, P.L.K.; Chen, A.S.H. Preliminary results of using inertial sensors to detect dementia-related wandering patterns. In Proceedings of the 2015 37th Annual International Conference of the IEEE Engineering in Medicine and Biology Society (EMBC), Milan, Italy, 25–29 August 2015; pp. 3703–3706. [CrossRef]
9. Labarrière, F.; Thomas, E.; Calistri, L.; Optasanu, V.; Gueugnon, M.; Ornetti, P.; Laroche, D. Machine Learning Approaches for Activity Recognition and/or Activity Prediction in Locomotion Assistive Devices—A Systematic Review. *Sensors* **2020**, *20*, 6345. [CrossRef]
10. Wang, K.; Cao, S.; Kaur, J.; Ghafurian, M.; Butt, Z.A.; Morita, P. Heart rate prediction with contactless active assisted living technology: A smart home approach for older adults. *Front. Artif. Intell.* **2024**, *6*, 1342427. [CrossRef]
11. Lu, L.; Zhang, J.; Xie, Y.; Gao, F.; Xu, S.; Wu, X.; Ye, Z. Wearable Health Devices in Health Care: Narrative Systematic Review. *Jmir Mhealth Uhealth* **2020**, *8*, e18907. [CrossRef]
12. De Vita, A.; Licciardo, G.D.; Benedetto, L.D.; Pau, D.; Plebani, E.; Bosco, A. Low-power Design of a Gravity Rotation Module for HAR Systems Based on Inertial Sensors. In Proceedings of the 2018 IEEE 29th International Conference on Application-Specific Systems, Architectures and Processors (ASAP), Milan, Italy, 10–12 July 2018; pp. 1–4. [CrossRef]
13. Yu, H.; Cang, S.; Wang, Y. A review of sensor selection, sensor devices and sensor deployment for wearable sensor-based human activity recognition systems. In Proceedings of the 2016 10th International Conference on Software, Knowledge, Information Management & Applications (SKIMA), Chengdu, China, 15–17 December 2016; pp. 250–257. [CrossRef]
14. Biagetti, G.; Crippa, P.; Curzi, A.; Orcioni, S.; Turchetti, C. Analysis of the EMG Signal During Cyclic Movements Using Multicomponent AM-FM Decomposition. *IEEE J. Biomed. Health Inform.* **2015**, *19*, 1672–1681. [CrossRef]
15. Lai, Z.; Xu, J.; Bowen, C.R.; Zhou, S. Self-powered and self-sensing devices based on human motion. *Joule* **2022**, *6*, 1501–1565. [CrossRef]
16. Rodriguez-Martin, D.; Samà, A.; Perez-Lopez, C.; Català, A.; Cabestany, J.; Rodriguez-Molinero, A. SVM-based posture identification with a single waist-located triaxial accelerometer. *Expert Syst. Appl.* **2013**, *40*, 7203–7211. [CrossRef]

17. Vakacherla, S.S.; Kantharaju, P.; Mevada, M.; Kim, M. Single Accelerometer to Recognize Human Activities Using Neural Networks. *J. Biomech. Eng.* **2023**, *145*, 061005. [CrossRef]
18. Tian, Y.; Zhang, J.; Wang, J.; Geng, Y.; Wang, X. Robust human activity recognition using single accelerometer via wavelet energy spectrum features and ensemble feature selection. *Syst. Sci. Control Eng.* **2020**, *8*, 83–96. [CrossRef]
19. Meng, L.; Zhang, A.; Chen, C.; Wang, X.; Jiang, X.; Tao, L.; Fan, J.; Wu, X.; Dai, C.; Zhang, Y.; et al. Exploration of Human Activity Recognition Using a Single Sensor for Stroke Survivors and Able-Bodied People. *Sensors* **2021**, *21*, 799. [CrossRef]
20. Mannini, A.; Rosenberger, M.; Haskell, W.; Sabatini, A.; Intille, S. Activity recognition in youth using single accelerometer placed at wrist or ankle. *Med. Sci. Sport. Exerc.* **2017**, *49*, 801–812. [CrossRef]
21. Mannini, A.; Sabatini, A.M. Machine Learning Methods for Classifying Human Physical Activity from On-Body Accelerometers. *Sensors* **2010**, *10*, 1154–1175. [CrossRef]
22. Iadarola, G.; Meletani, S.; Di Nardo, F.; Spinsante, S. A New Method for sEMG Envelope Detection from Reduced Measurements. In Proceedings of the 2022 IEEE International Symposium on Medical Measurements and Applications (MeMeA), Messina, Italy, 22–24 June 2022; pp. 1–6. [CrossRef]
23. Biagetti, G.; Crippa, P.; Falaschetti, L.; Turchetti, C. Classifier Level Fusion of Accelerometer and sEMG Signals for Automatic Fitness Activity Diarization. *Sensors* **2018**, *18*, 2850. [CrossRef]
24. Avila, E.; Junker, E.; Disselhorst-Klug, C. Introduction of a sEMG sensor system for autonomous use by inexperienced users. *Sensors* **2020**, *20*, 7348. [CrossRef]
25. Sun, J.; Liu, G.; Sun, Y.; Lin, K.; Zhou, Z.; Cai, J. Application of Surface Electromyography in Exercise Fatigue: A Review. *Front. Syst. Neurosci.* **2022**, *16*, 893275. [CrossRef]
26. Chang, K.M.; Liu, S.H.; Wu, X.H. A wireless sEMG recording system and its application to muscle fatigue detection. *Sensors* **2012**, *12*, 489–499. [CrossRef] [PubMed]
27. Doheny, E.; Goulding, C.; Flood, M.; McManus, L.; Lowery, M. Feature-Based Evaluation of a Wearable Surface EMG Sensor against Laboratory Standard EMG during Force-Varying and Fatiguing Contractions. *IEEE Sens. J.* **2020**, *20*, 2757–2765. [CrossRef]
28. Biagetti, G.; Crippa, P.; Falaschetti, L.; Orcioni, S.; Turchetti, C. Wireless surface electromyograph and electrocardiograph system on 802.15.4. *IEEE Trans. Consum. Electron.* **2016**, *62*, 258–266. [CrossRef]
29. Kobayashi, H. EMG/ECG acquisition system with online adjustable parameters using ZigBee wireless technology. *Electron. Commun. Jpn.* **2013**, *96*, 1–10. [CrossRef]
30. Burns, A.; Doheny, E.P.; Greene, B.R.; Foran, T.; Leahy, D.; O'Donovan, K.; McGrath, M.J. SHIMMER: An extensible platform for physiological signal capture. In Proceedings of the 2010 Annual International Conference of the IEEE Engineering in Medicine and Biology, Buenos Aires, Argentina, 31 August–4 September 2010; pp. 3759–3762. [CrossRef]
31. Wu, J.; Li, X.; Liu, W.; Wang, Z.J. sEMG Signal Processing Methods: A Review. *J. Phys. Conf. Ser.* **2019**, *1237*, 032008. [CrossRef]
32. Jeong, J.W.; Lee, W.; Kim, Y.J. A Real-Time Wearable Physiological Monitoring System for Home-Based Healthcare Applications. *Sensors* **2022**, *22*, 104. [CrossRef]
33. Yuk, S.W.; Hwang, I.H.; Cho, H.R.; Park, S.G. A study on an EMG sensor with high gain and low noise for measuring human muscular movement patterns for smart healthcare. *Micromachines* **2018**, *9*, 555. [CrossRef]
34. Orguc, S.; Khurana, H.; Lee, H.S.; Chandrakasan, A. 0.3 V ultra-low power sensor interface for EMG. In Proceedings of the ESSCIRC 2017—43rd IEEE European Solid State Circuits Conference, Leuven, Belgium, 11–14 September 2017; pp. 219–222. [CrossRef]
35. De Luca, C.J.; Donald Gilmore, L.; Kuznetsov, M.; Roy, S.H. Filtering the surface EMG signal: Movement artifact and baseline noise contamination. *J. Biomech.* **2010**, *43*, 1573–1579. [CrossRef] [PubMed]
36. Dow, H.K.; Huang, I.J.; Rieger, R.; Kuo, K.C.; Guo, L.Y.; Pao, S.J. A Bio-Sensing System-on-Chip and Software for Smart Clothes. In Proceedings of the 2019 IEEE International Conference on Consumer Electronics, (ICCE 2019), Las Vegas, NV, USA, 11–13 January 2019. [CrossRef]
37. Liu, C.; Li, J.; Zhang, S.; Yang, H.; Guo, K. Study on Flexible sEMG Acquisition System and Its Application in Muscle Strength Evaluation and Hand Rehabilitation. *Micromachines* **2022**, *13*, 2047. [CrossRef]
38. Wang, Y.Z.; Wang, Y.P.; Wu, Y.C.; Yang, C.H. A 12.6 mW, 573–2901 kS/s Reconfigurable Processor for Reconstruction of Compressively Sensed Physiological Signals. *IEEE J.-Solid-State Circuits* **2019**, *54*, 2907–2916. [CrossRef]
39. Vijayvargiya, A.; Singh, P.; Kumar, R.; Dey, N. Hardware Implementation for Lower Limb Surface EMG Measurement and Analysis Using Explainable AI for Activity Recognition. *IEEE Trans. Instrum. Meas.* **2022**, *71*, 2004909. [CrossRef]
40. Tu, Y.H.; Yao, K.W.; Huang, M.H.; Lin, Y.Y.; Chi, H.Y.; Cheng, P.M.; Tsai, P.Y.; Shiue, M.T.; Liu, C.N.; Cheng, K.H.; et al. A body sensor node SoC for ECG/EMG applications with compressed sensing and wireless powering. In Proceedings of the 017 International Symposium on VLSI Design, Automation and Test (VLSI-DAT), Hsinchu, Taiwan, 24–27 April 2017; pp. 1–4. [CrossRef]
41. Di Giminiani, R.; Cardinale, M.; Ferrari, M.; Quaresima, V. Validation of fabric-based thigh-wearable EMG sensors and oximetry for monitoring quadricep activity during strength and endurance exercises. *Sensors* **2020**, *20*, 4664. [CrossRef]
42. Rescio, G.; Leone, A.; Giampetruzzi, L.; Siciliano, P. Fall Risk Assessment Using New sEMG-Based Smart Socks. In *Advances in Data Science: Methodologies and Applications*; Phillips-Wren, G., Esposito, A., Jain, L.C., Eds.; Springer International Publishing: Cham, Switzerland, 2021; pp. 147–166. [CrossRef]

43. Liu, S.; Liu, X.; Jiang, Y.; Wang, X.; Huang, P.; Wang, H.; Zhu, M.; Tan, J.; Li, P.; Lin, C.; et al. Flexible Non-contact Electrodes for Bioelectrical Signal Monitoring. In Proceedings of the Annual International Conference of the IEEE Engineering in Medicine and Biology Society, (EMBS), Honolulu, HI, USA, 18–21 July 2018; pp. 4305–4308. [CrossRef]
44. Dandumahanti, B.P.; Subramaniyam, M. Wireless sEMG Sensor for Neck Muscle Activity Measurement and Posture Classification Using Machine Learning. *IEEE Sens. J.* **2023**, *23*, 31220–31228. [CrossRef]
45. Martins, J.; Cerqueira, S.M.; Catarino, A.W.; da Silva, A.F.; Rocha, A.M.; Vale, J.; Ângelo, M.; Santos, C.P. Integrating sEMG and IMU Sensors in an e-Textile Smart Vest for Forward Posture Monitoring: First Steps. *Sensors* **2024**, *24*, 4717. [CrossRef]
46. Jani, A.; Bagree, R.; Roy, A. Design of a low-power, low-cost ECG & EMG sensor for wearable biometric and medical application. In Proceedings of the IEEE Sensors, Glasgow, UK, 29 October–1 November 2017; pp. 1–3. [CrossRef]
47. Biagetti, G.; Crippa, P.; Falaschetti, L.; Turchetti, C. A multi-channel electromyography, electrocardiography and inertial wireless sensor module using bluetooth low-energy. *Electronics* **2020**, *9*, 934. [CrossRef]
48. Said, S.; AlKork, S.; Beyrouthy, T.; Abdrabbo, M.F. Wearable bio-sensors bracelet for driveras health emergency detection. In Proceedings of the 2017 2nd International Conference on Bio-engineering for Smart Technologies (BioSMART), Paris, France, 30 August–1 September 2017; pp. 1–4.
49. Pinto, A.G.; Dias, G.; Felizardo, V.; Pombo, N.; Silva, H.; Fazendeiro, P.; Crisóstomo, R.; Garcia, N. Electrocardiography, electromyography, and accelerometry signals collected with BITalino while swimming: Device assembly and preliminary results. In Proceedings of the 2016 IEEE 12th International Conference on Intelligent Computer Communication and Processing (ICCP), Cluj-Napoca, Romania, 8–10 September 2016; pp. 37–41.
50. Zhang, X.; Chen, X.; Li, Y.; Lantz, V.; Wang, K.; Yang, J. A framework for hand gesture recognition based on accelerometer and EMG sensors. *IEEE Trans. Syst. Man. Cybern. Part A Systems Humans* **2011**, *41*, 1064–1076. [CrossRef]
51. Wu, J.; Sun, L.; Jafari, R. A Wearable System for Recognizing American Sign Language in Real-Time Using IMU and Surface EMG Sensors. *IEEE J. Biomed. Health Inform.* **2016**, *20*, 1281–1290. [CrossRef]
52. Tanweer, M.; Halonen, K.A.I. Development of wearable hardware platform to measure the ECG and EMG with IMU to detect motion artifacts. In Proceedings of the 2019 IEEE 22nd International Symposium on Design and Diagnostics of Electronic Circuits & Systems (DDECS), Cluj-Napoca, Romania, 24–26 April 2019; pp. 1–4. [CrossRef]
53. Das, D.M.; Vidwans, A.; Srivastava, A.; Ahmad, M.; Vaishnav, S.; Dewan, S.; Baghini, M.S. Design and development of an Internet-of-Things enabled wearable ExG measuring system with a novel signal processing algorithm for electrocardiogram. *IET Circuits Devices Syst.* **2019**, *13*, 903–907. [CrossRef]
54. Ding, X.; Wang, M.; Guo, W.; Sheng, X.; Zhu, X. Hybrid sEMG, NIRS and MMG Sensor System. In Proceedings of the 2018 25th International Conference on Mechatronics and Machine Vision in Practice (M2VIP), Stuttgart, Germany, 20–22 November 2018; pp. 1–6. [CrossRef]
55. Ke, A.; Huang, J.; Chen, L.; Gao, Z.; He, J. An ultra-sensitive modular hybrid EMG–FMG sensor with floating electrodes. *Sensors* **2020**, *20*, 4775. [CrossRef]
56. *IEEE Std 802.15.4-2020 (Revision of IEEE Std 802.15.4-2015)*; IEEE Standard for Low-Rate Wireless Networks. IEEE: Piscataway, NJ, USA, 2020; pp. 1–800. [CrossRef]
57. Kobayashi, H. A ZigBee based wireless EMG/ECG streaming system for the universal interface. In Proceedings of the IECON 2011—37th Annual Conference IEEE Industrial Electronics Society, Melbourne, VIC, Australia, 7–10 November 2011; pp. 2094–2099. [CrossRef]
58. Kobayashi, H. Intelligent wireless EMG/ECG electrode employing ZigBee technology. In Proceedings of the SICE Annual Conference 2011, Tokyo, Japan, 13–18 September 2011; pp. 2856–2861.
59. Biagetti, G.; Crippa, P.; Falaschetti, L.; Mansour, A.; Turchetti, C. Energy and Performance Analysis of Lossless Compression Algorithms for Wireless EMG Sensors. *Sensors* **2021**, *21*, 5160. [CrossRef]
60. Xing, K.; Zhong, J.; Zhu, W.; Ma, G.; Tong, T.; Wang, L. A Wearable High-Precision Skin Resistance Acquisition System. In Proceedings of the 2018 IEEE SmartWorld, Ubiquitous Intelligence & Computing, Advanced & Trusted Computing, Scalable Computing & Communications, Cloud & Big Data Computing, Internet of People and Smart City Innovation (SmartWorld/SCALCOM/UIC/ATC/CBDCom/IOP/SCI), Guangzhou, China, 8–12 October 2018; pp. 123–127. [CrossRef]
61. Boucsein, W. *Electrodermal Activity*, 2nd ed.; Springer: New York, NY, USA, 2012; pp. 1–618. [CrossRef]
62. Kalinkov, K. Algorithm for peak detection in the Skin Conductance Response component of the EDA signals. In Proceedings of the 2020 International Conference on Biomedical Innovations and Applications (BIA), Varna, Bulgaria, 24–27 September 2020; pp. 89–91. [CrossRef]
63. Can, Y.S.; Chalabianloo, N.; Ekiz, D.; Ersoy, C. Continuous Stress Detection Using Wearable Sensors in Real Life: Algorithmic Programming Contest Case Study. *Sensors* **2019**, *19*, 1849. [CrossRef]
64. Johnson, K.T.; Picard, R.W. Advancing Neuroscience through Wearable Devices. *Neuron* **2020**, *108*, 8–12. [CrossRef]
65. Kleckner, I.R.; Wormwood, J.B.; Jones, R.M.; Culakova, E.; Barrett, L.F.; Lord, C.; Quigley, K.S.; Goodwin, M.S. Adaptive thresholding increases sensitivity to detect changes in the rate of skin conductance responses to psychologically arousing stimuli in both laboratory and ambulatory settings. *Int. J. Psychophysiol.* **2024**, *196*, 112280. [CrossRef]
66. Christopoulos, G.I.; Uy, M.A.; Yap, W.J. The Body and the Brain: Measuring Skin Conductance Responses to Understand the Emotional Experience. *Organ. Res. Methods* **2019**, *22*, 394–420. [CrossRef]

67. Iadarola, G.; Poli, A.; Spinsante, S. Reconstruction of Galvanic Skin Response Peaks via Sparse Representation. In Proceedings of the 2021 IEEE International Instrumentation and Measurement Technology Conference (I2MTC), Glasgow, UK, 17–20 May 2021; pp. 1–6. [CrossRef]
68. Iadarola, G.; Poli, A.; Spinsante, S. Analysis of Galvanic Skin Response to Acoustic Stimuli by Wearable Devices. In Proceedings of the 2021 IEEE International Symposium on Medical Measurements and Applications (MeMeA), Lausanne, Switzerland, 23–25 June 2021; pp. 1–6. [CrossRef]
69. Iadarola, G.; Poli, A.; Spinsante, S. Compressed Sensing of Skin Conductance Level for IoT-based wearable sensors. In Proceedings of the 2022 IEEE International Instrumentation and Measurement Technology Conference (I2MTC), Ottawa, ON, Canada, 16–19 May 2022; pp. 1–6. [CrossRef]
70. Zhang, T.; El Ali, A.; Wang, C.; Hanjalic, A.; Cesar, P. CorrNet: Fine-Grained Emotion Recognition for Video Watching Using Wearable Physiological Sensors. *Sensors* **2021**, *21*, 52. [CrossRef]
71. Montgomery, S.M.; Nair, N.; Chen, P.; Dikker, S. Introducing EmotiBit, an open-source multi-modal sensor for measuring research-grade physiological signals. *Sci. Talks* **2023**, *6*, 100181. [CrossRef]
72. Klimek, A.; Mannheim, I.; Schouten, G.; Wouters, E.J.M.; Peeters, M.W.H. Wearables measuring electrodermal activity to assess perceived stress in care: A scoping review. *Acta Neuropsychiatr.* **2023**, 1–11. [CrossRef]
73. Betancourt, M.A.; Dethorne, L.S.; Karahalios, K.; Kim, J.G. Skin Conductance as an In Situ Marker for Emotional Arousal in Children with Neurodevelopmental Communication Impairments: Methodological Considerations and Clinical Implications. *ACM Trans. Access. Comput.* **2017**, *9*, 1–29. [CrossRef]
74. Khreis, S.; Ge, D.; Rahman, H.; Carrault, G. Breathing Rate Estimation Using Kalman Smoother with Electrocardiogram and Photoplethysmogram. *IEEE Trans. Biomed. Eng.* **2020**, *67*, 893–904. [CrossRef]
75. Oliveira, A.; Aguiar, J.; Silva, E.; Faria, B.; Gonçalves, H.; Teófilo, L.; Gonçalves, J.; Carvalho, V.; Cardoso, H.; Reis, L. Assessing Daily Activities Using a PPG Sensor Embedded in a Wristband-Type Activity Tracker. In *Advances in Intelligent Systems and Computing*; Springer: Cham, Switzerland, 2020; Volume 1161, pp. 108–119. [CrossRef]
76. Elgendi, M.; Jonkman, M.; DeBoer, F. Heart rate variability measurement using the second derivative photoplethysmogram. In Proceedings of the Third International Conference on Bio-inspired Systems and Signal Processing (BIOSIGNALS 2010), Valencia, Spain, 20–23 January 2010; pp. 82–87.
77. Longmore, S.K.; Lui, G.Y.; Naik, G.; Breen, P.P.; Jalaludin, B.; Gargiulo, G.D. A Comparison of Reflective Photoplethysmography for Detection of Heart Rate, Blood Oxygen Saturation, and Respiration Rate at Various Anatomical Locations. *Sensors* **2019**, *19*, 1874. [CrossRef]
78. Allen, J. Photoplethysmography and its application in clinical physiological measurement. *Physiol. Meas.* **2007**, *28*, R1–R39. [CrossRef]
79. Gil, E.; Orini, M.; Bailón, R.; Vergara, J.; Mainardi, L.; Laguna, P. Photoplethysmography pulse rate variability as a surrogate measurement of heart rate variability during non-stationary conditions. *Physiol. Meas.* **2010**, *31*, 1271–1290. [CrossRef]
80. Gil, E.; Laguna, P.; Martínez, J.P.; Barquero-Pérez, O.; García-Alberola, A.; Sörnmo, L. Heart Rate Turbulence Analysis Based on Photoplethysmography. *IEEE Trans. Biomed. Eng.* **2013**, *60*, 3149–3155. [CrossRef]
81. Raghuram, M.; Madhav, K.V.; Krishna, E.H.; Komalla, N.R.; Sivani, K.; Reddy, K.A. HHT based signal decomposition for reduction of motion artifacts in photoplethysmographic signals. In Proceedings of the 2012 IEEE International Instrumentation and Measurement Technology Conference Proceedings (I2MTC), Graz, Austria, 13–16 May 2012; pp. 1730–1734. [CrossRef]
82. Raghuram, M.; Sivani, K.; Reddy, K.A. E2MD for reduction of motion artifacts from photoplethysmographic signals. In Proceedings of the International Conference Electronics and Communication Systems (ICECS), Coimbatore, India, 13–14 February 2014; pp. 1–6. [CrossRef]
83. Raghuram, M.; Madhav, K.V.; Krishna, E.H.; Reddy, K.A. Evaluation of wavelets for reduction of motion artifacts in photoplethysmographic signals. In Proceedings of the 10th International Conference Information Sciences Signal Processing and their Applications (ISSPA), Kuala Lumpur, Malaysia, 10–13 May 2010; pp. 460–463. [CrossRef]
84. Ram, M.R.; Madhav, K.V.; Krishna, E.H.; Komalla, N.R.; Reddy, K.A. A Novel Approach for Motion Artifact Reduction in PPG Signals Based on AS-LMS Adaptive Filter. *IEEE Trans. Instrum. Meas.* **2012**, *61*, 1445–1457. [CrossRef]
85. Lin, Q.; Song, S.; Castro, I.; Jiang, H.; Konijnenburg, M.; Van Wegberg, R.; Biswas, D.; Stanzione, S.; Sijbers, W.; Van Hoof, C.; et al. Wearable Multiple Modality Bio-Signal Recording and Processing on Chip: A Review. *IEEE Sens. J.* **2021**, *21*, 1108–1123. [CrossRef]
86. Pribadi, E.; Pandey, R.; Chao, P.P. Design and implementation of a new light to digital converter for the PPG sensor. *Microsyst. Technol.* **2021**, *27*, 2461–2472. [CrossRef]
87. Biagetti, G.; Crippa, P.; Falaschetti, L.; Saraceni, L.; Tiranti, A.; Turchetti, C. Dataset from PPG wireless sensor for activity monitoring. *Data Brief* **2020**, *29*, 105044. [CrossRef]
88. Aydemir, T.; Şahin, M.; Aydemir, O. A New Method for Activity Monitoring Using Photoplethysmography Signals Recorded by Wireless Sensor. *J. Med. Biol. Eng.* **2020**, *40*, 934–942. [CrossRef]
89. Biagetti, G.; Crippa, P.; Falaschetti, L.; Focante, E.; Madrid, N.M.; Seepold, R.; Turchetti, C. Machine Learning and Data Fusion Techniques Applied to Physical Activity Classification Using Photoplethysmographic and Accelerometric Signals. *Procedia Comput. Sci.* **2020**, *176*, 3103–3111. [CrossRef]

90. Khushaba, R.N.; Scheme, E.; Al-Timemy, A.H.; Phinyomark, A.; Al-Taee, A.; Al-Jumaily, A. A long short-term recurrent spatial-temporal fusion for myoelectric pattern recognition. *Expert Syst. Appl.* **2021**, *178*, 114977. [CrossRef]
91. Kristoffersson, A.; Lindén, M. A systematic review of wearable sensors for monitoring physical activity. *Sensors* **2022**, *22*, 573. [CrossRef]
92. Mengarelli, A.; Tigrini, A.; Fioretti, S.; Cardarelli, S.; Verdini, F. On the use of fuzzy and permutation entropy in hand gesture characterization from EMG signals: Parameters selection and comparison. *Appl. Sci.* **2020**, *10*, 7144. [CrossRef]
93. Leelaarporn, P.; Wachiraphan, P.; Kaewlee, T.; Udsa, T.; Chaisaen, R.; Choksatchawathi, T.; Laosirirat, R.; Lakhan, P.; Natnithikarat, P.; Thanontip, K.; et al. Sensor-driven achieving of smart living: A review. *IEEE Sens. J.* **2021**, *21*, 10369–10391. [CrossRef]
94. Totty, M.S.; Wade, E. Muscle activation and inertial motion data for noninvasive classification of activities of daily living. *IEEE Trans. Biomed. Eng.* **2017**, *65*, 1069–1076. [CrossRef]
95. Perusquia-Hernandez, M.; Hirokawa, M.; Suzuki, K. A wearable device for fast and subtle spontaneous smile recognition. *IEEE Trans. Affect. Comput.* **2017**, *8*, 522–533. [CrossRef]
96. Huang, Q.; Wang, W.; Zhang, Q. Your glasses know your diet: Dietary monitoring using electromyography sensors. *IEEE Internet Things J.* **2017**, *4*, 705–712. [CrossRef]
97. Zhang, R.; Amft, O. Monitoring chewing and eating in free-living using smart eyeglasses. *IEEE J. Biomed. Health Inform.* **2017**, *22*, 23–32. [CrossRef]
98. Hossain, D.; Imtiaz, M.H.; Sazonov, E. Comparison of wearable sensors for estimation of chewing strength. *IEEE Sens. J.* **2020**, *20*, 5379–5388. [CrossRef]
99. Reis Carneiro, M.; Majidi, C.; Tavakoli, M. Multi-Electrode Printed Bioelectronic Patches for Long-Term Electrophysiological Monitoring. *Adv. Funct. Mater.* **2022**, *32*, 2205956. [CrossRef]
100. Shahzad, W.; Ayaz, Y.; Khan, M.J.; Naseer, N.; Khan, M. Enhanced performance for multi-forearm movement decoding using hybrid IMU–SEMG interface. *Front. Neurorobot.* **2019**, *13*, 43. [CrossRef]
101. Song, X.; van de Ven, S.S.; Chen, S.; Kang, P.; Gao, Q.; Jia, J.; Shull, P.B. Proposal of a wearable multimodal sensing-based serious games approach for hand movement training after stroke. *Front. Physiol.* **2022**, *13*, 811950. [CrossRef]
102. Song, X.; Van De Ven, S.S.; Liu, L.; Wouda, F.J.; Wang, H.; Shull, P.B. Activities of daily living-based rehabilitation system for arm and hand motor function retraining after stroke. *IEEE Trans. Neural Syst. Rehabil. Eng.* **2022**, *30*, 621–631. [CrossRef]
103. Chang, W.; Dai, L.; Sheng, S.; Tan, J.T.C.; Zhu, C.; Duan, F. A hierarchical hand motions recognition method based on IMU and sEMG sensors. In Proceedings of the 2015 IEEE International Conference on Robotics and Biomimetics (ROBIO), Zhuhai, China, 6–9 December 2015; IEEE: Piscataway, NJ, USA, 2015; pp. 1024–1029. [CrossRef]
104. Bassani, G.; Filippeschi, A.; Avizzano, C.A. A dataset of human motion and muscular activities in manual material handling tasks for biomechanical and ergonomic analyses. *IEEE Sens. J.* **2021**, *21*, 24731–24739. [CrossRef]
105. Mekruksavanich, S.; Jantawong, P.; Jitpattanakul, A. A deep learning-based model for human activity recognition using biosensors embedded into a smart knee bandage. *Procedia Comput. Sci.* **2022**, *214*, 621–627. [CrossRef]
106. Liu, H.; Hartmann, Y.; Schultz, T. CSL-SHARE: A multimodal wearable sensor-based human activity dataset. *Front. Comput. Sci.* **2021**, *3*, 800056. [CrossRef]
107. Zhou, B.; Wang, H.; Hu, F.; Feng, N.; Xi, H.; Zhang, Z.; Tang, H. Accurate recognition of lower limb ambulation mode based on surface electromyography and motion data using machine learning. *Comput. Methods Programs Biomed.* **2020**, *193*, 105486. [CrossRef]
108. Nouredanesh, M.; Tung, J. IMU, sEMG, or their cross-correlation and temporal similarities: Which signal features detect lateral compensatory balance reactions more accurately? *Comput. Methods Programs Biomed.* **2019**, *182*, 105003. [CrossRef]
109. Esaa, R.R.; Jaber, H.A.; Ameer, A.A. Hand movements classification based on Myo armband signals. In Proceedings of the 2022 4th International Conference on Electrical, Control and Instrumentation Engineering (ICECIE), KualaLumpur, Malaysia, 26 November 2022; IEEE: Piscataway, NJ, USA, 2022; pp. 1–5. [CrossRef]
110. Schabron, B.; Reust, A.; Desai, J.; Yihun, Y. Integration of forearm sEMG signals with IMU sensors for trajectory planning and control of assistive robotic arm. In Proceedings of the 2019 41st Annual International Conference of the IEEE Engineering in Medicine and Biology Society (EMBC), Berlin, Germany, 23–27 July 2019; IEEE: Piscataway, NJ, USA, 2019; pp. 5274–5277. [CrossRef]
111. Colli Alfaro, J.G.; Trejos, A.L. User-independent hand gesture recognition classification models using sensor fusion. *Sensors* **2022**, *22*, 1321. [CrossRef]
112. Ertuğrul, Ö.F.; Dal, S.; Hazar, Y.; Aldemir, E. Determining relevant features in activity recognition via wearable sensors on the MYO Armband. *Arab. J. Sci. Eng.* **2020**, *45*, 10097–10113. [CrossRef]
113. Uswatte, G.; Hobbs Qadri, L. A behavioral observation system for quantifying arm activity in daily life after stroke. *Rehabil. Psychol.* **2009**, *54*, 398. [CrossRef]
114. Karnam, N.K.; Turlapaty, A.C.; Dubey, S.R.; Gokaraju, B. EMAHA-DB1: A new upper limb sEMG dataset for classification of activities of daily living. *IEEE Trans. Instrum. Meas.* **2023**, *72*, 1–11. [CrossRef]
115. Bangaru, S.S.; Wang, C.; Aghazadeh, F. Automated and continuous fatigue monitoring in construction workers using forearm EMG and IMU wearable sensors and recurrent neural network. *Sensors* **2022**, *22*, 9729. [CrossRef]
116. Lee, Y.J.; Wei, M.Y.; Chen, Y.J. Multiple inertial measurement unit combination and location for recognizing general, fatigue, and simulated-fatigue gait. *Gait Posture* **2022**, *96*, 330–337. [CrossRef]

117. Biagetti, G.; Crippa, P.; Falaschetti, L.; Orcioni, S.; Turchetti, C. Human activity monitoring system based on wearable sEMG and accelerometer wireless sensor nodes. *Biomed. Eng. Online* **2018**, *17*, 132. [CrossRef]
118. Payne, A.F.H.; Dawson, M.E.; Schell, A.M.; Singh, K.; Courtney, C.G. Can you give me a hand? A comparison of hands and feet as optimal anatomical sites for skin conductance recording. *Psychophysiology* **2013**, *50*, 1065–1069. [CrossRef]
119. van Dooren, M.; de Vries, J.G.J.; Janssen, J.H. Emotional sweating across the body: Comparing 16 different skin conductance measurement locations. *Physiol. Behav.* **2012**, *106*, 298–304. [CrossRef]
120. Sim, D.J.K.; Kim, S.M.; Kim, S.S.; Doh, I. Portable Skin Analyzers with Simultaneous Measurements of Transepidermal Water Loss, Skin Conductance and Skin Hardness. *Sensors* **2019**, *19*, 3857. [CrossRef]
121. Iadarola, G.; Bruschi, V.; Cecchi, S.; Dourou, N.; Spinsante, S. Low-cost monitoring for stimulus detection in skin conductance. *Acta IMEKO* **2023**, *12*, 1–6. [CrossRef]
122. Ebrahimi, Z.; Gosselin, B. Ultralow-Power Photoplethysmography (PPG) Sensors: A Methodological Review. *IEEE Sens. J.* **2023**, *23*, 16467–16480. [CrossRef]
123. Ismail, S.; Akram, U.; Siddiqi, I. Heart rate tracking in photoplethysmography signals affected by motion artifacts: A review. *Eurasip J. Adv. Signal Process.* **2021**, *2021*, 5. [CrossRef]
124. Kim, B.S.; Yoo, S.K. Motion artifact reduction in photoplethysmography using independent component analysis. *IEEE Trans. Biomed. Eng.* **2006**, *53*, 566–568. [CrossRef]
125. Foo, J.Y.A. Comparison of wavelet transformation and adaptive filtering in restoring artefact-induced time-related measurement. *Biomed. Signal Process. Control* **2006**, *1*, 93–98. [CrossRef]
126. Gibbs, P.T.; Wood, L.B.; Asada, H.H. Active motion artifact cancellation for wearable health monitoring sensors using collocated MEMS accelerometers. In Proceedings of the Smart Structures and Materials. International Society for Optics and Photonics, San Diego, CA, USA, 7–10 March 2005; Volume 5765, pp. 811–819. [CrossRef]
127. Tanweer, K.T.; Hasan, S.R.; Kamboh, A.M. Motion artifact reduction from PPG signals during intense exercise using filtered X-LMS. In Proceedings of the 2017 IEEE International Symposium on Circuits and Systems (ISCAS), Baltimore, MD, USA, 28–31 May 2017; pp. 1–4. [CrossRef]
128. Chowdhury, S.S.; Hyder, R.; Hafiz, M.S.B.; Haque, M. Real-Time Robust Heart Rate Estimation from Wrist-Type PPG Signals Using Multiple Reference Adaptive Noise Cancellation. *IEEE J. Biomed. Health Inform.* **2018**, *22*, 450–459. [CrossRef]
129. Galli, A.; Frigo, G.; Narduzzi, C.; Giorgi, G. Robust estimation and tracking of heart rate by PPG signal analysis. In Proceedings of the IEEE International Instrumentation and Measurement Technology Conference (I2MTC), Torino, Italy, 22–25 May 2017; pp. 1–6. [CrossRef]
130. Lee, B.; Han, J.; Baek, H.J.; Shin, J.H.; Park, K.S.; Yi, W.J. Improved elimination of motion artifacts from a photoplethysmographic signal using a Kalman smoother with simultaneous accelerometry. *Physiol. Meas.* **2010**, *31*, 1585. [CrossRef]
131. Zhang, Z.; Pi, Z.; Liu, B. TROIKA: A General Framework for Heart Rate Monitoring Using Wrist-Type Photoplethysmographic Signals During Intensive Physical Exercise. *IEEE Trans. Biomed. Eng.* **2015**, *62*, 522–531. [CrossRef]
132. Zhang, Z. Photoplethysmography-Based Heart Rate Monitoring in Physical Activities via Joint Sparse Spectrum Reconstruction. *IEEE Trans. Biomed. Eng.* **2015**, *62*, 1902–1910. [CrossRef]
133. Islam, M.T.; Zabir, I.; Ahamed, S.T.; Yasar, M.T.; Shahnaz, C.; Fattah, S.A. A time-frequency domain approach of heart rate estimation from photoplethysmographic (PPG) signal. *Biomed. Signal Process. Control* **2017**, *36*, 146–154. [CrossRef]
134. Zhang, Y.; Liu, B.; Zhang, Z. Combining ensemble empirical mode decomposition with spectrum subtraction technique for heart rate monitoring using wrist-type photoplethysmography. *Biomed. Signal Process. Control* **2015**, *21*, 119–125. [CrossRef]
135. Islam, M.S.; Shifat-E-Rabbi, M.; Dobaie, A.M.A.; Hasan, M.K. PREHEAT: Precision heart rate monitoring from intense motion artifact corrupted PPG signals using constrained RLS and wavelets. *Biomed. Signal Process. Control* **2017**, *38*, 212–223. [CrossRef]
136. Biagetti, G.; Crippa, P.; Falaschetti, L.; Orcioni, S.; Turchetti, C. Reduced complexity algorithm for heart rate monitoring from PPG signals using automatic activity intensity classifier. *Biomed. Signal Process. Control* **2019**, *52*, 293–301. [CrossRef]
137. Zhao, D.; Sun, Y.; Wan, S.; Wang, F. SFST: A robust framework for heart rate monitoring from photoplethysmography signals during physical activities. *Biomed. Signal Process. Control* **2016**, *33*, 316–324. [CrossRef]
138. Brophy, E.; Muehlhausen, W.; Smeaton, A.; Ward, T. CNNs for Heart Rate Estimation and Human Activity Recognition in Wrist Worn Sensing Applications. In Proceedings of the 2020 IEEE International Conference on Pervasive Computing and Communications Workshops, PerCom Workshops 2020, Austin, TX, USA, 23–27 March 2020. [CrossRef]
139. Dubey, H.; Kumaresan, R.; Mankodiya, K. Harmonic sum-based method for heart rate estimation using PPG signals affected with motion artifacts. *J. Ambient. Intell. Humaniz. Comput.* **2018**, *9*, 137–150. [CrossRef]
140. Wójcikowski, M.; Pankiewicz, B. Photoplethysmographic time-domain heart rate measurement algorithm for resource-constrained wearable devices and its implementation. *Sensors* **2020**, *20*, 1783. [CrossRef]
141. Temko, A. Accurate Heart Rate Monitoring during Physical Exercises Using PPG. *IEEE Trans. Biomed. Eng.* **2017**, *64*, 2016–2024. [CrossRef]
142. Cajas, S.; Landínez, M.; López, D. Modeling of motion artifacts on PPG signals for heart-monitoring using wearable devices. In Proceedings of the SPIE—The International Society for Optical Engineering, Medelin, Colombia, 6–8 November 2019; Volume 11330. [CrossRef]
143. Badeau, R.; Richard, G.; David, B. Sliding window adaptive SVD algorithms. *IEEE Trans. Signal Process.* **2004**, *52*, 1–10. [CrossRef]

144. Strobach, P. Sliding window adaptive SVD using the unsymmetric householder partial compressor. *Signal Process.* **2010**, *90*, 352–362. [CrossRef]
145. Bianchi, V.; Bassoli, M.; Lombardo, G.; Fornacciari, P.; Mordonini, M.; De Munari, I. IoT wearable sensor and deep learning: An integrated approach for personalized human activity recognition in a smart home environment. *IEEE Internet Things J.* **2019**, *6*, 8553–8562. [CrossRef]
146. Amidei, A.; Spinsante, S.; Iadarola, G.; Benatti, S.; Tramarin, F.; Pavan, P.; Rovati, L. Driver Drowsiness Detection: A Machine Learning Approach on Skin Conductance. *Sensors* **2023**, *23*, 4004. [CrossRef]
147. Wang, C.; Jiang, W.; Yang, K.; Yu, D.; Newn, J.; Sarsenbayeva, Z.; Goncalves, J.; Kostakos, V. Electronic monitoring systems for hand hygiene: Systematic review of technology. *J. Med. Internet Res.* **2021**, *23*, e27880. [CrossRef]
148. Yadav, S.K.; Tiwari, K.; Pandey, H.M.; Akbar, S.A. A review of multimodal human activity recognition with special emphasis on classification, applications, challenges and future directions. *Knowl.-Based Syst.* **2021**, *223*, 106970. [CrossRef]
149. Chun, K.S.; Sanders, A.B.; Adaimi, R.; Streeper, N.; Conroy, D.E.; Thomaz, E. Towards a generalizable method for detecting fluid intake with wrist-mounted sensors and adaptive segmentation. In Proceedings of the 24th International Conference on Intelligent User Interfaces, Marina del Ray, CA, USA, 17–20 March 2019; pp. 80–85. [CrossRef]
150. Sabry, F.; Eltaras, T.; Labda, W.; Hamza, F.; Alzoubi, K.; Malluhi, Q. Towards on-device dehydration monitoring using machine learning from wearable device's data. *Sensors* **2022**, *22*, 1887. [CrossRef]
151. Moccia, S.; Solbiati, S.; Khornegah, M.; Bossi, F.F.; Caiani, E.G. Automated classification of hand gestures using a wristband and machine learning for possible application in pill intake monitoring. *Comput. Methods Programs Biomed.* **2022**, *219*, 106753. [CrossRef]
152. Moschetti, A.; Fiorini, L.; Esposito, D.; Dario, P.; Cavallo, F. Recognition of daily gestures with wearable inertial rings and bracelets. *Sensors* **2016**, *16*, 1341. [CrossRef]
153. Liang, Z.; Wang, X.; Guo, J.; Ye, Y.; Zhang, H.; Xie, L.; Tao, K.; Zeng, W.; Yin, E.; Ji, B. A Wireless, High-Quality, Soft and Portable Wrist-Worn System for SEMG Signal Detection. *Micromachines* **2023**, *14*, 1085. [CrossRef]
154. Madgwick, S.O.; Wilson, S.; Turk, R.; Burridge, J.; Kapatos, C.; Vaidyanathan, R. An extended complementary filter for full-body MARG orientation estimation. *IEEE/ASME Trans. Mechatron.* **2020**, *25*, 2054–2064. [CrossRef]
155. Thakur, D.; Biswas, S. Permutation importance based modified guided regularized random forest in human activity recognition with smartphone. *Eng. Appl. Artif. Intell.* **2024**, *129*, 107681. [CrossRef]
156. Luwe, Y.J.; Lee, C.P.; Lim, K.M. Wearable Sensor-Based Human Activity Recognition with Hybrid Deep Learning Model. *Informatics* **2022**, *9*, 56. [CrossRef]

Disclaimer/Publisher's Note: The statements, opinions and data contained in all publications are solely those of the individual author(s) and contributor(s) and not of MDPI and/or the editor(s). MDPI and/or the editor(s) disclaim responsibility for any injury to people or property resulting from any ideas, methods, instructions or products referred to in the content.

sensors

MDPI

Article

Measuring CO_2 Concentration and Thermal Comfort in Italian University Classrooms: A Seasonal Analysis

Alessia Fedele [1], Andrea Colantoni [2], Giuseppe Calabrò [1], Mauro Scungio [1], Stefano Rossi [1] and Juri Taborri [1,*]

[1] Department of Economics, Engineering, Society and Business Organization, University of Tuscia, 01100 Viterbo, Italy; alessiafedele96@gmail.com (A.F.); giuseppe.calabro@unitus.it (G.C.); mauro.scungio@unitus.it (M.S.); stefano.rossi@unitus.it (S.R.)
[2] Department of Agriculture and Forest Sciences, University of Tuscia, 01100 Viterbo, Italy; colantoni@unitus.it
* Correspondence: juri.taborri@unitus.it

Abstract: This study investigates indoor air quality (IAQ) and thermal comfort in Italian university classrooms, considering seasonal variations. Poor IAQ can impair students' cognitive performance and well-being, making ventilation strategies crucial. The aim is to assess CO_2 levels, temperature, and humidity, analyzing their interactions across seasons. A monitoring protocol was applied in three classrooms using NDIR sensors and a microclimate assessment system. Sensors were placed strategically to capture representative data in 20 days in spring and autumn. Results indicate that CO_2 levels peaked at 2324.2 ppm in autumn, significantly exceeding the 1000 ppm threshold, whereas spring levels remained below 953.4 ppm. Relative humidity ranged from 32.7% to 55.6%, with higher values in autumn. Temperatures varied from 19.1 °C to 27.5 °C, with warmer conditions in spring. Strong positive correlations (always greater than 0.70) between CO_2 and humidity suggest inadequate air exchange reduces IAQ, potentially affecting cognitive performance. This research provides valuable insights for improving student well-being through better air quality management. This research provides valuable insights for optimizing classroom environments, supporting cognitive performance, and improving student well-being through better air quality management.

Keywords: thermal comfort; CO_2 concentration; sensor systems

1. Introduction

Indoor air quality (IAQ), mainly CO_2 concentration, and thermal comfort are increasingly recognized as critical elements influencing health and well-being [1,2]. IAQ and thermal comfort are also demonstrated to be essential components to guarantee a healthy learning environment, reducing absence rate and improving test scores, as well as cognitive performance, in educational settings [3,4]. Classrooms are often subject to fluctuating and elevated levels of carbon dioxide (CO_2) due to high occupancy, limited space, and varying ventilation quality, which can lead to suboptimal conditions for learning [5,6]. Research underscores the significance of maintaining CO_2 concentrations below 1000 parts per million (ppm), as higher levels have been shown to impair cognitive abilities essential for academic tasks [7]. For instance, students exposed to environments with high CO_2 concentrations display reduced problem-solving skills, memory retention, and information processing, which collectively diminish academic performance [8].

Studies on indoor air quality in educational environments indicate that CO_2 concentration serves as a reliable marker for assessing ventilation adequacy and occupancy levels. For example, Hui et al. found that indoor CO_2 levels could serve as a proxy for ventilation efficiency, as higher concentrations indicate insufficient fresh air exchange [9]. As a direct product of human respiration, CO_2 accumulates rapidly in enclosed spaces with inadequate air exchange, creating environments with lower oxygen availability and higher levels of contaminants. Several investigations into the CO_2 concentration of schools and universities show that the concentration of CO_2 tends to rise considerably during class hours, peaking at levels that can exceed recommended standards, particularly in colder seasons when natural ventilation (e.g., open windows) is minimized to maintain thermal comfort [10]. Previous studies have consistently highlighted the seasonal impact on CO_2 concentration in classrooms, with higher levels typically recorded during the winter months due to reduced ventilation practices. Investigations conducted in various educational settings, including naturally ventilated classrooms in different climates, have shown that CO_2 accumulation is significantly influenced by outdoor temperatures, as lower temperatures discourage window opening and natural ventilation [11,12]. In particular, a comparison study between Polish and Spanish classrooms reported consistently higher CO_2 levels in the coolest region, i.e., Poland, with a pronounced effect in the early morning hours, suggesting a strong correlation between seasonal ventilation habits and indoor air quality [12]. A similar trend has been observed in Italian primary schools, where research has demonstrated that manual airing techniques alone are not sufficient to maintain acceptable indoor air quality during the heating season [13]. Furthermore, the study emphasized that while spring conditions may allow for more effective natural ventilation, wintertime air exchange remains inadequate, requiring alternative solutions to maintain CO_2 concentrations within recommended thresholds. In addition to CO_2 accumulation, research also suggests that high levels of indoor pollutants, such as particulate matter (PM), can coincide with increased CO_2 concentrations, particularly in schools where manual airing strategies are employed inconsistently [14].

In addition to IAQ, there has been a growing interest in research analyzing the role of thermal comfort parameters—namely temperature and relative humidity—on both physiological and cognitive functions within classroom environments [15,16]. Temperature fluctuations, even within mild ranges, can influence students' heart rates and thermal perceptions, potentially inducing discomfort that affects their focus and task performance [17]. Studies by Brink et al. indicate that both high and low extremes of thermal conditions disrupt cognitive tasks such as memory recall and vigilance [18], even if it is worth underlining that, regarding the thermal perception, female students tend to feel colder than their male counterparts. Similarly, Barbic et al. observed a decrease of up to 24% in cognitive performance among students experiencing thermal discomfort due to high temperature [19]. However, not all thermal discomfort sensations lead to a deterioration in cognitive performance, and the impact is most likely activity-dependent; in fact, a cooler environment has been demonstrated to positively influence cognitive performance in comparison to the colder and warmer ones. Moreover, the thermal sensation 'warm' influences cognitive performance in tasks related to vigilance, memory and learning more than the thermal sensation 'cold'. Evidence suggests a consistent relationship between CO_2 levels and other microclimate variables. Specifically, research by Lazović et al. highlighted that the correlation between CO_2 concentration and relative humidity can serve as a key indicator of air quality and ventilation efficiency, especially in naturally ventilated spaces [20]. For example, high occupancy without adequate air exchange often results in both increased CO_2 levels and higher relative humidity due to human exhalation and respiration processes [21]. These interconnected variables suggest that managing one factor can often influence the others, underscoring the importance of a comprehensive IAQ-monitoring approach that encompasses CO_2, temperature, and humidity [22].

From this perspective, it is clear how a sensor-based measurement of not only CO_2 levels but also temperature and humidity is mandatory in classrooms to sustain a productive learning atmosphere and to properly design and manage classroom ventilation strategies, moreso than in other types of buildings, such as the ones dedicated for office work [17,23]. Even if a plethora of studies have been already proposed in the literature, to the best of the authors' knowledge, analyses are limited to specific seasons and specific geographical areas. In addition, university classrooms, characterized by high occupancy levels and variable ventilation conditions, are particularly susceptible to fluctuations in CO_2 concentration, temperature, and humidity. While extensive research has been conducted on indoor air quality in primary and secondary schools, fewer studies have focused on higher education environments, where students typically spend extended periods in enclosed spaces with limited ventilation control. In light of these findings, the present study investigates the seasonal variations in CO_2 concentration and thermal comfort in university classrooms, aiming to assess how different ventilation practices impact air quality throughout the academic year, especially in the seasons with higher occupancy due to lectures rather than exams, which are spring and autumn. Our central hypothesis is that CO_2 levels significantly increase during autumn due to reduced natural ventilation, leading to higher relative humidity and potential discomfort for students, whereas spring conditions allow for improved air exchange, maintaining better indoor air quality. To test this hypothesis, we conducted continuous environmental monitoring across different classroom settings, analyzing the relationship between CO_2 concentration, humidity, and thermal comfort. By focusing on university classrooms, this study addresses a critical gap in the literature and provides practical insights for optimizing ventilation strategies in higher education settings. These findings contribute to a growing body of research on indoor air quality management and offer evidence-based recommendations for improving student well-being and learning conditions in academic institutions.

2. Materials and Methods

2.1. Experimental Setup

For the objective of measuring carbon dioxide concentration, temperature and humidity, a commercially available Non-Dispersive Infrared (NDIR) sensor produced by Fybra (model: Fybra School, Milan, Italy) has been used (Figure 1).

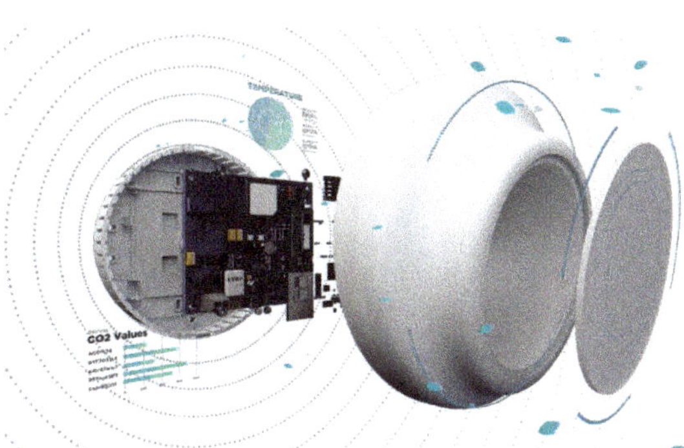

Figure 1. Structure of the NDIR sensor.

The working principle of the sensor for the measurement of the carbon dioxide concentration is based on spectroscopy, specifically on the energy absorption characteristics of CO_2 in the infrared zone. The NDIR sensor has an infrared light source that emits through a measuring chamber. The gas to be measured flows within this chamber. The infrared light traverses the gas sample, and the gas molecules absorb energy at specific infrared wavelengths. Each gas possesses its own unique infrared absorption spectrum. For instance, in the case of CO_2, absorptions of radiation with wavelengths at 2.7 μm, 4.7 μm and 15 μm are detected in the infrared region. Following the passage of infrared light through the gas sample, it reaches a detector responsible for measuring the intensity of the light striking it. The difference between the intensity of the emitted infrared light and the intensity of the detected light offers insights into the gas absorption within the sample. The same sensor is also able to gather measurements on temperature (T) and relative humidity (RH). Metrological characteristics of Fybra sensors are reported in Table 1.

Table 1. Metrological characteristics of Fybra sensors.

Variable	Measurement Range	Accuracy	Repeatability	Response Time
CO_2	0–40,000 ppm	±(40 ppm + 5% reading error)	±10 ppm	60 s
RH	0–100%RH	±9%RH	±0.4%RH	90 s
T	−10–60 °C	±1.5 °C	±0.4 °C	120 s

For the microclimate assessment, the BABUC (LSI Lastem srl, Milan, Italy) system has been adopted. The system consists of an instrument assembly (data logger and sensors) mounted on a tripod, as depicted in Figure 2. Depending on the specific environment and the related microclimatic survey (moderate, hot and cold environments and localized discomforts), it is possible to choose different types of sensors. For our application, a moderate indoor environment was analyzed, and therefore the utilized probes consisted of a *hot-wire anemometer* (model ESV126, LSI Lastem srl, Milan, Italy), a *psychrometer* (model MINI-DIN ESU102A, LSI Lastem srl, Milan, Italy) and a *globe thermometer* (model DMA131A, LSI Lastem srl, Milan, Italy).

Through the sensor system, we gathered the following measures: (i) flow velocity; (ii) dry air temperature; (iii) wet air temperature; (iv) relative humidity; and (v) mean radiant temperature. The BABUC system has been already used in the literature for microclimatic analysis in indoor environments [24,25]. Metrological characteristics of the BABUC system are reported in Table 2.

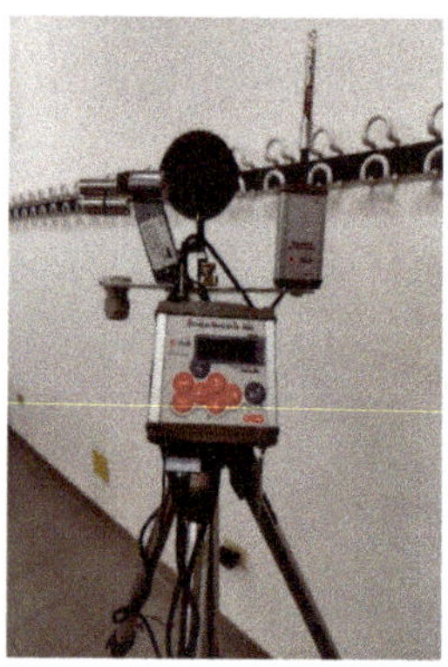

(a)

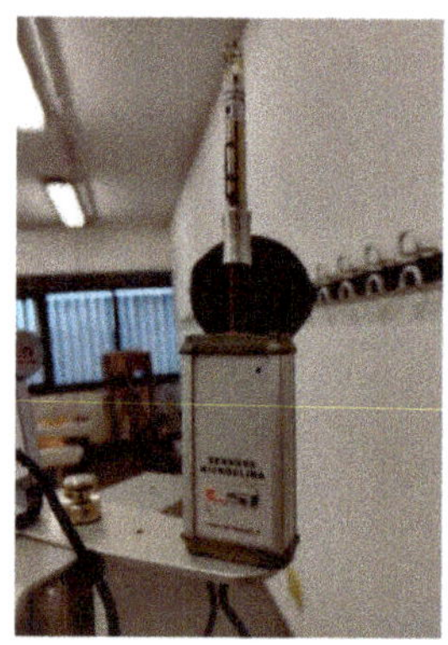

(b)

Figure 2. *Cont.*

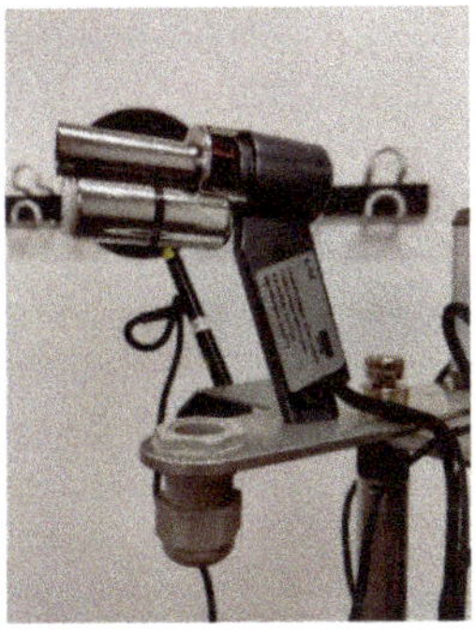

(c)

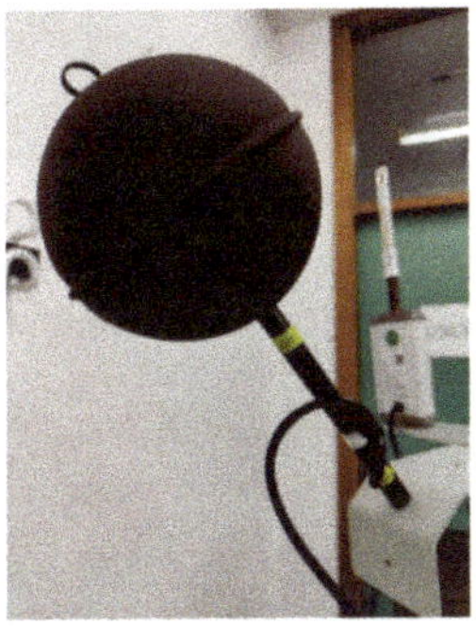

(d)

Figure 2. (**a**) overall BABUC system for microclimate assessment; (**b**) hot-wire anemometer; (**c**) psychrometer; and (**d**) globe thermometer.

Table 2. Metrological characteristics of the BABUC system.

Sensor	Operative Range	Accuracy
Hot-wire anemometer	0–50 m/s	±0.05 m/s
Globe thermometer	−50–100 °C	±0.17 °C
Psychrometer	−50–150 °C	±0.13 °C
	40–100%RH	±2%RH

2.2. Experimental Protocol

The monitoring and evaluation of CO_2 concentration and thermal comfort was conducted in three classrooms of varying size at the University of Tuscia, located in the School of Engineering. The selected classrooms included the following: (i) F8-97 m^2, volume 291 m^3, maximum capacity of 70 students; (ii) F9-78.23 m^2, volume 234.69 m^3, maximum capacity of 54 students; and (iii) B1-112.40 m^2, volume 337.20 m^3, maximum capacity of 40 students.

Each classroom is characterized by the following:

- Ventilation system—Natural ventilation through windows;
- Windows—B1 and F8 are equipped with 4 single-glazed aluminum inward-opening windows; whereas F9 with 2 single-glazed aluminum inward-opening windows.
- Equipment—Desks, chairs, projector, whiteboard;
- Heating system—Central heating with radiators along walls;
- Location and orientation: low-traffic area surrounded by open fields and east-to-northeast orientation; the direct sunlight exposure is primarily limited to the morning hours.

Given the sizes of the classrooms, sensor placement was optimized to capture representative data for each environment. Classrooms F8 and F9 were equipped with one sensor each, positioned at the front of the room near the desk and front rows. In Classroom B1, due to its larger size, two sensors were installed: one near the desk and front rows, and another at the rear of the room. The placement of sensors within the classrooms adhered strictly to the following criteria, as recommended by ISO 16000:26 and GdS Indoor Pollution guidelines:

- *Height*: Sensors were installed at approximately 1.5 m above the floor, which corresponds to the breathing zone of seated occupants. This placement minimized measurement bias from non-representative air strata.

- *Distance from Occupants*: Sensors were placed 1.5–2 m away from occupants to reduce direct interference from exhaled air.
- *Distance from Obstacles*: A minimum distance of 1 m was maintained from walls, bookshelves, and other potential obstructions to ensure an unobstructed measurement of the indoor air.
- *Ventilation Interference*: Sensors were positioned away from direct airflow caused by windows, doors, or air conditioning units to avoid artificial fluctuations in readings.
- *Heat Sources*: Care was taken to avoid proximity to radiators, spotlights, or other heat sources that could influence temperature and relative humidity measurements.

In addition to continuous monitoring, the sensors were equipped with a visual alert system to provide real-time feedback to occupants: (i) blue, indicating good air quality; (ii) red, indicating poor air quality, prompting immediate ventilation; (iii) and violet, denoting improving air quality. Specifically, the red light starts when CO_2 overcomes the limit of 1000 ppm, whereas the violet starts at 900 ppm. Values lower than 900 are associated with blue light.

Data acquisition through NDIR sensors was performed in two distinct periods, spring and autumn, to capture seasonal variations in indoor air quality and thermal comfort conditions. The data collection schedule comprised 50 days between May 2022 and December 2022, twenty-five per each season, for the three above-mentioned classrooms. Details of students' occupancy are reported in Table 3.

Table 3. Students' occupancy during case study and demographic information.

	Season	Avg. Occupancy	Max Occupancy	Gender (%Male/ %Female)	Avg. Height (m)	Avg. Body Mass (kg)	Avg Age (y.o.)
F8	Spring	27	30	70/30	174 ± 7	71 ± 12	22 ± 3
	Autumn	26	30	70/30	174 ± 7	71 ± 12	22 ± 2
F9	Spring	23	27	75/25	173 ± 7	72 ± 10	22 ± 2
	Autumn	23	28	77/23	172 ± 6	73 ± 10	22 ± 2
B1	Spring	29	30	65/35	175 ± 7	68 ± 11	25 ± 1
	Autumn	28	30	64/36	176 ± 8	71 ± 11	25 ± 2

In all classes of the case study, the lecture starts at 9:00 a.m. Depending on the type of activities, they last 60 or 120 min, and after that a 15 min break takes place. A long break is scheduled from 1 to 2 pm. During both autumn and spring, windows were kept closed before the school start time, whereas the classroom's door was kept open till the lesson start. During the lecture time, a different behavior was recognized in autumn and spring: (i) in the autumn, windows and doors were kept ordinarily closed (aside from the previously mentioned break time, during which both windows and doors were opened to let fresh air enter the classroom); (ii) in spring, windows are mainly kept open, whereas the door was kept closed (expect during the break time). The specifications of this case study are in line with other previously available case studies in the literature [11–13,26].

These timeframes were chosen to reflect different external environmental conditions, influencing natural ventilation behavior, and occupancy patterns. Outputs of the BABUC system were collected for a complete microclimate assessment only in classroom F8 due to its higher student occupancy, making it a critical space for IAQ evaluation.

On each survey day, sensors were calibrated and the actual data acquisition started after 30 min to guarantee the stabilization of the signals. Measurements were recorded every 120 s to ensure consistent temporal resolution across all monitored variables. This interval was selected to align with the response times of the sensors. The measurement duration for each classroom spanned the full teaching schedule, capturing occupancy dynamics and ventilation practices throughout the day.

For the microclimate analysis, subjective thermal comfort was assessed through questionnaires distributed to students present during the survey to assess the Predicted Mean Vote (PMV). Participants rated their thermal sensation on a 7-point scale ranging from −3 (very cold) to +3 (very warm), as in the following Table 4, as requested in the norm ISO 7730:2006.

Table 4. Seven-point thermal sensation scale.

PMV	Thermal Sensation
+3	Very warm
+2	Warm
+1	Slightly warm
0	Neutral
−1	Slightly cold
−2	Cold
−3	Very cold

2.3. Data Analysis

The collected data were exported in CSV format, with separate files created for each variable and classroom. The data analysis was conducted using MATLAB software (version R2023a) to process, evaluate, and interpret the datasets collected from the NDIR sensors and the BABUC microclimate station.

Concerning NDIR sensors, data were firstly structured into matrices, categorized by four daily intervals, which were as follows: (i) 9–11 a.m.; (ii) 11 a.m.–1 p.m.; (iii) 2–4 p.m.; and (iv) 4–6 p.m.. It is worth noticing that the interval 1–2 p.m. has been excluded by the successive analysis for the absence of occupants in the classroom due to the lunch break. Successively, descriptive statistics, including maximum, minimum, mean, and standard deviation, were computed for CO_2 concentration, temperature and relative humidity per each daily interval. It is worth highlighting that the results for the classroom B1, in which two NDIR sensors were placed, were obtained by firstly averaging the measurements of the two sensors. Apposite graphs were realized to monitor when CO_2 concentration overcame the norm threshold of 1000 ppm.

Moving to the microclimate analysis through the BABUC system, performed only in classroom F8, we applied the approach based on ISO 7730:2006, which defines thermal comfort using the Predicted Mean Vote (PMV) and Predicted Percentage of Dissatisfied (PPD) indices. Using the replies to the questionnaire, the PPD index was calculated by using the following formula:

$$\mathrm{PPD} = 100 - 95e^{(-0.03353PMV^4 - 0.2179PMV^2)}$$

The previous equation can be applied only if the following values are met: (i) indoor air temperature in the range 10–30 °C; (ii) average radiant temperature in the range 10–40 °C; and (iii) air velocity in the range 0–1 m/s. The coherence with the requirement has been verified by using data gathered from the BABUC system.

2.4. Statistical Analysis

Correlation between CO_2 concentration and relative humidity was assessed. Scatter plots were created with relative humidity values on the x-axis and CO_2 concentrations on the y-axis, and a trend line was fitted to visualize the relationship. Pearson's correlation coefficient (r) was calculated to quantify the strength of the relationship. Correlation coefficients approaching +1 indicated a strong positive correlation, values near −1 indicated a strong negative correlation, and values close to 0 suggested no significant relationship. Furthermore, to determine whether significant differences in mean CO_2 concentrations existed across different time intervals, a one-way ANOVA test was conducted. Bonferroni's multiple comparison test was performed in case of significant ANOVA results. Seasonal differences in CO_2 concentrations and PMV between spring and autumn were further examined using a two-sample *t*-test, independently per each variable.

3. Results and Discussions

3.1. *Temperature, Relative Humidity and CO_2 Concentration: Seasoning Variations and Correlation*

Figure 3a,b reports an example of the behavior of the CO_2 concentration in one classroom in relation to the time intervals and the number of occupants. Qualitatively, the graphs compare CO_2 concentration and student presence in a classroom during spring and autumn. In both cases, CO_2 levels increase when students are present and decrease after they leave, highlighting the direct relationship between occupancy and indoor air quality. Anyway, even if it is true that when windows and doors were opened the CO_2 concentration fell (in the break 1–2 pm), the concentration suddenly started increasing again when the openings were fully closed. Thus, the short-term manual airing periods were insufficient to maintain average CO_2 concentrations below 1000 ppm, especially in autumn. This qualitative example is in line with the results reported in [13], also considering similar climate regions.

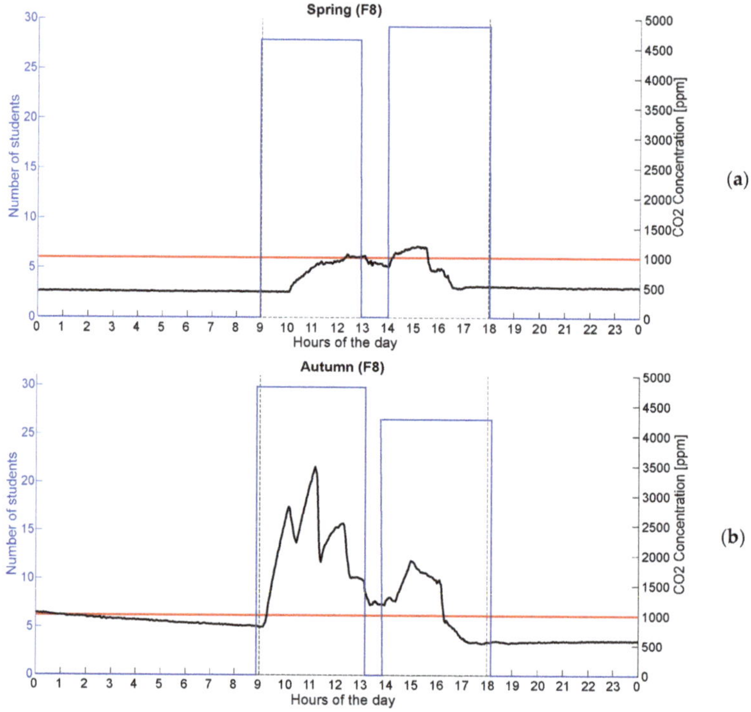

Figure 3. Example of the trend of CO_2 concentration in F8 during spring (**a**) and autumn (**b**). Solid red, black and blue lines represent the CO_2 threshold, the trend of CO_2 and the number of occupants, respectively. Dotted black line indicates the tested time intervals.

The red line represents a threshold beyond which CO_2 levels may become critical, and in autumn, the concentration exceeds this limit more significantly than in spring.

Table 5 summarizes the mean, maximum, and minimum values (with standard deviations) for T, RH, and CO_2 concentration in three rooms (F8, F9, and B1) across two seasons (spring and autumn), considering whole-day measurements.

Table 5. Mean (SD) of the mean, maximum and minimum value of temperature (T), relative humidity (RH) and CO_2 concentration for each room, considering the whole day, divided for the two seasons. * indicates statistical differences between seasons.

		F8			F9			B1		
		Mean	*Max*	*Min*	*Mean*	*Max*	*Min*	*Mean*	*Max*	*Min*
SPRING	**T (°C)**	23.5 * (0.8)	24.6 (1.3)	22.6 (1.1)	23.9 * (0.9)	27.5 (4.0)	22.4 (1.1)	24.5 * (0.7)	25.6 (0.9)	23.4 (0.6)
	RH (%)	36.5 * (2.1)	39.4 (2.7)	32.7 (3.7)	32.7 * (1.8)	37.2 (2.6)	27.4 (3.7)	34.1 * (3.0)	38.1 (3.8)	30.3 (4.3)
	CO_2 (ppm)	492.0 * (62.4)	729.8 (280.1)	422.71 (33.8)	503.0 * (36.3)	953.4 (286.7)	419.2 (12.6)	547.4 * (84.6)	934.6 (200.1)	425.5 (32.0)
AUTUMN	**T (°C)**	21.0 * (1.1)	24.0 (1.0)	19.1 (1.5)	21.5 * (0.9)	23.5 (0.5)	20.4 (1.0)	22.9 * (1.3)	24.5 (2.1)	20.9 (2.5)
	RH (%)	51.6 * (5.3)	55.6 (6.5)	47.1 (6.6)	50.0 * (4.0)	51.8 (4.8)	47.0 (4.4)	49.4 * (3.9)	54.2 (4.2)	46.0 (4.7)
	CO_2 (ppm)	1034.0 * (153.9)	2324.2 (524.0)	457.2 (51.7)	664.4 * (92.1)	1419.6 (224.5)	475.8 (84.5)	702.8 * (92.5)	1399.4 (91.9)	457.0 (22.1)

During spring, the mean temperature ranged from 23.5 °C in F8 to 24.5 °C in B1, with maximum values reaching 27.5 °C in F9 and minimum values as low as 22.4 °C in F9. RH was generally lower in spring, with mean values between 32.7% (F9) and 36.5% (F8). Maximum RH peaked at 39.4% in F8, while minimum RH values went as low as 27.4% in F9. CO_2 levels showed substantial variation, with mean concentrations highest in B1 (547.4 ppm) and lowest in F8 (492.0 ppm). Maximum CO_2 levels peaked at 953.4 ppm in F9, whereas minimum values were consistent across rooms, with F9 recording the lowest minimum at 419.2 ppm. In autumn, temperatures were slightly lower, with mean values ranging from 21.0 °C in F8 to 22.9 °C in B1. Maximum temperatures reached 24.5 °C in B1, while minimum temperatures dropped to 19.1 °C in F8. Relative humidity was notably higher in autumn compared to spring, with mean values ranging between 49.4% (B1) and 51.6% (F8). Maximum RH reached 55.6% in F8, while minimum values were lowest at 46.0% in B1. CO_2 concentrations exhibited a pronounced seasonal increase in autumn, particularly in F8, where the mean reached 1034.0 ppm with a maximum of 2324.2 ppm. In contrast, minimum CO_2 levels across rooms remained comparable, with values between 457.0 ppm (B1) and 475.8 ppm (F9). Statistical differences between the two seasons were found for each mean parameter and for all the rooms, with p-values ranging from <0.01 and 0.04. Moving to the time interval analysis, no statistical differences were found, with all the ANOVA tests associated with p-values ranging from 0.10 to 0.85.

Overall, the results highlight seasonal variations, with higher CO_2 concentrations and relative humidity observed in autumn and higher temperatures recorded during spring. The observed increase in RH and CO_2 concentrations during the autumn season, compared to spring, can be attributed to several environmental and seasonal factors, as the effects induced by occupancy can be considered negligible since, during the survey period, the occupancy of each room remained similar, ranging consistently between 25 to 30 students. This stable occupancy ensures that variations in CO_2 and RH levels are not influenced by differences in the number of occupants but rather by other external or environmental factors. Similar results have been reported in [11], where an increment of both CO_2 concentration and RH has been found in the winter semester in comparison to the summer one when seeking to evaluate the indoor environment quality in Slovakian classrooms. The results are also consistent in terms of students' occupancy that is comparable, as well as the use of only natural ventilation. Likewise, when compared different climates (Poland vs. Spain), the coolest region was found to be related to higher concentrations of carbon dioxide, revealing the influence of external temperature [12]. However, it is worth noting that in the reported study, the differences are significant, especially during the first hours of the day, whereas in our study no differences among hours were found. The increment in CO_2 concentration is also in line with the results proposed in [13], where the analysis of indoor air quality has been conducted in primary school classrooms with natural ventilation and manual airing, as in this study. Data clearly highlight that students were exposed to high CO_2 concentrations in the cooler season also in [13,14,26].

Focusing on the reasons, the rise in CO_2 levels during autumn may be explained primarily by changes in ventilation practices, confirming the results presented in [5,10]. Cooler outdoor temperatures in autumn often result in reduced natural ventilation as windows and doors are kept closed to maintain indoor thermal comfort. Limited ventilation diminishes the air exchange rate, causing CO_2 exhaled by occupants to accumulate indoors. Studies have shown that ventilation rates significantly influence indoor CO_2 concentrations, with insufficient ventilation resulting in higher indoor CO_2 levels, especially in occupied spaces. RH is also elevated during autumn due to a combination of seasonal and indoor factors. Cooler temperatures reduce the air's capacity to hold moisture, causing a higher relative humidity for the same amount of water vapor present [27]. Additionally, reduced ventilation practices may trap moisture indoors, further contributing to elevated RH levels. Indoor sources of humidity, such as human respiration, can exacerbate the rise in RH in enclosed spaces. This phenomenon is supported by research highlighting the importance of ventilation in controlling indoor humidity, particularly during cooler seasons when natural ventilation is restricted [28]. Seasonal climatic differences further compound these indoor changes. In many climates, autumn is characterized by increased precipitation and cooler weather, which contribute to higher outdoor humidity levels [29]. This outdoor air, when introduced indoors via mechanical ventilation or infiltration, can raise indoor RH levels. Furthermore, it is worth highlighting that, as previously reported, the buildings have an east-to-northeast orientation, meaning that direct sunlight exposure is primarily limited to the morning hours. This orientation influences indoor thermal conditions, as classrooms receive solar radiation in the early hours but remain largely shaded in the afternoon. Consequently, natural heating effects from solar exposure are reduced later in the day, potentially affecting indoor temperature and humidity dynamics, particularly during colder months when ventilation practices are more restricted.

In summary, it is evident that relying solely on manual airing based on subjective physical perception or during the break is not sufficient to ensure minimum indoor air quality in university environments, especially during autumn when severe meteoclimatic conditions can worsen indoor air stagnation. However, this method might still be effective in milder conditions. Based on the findings, even in cases where mechanical ventilation systems cannot be implemented due to cost and structural constraints, it is crucial to adopt alternative solutions that facilitate effective ventilation during critical indoor air quality situations. The simplest and most cost-effective approach could involve establishing detailed window-opening protocols.

Figure 4a,b report an example of correlation betwenn CO_2 and RH for classroom F8 in both tested seasons. Table 6 reports all the r values across seasons and classrooms.

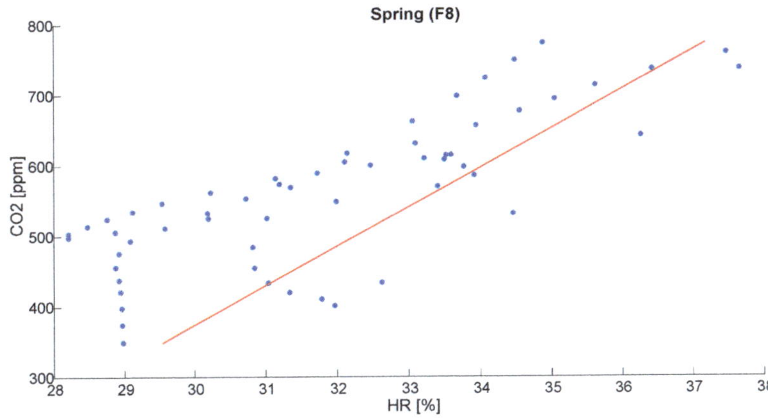

(a)

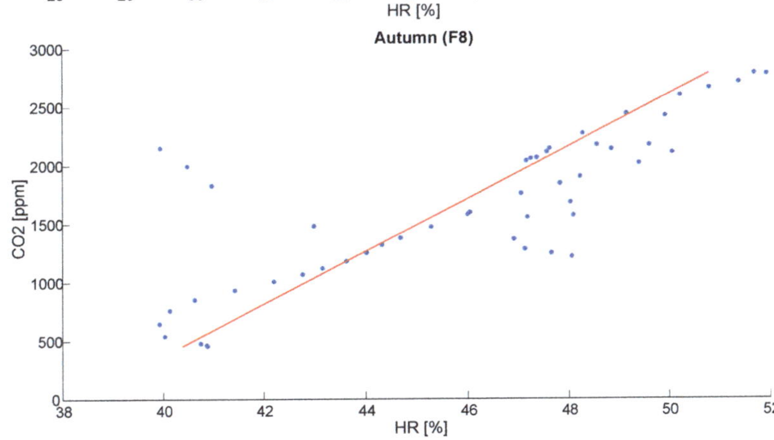

(b)

Figure 4. Example of correlation between CO_2 concentration and RH in F8 during spring (**a**) and autumn (**b**). Blue points represent the data, whereas the solid red line is the regression line.

Table 6. Correlation values between RH and CO_2 concentration.

		r-Value			
	Season	**Mean**	**STD**	**Min**	**Max**
F8	Spring	0.84	0.03	0.78	0.89
	Autumn	0.90	0.04	0.85	0.95
F9	Spring	0.85	0.02	0.80	0.88
	Autumn	0.91	0.03	0.87	0.94
B1	Spring	0.83	0.04	0.76	0.88
	Autumn	0.89	0.05	0.83	0.93

The results indicate a consistently strong positive correlation in both seasons, with higher values in autumn (average around 0.90) compared to spring (average r around 0.84). These correlations are also always statistically significant, with a p-value consistently below 0.02. The relatively low standard deviation values (ranging from 0.02 to 0.05) suggest that the correlation remains stable across different measurements, reinforcing the reliability of this trend. The maximum values of r reaching up to 0.95 in autumn and 0.89 in spring indicate periods where the relationship between CO_2 and RH is extremely strong, whereas the minimum values (down to 0.76 in spring) suggest occasional deviations, likely influenced by specific ventilation patterns or external environmental factors, mainly due to the opening of the windows.

By analyzing the overall results, there is a positive correlation between CO_2 and RH, as also demonostred for poorly ventilated indoor environments in [20,21]. Specifically, Lazovic et al. [20] found a very strong correlation (greater than 0.80) in aschool with the lowest ventilation rate, due to only manual ventilation. Additionally, the found correlation is greater than the 0.60 found in the study proposed in [30], where a similar case study has been conducted focusing on only natural ventilation. This relationship arises because both parameters share common influencing factors. Human respiration is a significant source of both CO_2 and water vapor, contributing simultaneously to increased concentrations of both variables in enclosed spaces. When ventilation is insufficient, these emissions accumulate, leading to a concurrent rise in CO_2 and RH levels. Research has shown that in such conditions, elevated CO_2 levels are often associated with higher RH due to the shared impact of limited air exchange [31]. This correlation underscores the importance of ventilation as a key factor in maintaining indoor air quality.

From a health and comfort perspective, the seasonal increase in both CO_2 and RH levels has important implications. Elevated CO_2 concentrations can lead to reduced cognitive performance, fatigue, and discomfort, particularly when exceeding 1000 ppm [7,8,32]. Similarly, higher RH levels can create favorable conditions for mold growth and dust mites, which can negatively impact indoor air quality and exacerbate respiratory conditions [33].

Therefore, the findings underscore the critical need for effective ventilation strategies and humidity control during autumn to maintain healthy and comfortable indoor environments. In fact, as also suggested by Norbäck et al. [21], the potential decline in learning capacity due to health issues triggered by poor indoor air quality in classrooms requires greater attention.

3.2. Thermal Comfort Analysis

Considering the data acquired by the BABUC, it was possible to conduct the thermal comfort analysis through the computation of PMV and PPD. Figure 5a,b show an example of the PMV curves obtained in spring and autumn in F8. Generally, in spring, the PMV values exhibit a relatively balanced distribution around neutrality, with the majority of responses concentrated between −1 and +1. The highest frequency is observed at PMV = 0, indicating that most students experienced neutral thermal conditions. This suggests that indoor environmental parameters, such as temperature, humidity, and ventilation, were relatively well regulated, maintaining comfort within acceptable limits. The overall distribution in spring appears symmetric, with no extreme deviations toward discomfort levels. In contrast, in autumn, the PMV distribution shifts, with a notable increase in responses at PMV = −1 and a reduction at PMV = 0 and +1. This shift implies that a larger proportion of occupants perceived the classroom as slightly cool, potentially indicating lower indoor temperatures or differences in thermal insulation due to seasonal variations. The asymmetry in the autumn distribution suggests that a subset of students may have experienced mild thermal discomfort, likely due to inadequate heating or ventilation adjustments to compensate for external temperature changes.

Such results align with previous research indicating that classroom environments often face challenges in maintaining optimal thermal comfort due to varying metabolic rates, clothing insulation, and localized air distribution [34,35]. Overall, these results imply that students perceive a slight thermal discomfort in the classroom, though it is not significant enough to adversely affect the learning environment. However, the persistence of dissatisfaction under certain conditions emphasizes the need for further optimization of HVAC systems, such as improved ventilation, individualized control systems, or adaptive comfort measures, to enhance thermal comfort and support cognitive performance [36].

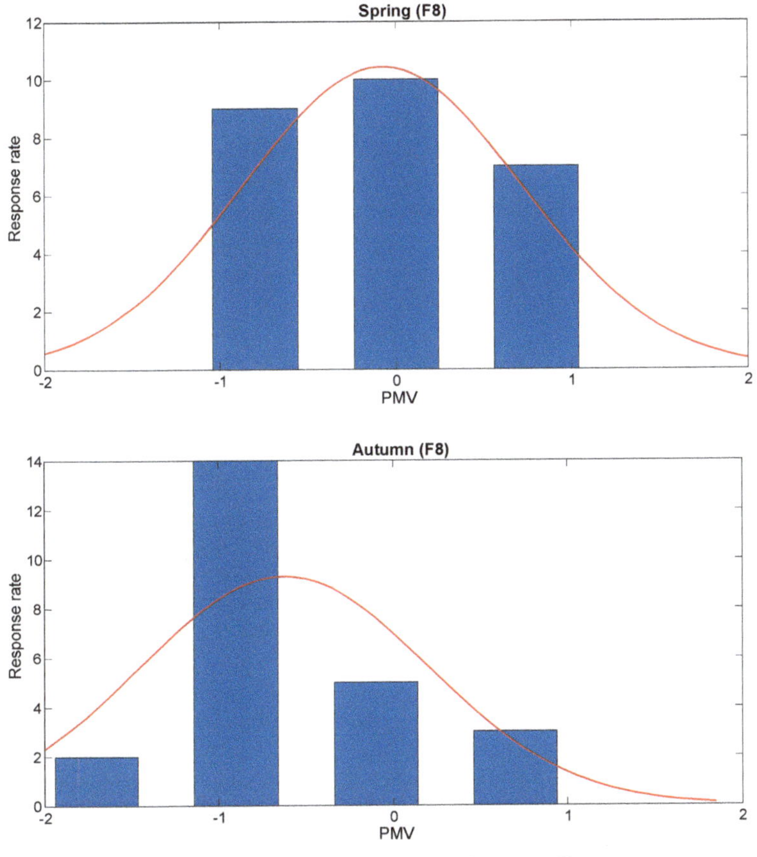

Figure 5. Example of PMV in F8 during spring (**a**) and autumn (**b**).

By considering both CO_2 concentration analysis and thermal comfort results, the comparative analysis between the two seasons highlights the importance of adaptive thermal management strategies in educational environments. While conditions in spring seem to align more closely with optimal CO_2 concentration and thermal comfort, autumn exhibits a trend toward cooler perceptions. Thus, the need is evident for targeted interventions, such as enhanced ventilation and humidity control, to mitigate indoor air quality challenges in cooler seasons.

3.3. Policy Implications and Recommendations

The findings of this study underscore the need for universities to adopt targeted strategies for maintaining optimal indoor environmental quality, particularly in response to seasonal variations in CO_2 concentration and thermal comfort. In fact, given the observed increase in CO_2 levels and relative humidity during autumn, institutions should prioritize enhanced ventilation strategies, such as a combination of natural and mechanical ventilation systems, to maintain air exchange rates within recommended thresholds. Implementing automated ventilation controls that adjust airflow based on real-time CO_2 levels could help mitigate periods of poor air quality, especially in high-occupancy classrooms.

Additionally, real-time air quality monitoring systems should be integrated into classroom management practices. Displaying CO_2 concentration levels in lecture halls through visual indicators can promote timely interventions, such as opening windows or adjusting mechanical ventilation settings. Universities should also establish policies that encourage regular ventilation breaks during lessons, particularly in colder months when natural ventilation is often reduced. From a long-term perspective, educational institutions should consider investing in advanced HVAC systems with demand-controlled ventilation (DCV) technology, which automatically adjusts airflow based on occupancy and air quality conditions. Future campus renovations and new building designs should incorporate these principles to ensure a healthier and more conducive learning environment. By implementing these recommendations, universities can significantly enhance indoor environmental quality, contributing to improved student concentration, well-being, and overall academic performance.

3.4. Limitations

While this study provides valuable insights into CO_2 concentration and thermal comfort in university classrooms, some limitations should be acknowledged. First, the analysis focused exclusively on CO_2 as an indicator of ventilation efficiency, without considering other air quality parameters such as particulate matter (PM), volatile organic compounds (VOCs), or other indoor pollutants that could further influence the learning environment. Additionally, this study was conducted within a single university in Italy, meaning the findings may not be fully generalizable to institutions in different climatic regions, building designs, or ventilation configurations. Future studies should explore similar analyses in a wider range of educational settings to enhance the applicability of the results.

Another limitation is the lack of direct cognitive performance assessments. While high CO_2 levels have been linked to impaired cognitive abilities in previous literature, this study did not include objective measurements such as student test performance, concentration levels, or physiological indicators like heart rate variability. Incorporating such data in future research could provide stronger evidence of the impact of CO_2 and thermal comfort on student well-being and academic performance. Lastly, the thermal comfort assessment was based on subjective self-reported data, which, despite being a widely used method, is inherently influenced by individual perception differences. A more comprehensive evaluation that integrates objective thermal indicators, such as skin temperature or metabolic rate, could offer a more accurate representation of students' thermal comfort. Despite these limitations, this study provides a strong foundation for understanding seasonal variations in CO_2 concentration and thermal comfort, highlighting the need for improved ventilation strategies in university classrooms.

4. Conclusions

This study provides valuable insights into the seasonal variations of CO_2 concentration and thermal comfort in university classrooms, emphasizing the importance setoff setting appropriate ventilation strategies for indoor environmental quality. The results indicate that during autumn, reduced ventilation leads to a significant increase in CO_2 levels, with peak values exceeding recommended thresholds. This, combined with higher relative humidity, suggests inadequate air exchange. In contrast, spring conditions allow for maintaining CO_2 concentrations within safer limits and contributing to a more comfortable learning environment. A strong positive correlation between CO_2 concentration and relative humidity was observed, reinforcing the role of ventilation in regulating indoor conditions. The findings demonstrate that ineffective ventilation not only leads to air stagnation but also increases humidity levels, which can further impact occupant comfort. The analysis of thermal comfort through the Predicted Mean Vote (PMV) index highlighted seasonal differences in perceived comfort, with students reporting slightly cooler sensations in autumn, likely due to temperature variations and heating system efficiency.

These findings underscore the need for targeted ventilation strategies that adapt to seasonal changes. Educational institutions should implement optimized ventilation protocols, combining natural and mechanical ventilation to maintain CO_2 levels within recommended thresholds. Real-time air quality monitoring systems could provide valuable data to support dynamic ventilation adjustments, improving indoor conditions and enhancing student performance. Future research should further investigate the impact of CO_2 exposure on cognitive function through direct performance assessments and physiological measurements. Additionally, expanding this study to different climatic regions and educational settings would provide a more comprehensive understanding of ventilation efficiency and its influence on indoor environmental quality. Exploring the integration of automated air exchange systems and advanced climate control technologies could lead to significant improvements in university classroom environments, fostering better health and learning outcomes for students.

Author Contributions: Conceptualization, G.C., M.S., S.R. and J.T.; methodology, A.C., M.S., S.R. and J.T.; software, A.F. and J.T.; formal analysis, A.F., M.S. and J.T.; investigation, A.F., A.C., M.S. and J.T.; resources, A.C. and G.C.; data curation, A.F., S.R. and J.T.; writing—original draft preparation, A.F. and J.T.; writing—review and editing, A.C., G.C., M.S., S.R. and J.T.; supervision, A.C., G.C., M.S., S.R. and J.T.; project administration, J.T.; funding acquisition, J.T. All authors have read and agreed to the published version of the manuscript.

Funding: This research was co-funded by European Union Committee within the Erasmus+ Programme 2023 Key Action 2: Cooperation Partnership in Higher Education (Agreement n. 2023-1-IT02-KA220-HED-000162181. P.I.: Dott. Juri Taborri). This publication reflects the views only of the authors, and the Commission cannot be held responsible for any use which may be made of the information contained therein.

Institutional Review Board Statement: Not applicable.

Informed Consent Statement: Not applicable.

Data Availability Statement: The dataset presented in this article are not readily available because the data are part of an ongoing European project. Requests to access the datasets should be directed to juri.taborri@unitus.it.

Conflicts of Interest: The authors declare no conflicts of interest.

References

1. Ma, N.; Aviv, D.; Guo, H.; Braham, W.W. Measuring the Right Factors: A Review of Variables and Models for Thermal Comfort and Indoor Air Quality. *Renew. Sustain. Energy Rev.* **2021**, *135*, 110436. [CrossRef]
2. Triantafyllou, A.G.; Zoras, S.; Evagelopoulos, V.; Garas, S. PM10, O3, CO Concentrations and Elemental Analysis of Airborne Particles in a School Building. *Water Air Soil Pollut. Focus* **2008**, *8*, 77–87. [CrossRef]
3. Sadrizadeh, S.; Yao, R.; Yuan, F.; Awbi, H.; Bahnfleth, W.; Bi, Y.; Cao, G.; Croitoru, C.; de Dear, R.; Haghighat, F.; et al. Indoor Air Quality and Health in Schools: A Critical Review for Developing the Roadmap for the Future School Environment. *J. Build. Eng.* **2022**, *57*, 104908. [CrossRef]

4. Brink, H.W.; Lechner, S.C.M.; Loomans, M.G.L.C.; Mobach, M.P.; Kort, H.S.M. Understanding How Indoor Environmental Classroom Conditions Influence Academic Performance in Higher Education. *Facilities* **2024**, *42*, 185–200. [CrossRef]
5. Deng, S.; Lau, J. Seasonal Variations of Indoor Air Quality and Thermal Conditions and Their Correlations in 220 Classrooms in the Midwestern United States. *Build. Environ.* **2019**, *157*, 79–88. [CrossRef]
6. Zhang, S.; Chen, Y.; Li, Y.; Yi, X.; Wu, J. Study on Air Quality and Its Annual Fluctuation in China Based on Cluster Analysis. *Int. J. Environ. Res. Public Health* **2022**, *19*, 4524. [CrossRef] [PubMed]
7. Guo, M.; Yu, W.; Sun, Y.; Wang, L.; Zhou, H.; Zhang, Y. Effects of Increasing Indoor Negative Air Ions on Cognitive Performance and Health of High Pure CO2 Level-Exposed College Students. *Indoor Air* **2023**, *2023*, 8298339. [CrossRef]
8. Cao, X.; Li, P.; Zhang, J.; Pang, L. Associations of Human Cognitive Abilities with Elevated Carbon Dioxide Concentrations in an Enclosed Chamber. *Atmosphere* **2022**, *13*, 891. [CrossRef]
9. Hui, P.S.; Wong, L.T.; Mui, K.W. Using Carbon Dioxide Concentration to Assess Indoor Air Quality in Offices. *Indoor Built Environ.* **2008**, *17*, 213–219. [CrossRef]
10. Korsavi, S.S.; Montazami, A.; Mumovic, D. Indoor Air Quality (IAQ) in Naturally-Ventilated Primary Schools in the UK: Occupant-Related Factors. *Build. Environ.* **2020**, *180*, 106992. [CrossRef]
11. Vilčeková, S.; Kapalo, P.; Mečiarová, L'.; Burdová, E.K.; Imreczeová, V. Investigation of Indoor Environment Quality in Classroom—Case Study. *Procedia Eng.* **2017**, *190*, 496–503. [CrossRef]
12. Krawczyk, D.A.; Rodero, A.; Gładyszewska-Fiedoruk, K.; Gajewski, A. CO_2 Concentration in Naturally Ventilated Classrooms Located in Different Climates—Measurements and Simulations. *Energy Build.* **2016**, *129*, 491–498. [CrossRef]
13. Stabile, L.; Dell'Isola, M.; Frattolillo, A.; Massimo, A.; Russi, A. Effect of Natural Ventilation and Manual Airing on Indoor Air Quality in Naturally Ventilated Italian Classrooms. *Build. Environ.* **2016**, *98*, 180–189. [CrossRef]
14. Heudorf, U.; Neitzert, V.; Spark, J. Particulate Matter and Carbon Dioxide in Classrooms—The Impact of Cleaning and Ventilation. *Int. J. Hyg. Environ. Health* **2009**, *212*, 45–55. [CrossRef] [PubMed]
15. Liu, C.; Zhang, Y.; Sun, L.; Gao, W.; Jing, X.; Ye, W. Influence of Indoor Air Temperature and Relative Humidity on Learning Performance of Undergraduates. *Case Stud. Therm. Eng.* **2021**, *28*, 101458. [CrossRef]
16. Wargocki, P.; Porras-Salazar, J.A.; Contreras-Espinoza, S. The Relationship between Classroom Temperature and Children's Performance in School. *Build. Environ.* **2019**, *157*, 197–204. [CrossRef]
17. Carvalho, J.P.; Barroso, B.I.L.; Da Silva, L.B.; Neves, A.I.A.; Torres, M.G.L.; Falcão, C.A.; Siqueira, J.C.F.; Souza, A.G.L.; Souza, E.L.; Da Silva, J.F. Students' Blood Pressure and Heart Rate in Learning Environments with Thermal Changes. *Int. J. Occup. Environ. Saf.* **2018**, *2*, 29–37. [CrossRef]
18. Brink, H.W.; Loomans, M.G.L.C.; Mobach, M.P.; Kort, H.S.M. Classrooms' Indoor Environmental Conditions Affecting the Academic Achievement of Students and Teachers in Higher Education: A Systematic Literature Review. *Indoor Air* **2021**, *31*, 405–425. [CrossRef]
19. Barbic, F.; Minonzio, M.; Cairo, B.; Shiffer, D.; Dipasquale, A.; Cerina, L.; Vatteroni, A.; Urechie, V.; Verzeletti, P.; Badilini, F.; et al. Effects of Different Classroom Temperatures on Cardiac Autonomic Control and Cognitive Performances in Undergraduate Students. *Physiol. Meas.* **2019**, *40*, 054005. [CrossRef]
20. Lazovic, I.; Stevanovic, Z.; Jovasevic-Stojanovic, M.; Zivkovic, M.; Banjac, M. Impact of CO2 Concentration on Indoor Air Quality and Correlation with Relative Humidity and Indoor Air Temperature in School Buildings in Serbia. *Therm. Sci.* **2016**, *20*, 297–307. [CrossRef]
21. Norbäck, D.; Nordström, K. Sick Building Syndrome in Relation to Air Exchange Rate, CO_2, Room Temperature and Relative Air Humidity in University Computer Classrooms: An Experimental Study. *Int. Arch. Occup. Environ. Health* **2008**, *82*, 21–30. [CrossRef] [PubMed]
22. Szczurek, A.; Maciejewska, M.; Pietrucha, T. Occupancy Determination Based on Time Series of CO2 Concentration, Temperature and Relative Humidity. *Energy Build.* **2017**, *147*, 142–154. [CrossRef]
23. Wargocki, P.; Wyon, D.P. The Effects of Outdoor Air Supply Rate and Supply Air Filter Condition in Classrooms on the Performance of Schoolwork by Children (RP-1257). *HVAC&R Res.* **2007**, *13*, 165–191. [CrossRef]
24. Zelinka, J. Microclimatic Research in the Slovakian Show Caves. *Acta Carsolog.* **2016**, *31*, 163. [CrossRef]
25. Buratti, C.; Ricciardi, P. Adaptive Analysis of Thermal Comfort in University Classrooms: Correlation between Experimental Data and Mathematical Models. *Build. Environ.* **2009**, *44*, 674–687. [CrossRef]
26. Stabile, L.; Frattolillo, A.; Dell'Isola, M.; Massimo, A.; Russi, A. Air Permeability of Naturally Ventilated Italian Classrooms. *Energy Procedia* **2015**, *78*, 3150–3155. [CrossRef]
27. Paull, R.E. Effect of Temperature and Relative Humidity on Fresh Commodity Quality. *Postharvest Biol. Technol.* **1999**, *15*, 263–277. [CrossRef]

28. Kuang, W. Seasonal Variation in Air Temperature and Relative Humidity on Building Areas and in Green Spaces in Beijing, China. *Chin. Geogr. Sci.* **2020**, *30*, 75–88. [CrossRef]
29. Lu, E.; Zeng, X. Understanding Different Precipitation Seasonality Regimes from Water Vapor and Temperature Fields: Case Studies. *Geophys. Res. Lett.* **2005**, *32*, L22707. [CrossRef]
30. Gładyszewska-Fiedoruk, K. Correlations of Air Humidity and Carbon Dioxide Concentration in the Kindergarten. *Energy Build.* **2013**, *62*, 45–50. [CrossRef]
31. Hou, J.; Sun, Y.; Dai, X.; Liu, J.; Shen, X.; Tan, H.; Yin, H.; Huang, K.; Gao, Y.; Lai, D.; et al. Associations of Indoor Carbon Dioxide Concentrations, Air Temperature, and Humidity with Perceived Air Quality and Sick Building Syndrome Symptoms in Chinese Homes. *Indoor Air* **2021**, *31*, 1018–1028. [CrossRef] [PubMed]
32. Wolkoff, P.; Azuma, K.; Carrer, P. Health, Work Performance, and Risk of Infection in Office-like Environments: The Role of Indoor Temperature, Air Humidity, and Ventilation. *Int. J. Hyg. Environ. Health* **2021**, *233*, 113709. [CrossRef] [PubMed]
33. Hurraß, J.; Nowak, D.; Heinzow, B.; Joest, M.; Stemler, J.; Wiesmüller, G.A. Indoor Mold. *Dtsch. Arztebl. Int.* **2024**, *121*, 265–271. [CrossRef] [PubMed]
34. Luo, M.; Wang, Z.; Ke, K.; Cao, B.; Zhai, Y.; Zhou, X. Human Metabolic Rate and Thermal Comfort in Buildings: The Problem and Challenge. *Build. Environ.* **2018**, *131*, 44–52. [CrossRef]
35. Corgnati, S.P.; Filippi, M.; Viazzo, S. Perception of the Thermal Environment in High School and University Classrooms: Subjective Preferences and Thermal Comfort. *Build. Environ.* **2007**, *42*, 951–959. [CrossRef]
36. Guevara, G.; Soriano, G.; Mino-Rodriguez, I. Thermal Comfort in University Classrooms: An Experimental Study in the Tropics. *Build. Environ.* **2021**, *187*, 107430. [CrossRef]

Disclaimer/Publisher's Note: The statements, opinions and data contained in all publications are solely those of the individual author(s) and contributor(s) and not of MDPI and/or the editor(s). MDPI and/or the editor(s) disclaim responsibility for any injury to people or property resulting from any ideas, methods, instructions or products referred to in the content.

sensors

MDPI

Article

A Wearable Sensor Node for Measuring Air Quality Through Citizen Science Approach: Insights from the SOCIO-BEE Project

Nicole Morresi [1,*], Maite Puerta-Beldarrain [2], Diego López-de-Ipiña [2], Alex Barco [2], Oihane Gómez-Carmona [2], Carlos López-Gomollon [3], Diego Casado-Mansilla [2], Maria Kotzagianni [4], Sara Casaccia [1], Sergi Udina [5] and Gian Marco Revel [1]

[1] Department of Industrial Engineering and Mathematical Sciences, Polytechnic University of Marche, 60131 Ancona, Italy; s.casaccia@univpm.it (S.C.); gm.revel@univpm.it (G.M.R.)
[2] DeustoTech, University of Deusto, 48007 Bilbao, Spain; mpuerta004@deusto.es (M.P.-B.); dipina@deusto.es (D.L.-d.-I.); a.barco@deusto.es (A.B.); oihane.gomezc@deusto.es (O.G.-C.); dcasado@deusto.es (D.C.-M.)
[3] Ayuntamiento de Zaragoza, 50003 Zaragoza, Spain; clopezg@zaragoza.es
[4] Municipality of Amaroussion, 15100 Athens, Greece; mkotzagianni@maroussi.gr
[5] Data Science Team, Bettair Cities SL, 08907 Barcelona, Spain; sudina@bettaircities.com
* Correspondence: n.morresi@pm.univpm.it

Abstract: Air pollution is a major environmental and public health challenge, especially in urban areas where fine-grained air quality data are essential to effective interventions. Traditional monitoring networks, while accurate, often lack spatial resolution and public engagement. This study presents a novel wearable wireless sensor node (WSN) that was developed within the Horizon Europe SOCIO-BEE project to support air quality monitoring through citizen science (CS). The low-cost, body-mounted WSN measures NO_2, O_3, and $PM_{2.5}$. Three pilot campaigns were conducted in Ancona (Italy), Maroussi (Greece), and Zaragoza (Spain), and involved diverse user groups—seniors, commuters, and students, respectively. $PM_{2.5}$ sensor data were validated through two approaches: direct comparison with reference stations and spatial clustering analysis using K-means. The results show strong correlation with official $PM_{2.5}$ data ($R^2 = 0.75$), with an average absolute error of 0.54 $\mu g/m^3$ and a statistical confidence interval of ± 3.3 $\mu g/m^3$. In Maroussi and Zaragoza, where no reference stations were available, the clustering approach yielded low intra-cluster coefficients of variation (CV = 0.50 $\pm$ 0.40 in Maroussi, CV = 0.28 $\pm$ 0.30 in Zaragoza), indicating that the measurements had high internal consistency and spatial homogeneity. Beyond technical validation, user engagement and perceptions were evaluated through pre-/post-campaign surveys. Across all pilots, over 70% of participants reported satisfaction with the system's usability and inclusiveness. The findings demonstrate that wearable low-cost sensors, when supported by a structured engagement and data validation framework, can provide reliable, actionable air quality data, empowering citizens and informing evidence-based environmental policy.

Keywords: citizen science; air quality; wearable devices

1. Introduction

One of the most pressing environmental challenges faced by citizens, especially in densely populated urban areas, is air pollution. Rising pollution levels not only threaten public health but also have severe environmental consequences (e.g., climate change), and this makes air quality a critical concern in the European Union (EU) and at the global level. In the EU alone, poor air quality is linked to approximately 400,000 premature deaths each year. Additionally, it is a major factor contributing to biodiversity loss, which further exacerbates ecological imbalances [1]. The COVID-19 pandemic, along with its relationship to climate change, further highlighted the need for a healthier planet. During the pandemic, lockdown policies drastically altered energy demand patterns, resulting in a significant drop in CO_2 emissions (up to 17%) due to reduced surface transportation [2]. This reduction underscores the potential impact of both policy measures and individual and collective societal actions in mitigating air pollution. Measuring the air quality in urban settings is the most fundamental step in raising awareness, in a sustainable and informed way, of the necessity of reducing CO_2 emissions and air pollution. Although most cities are equipped with standardized reference station for measuring air quality, the capillarity of these networks is reduced due to the high costs that are involved. To face this challenge, low-cost sensors (LCS) for measuring air quality are being used more and more frequently, and are typically combined with social approaches that aims to increase responsibility and citizen engagement. In fact, citizen science (CS) has emerged as a scalable strategy to complements conventional air quality monitoring networks through the deployment of LCS technologies operated by non-expert users in order to increase the spatial resolution of air quality measurements [3,4]. When properly designed, CS measurement campaigns not only support data-driven research but also promote awareness, community engagement, and participatory environmental governance. Many CS initiatives have emerged from community efforts aimed at resolving issues that directly impact local communities. This collaboration expands the scope and reaches of scientific endeavour, fosters community engagement, enhances scientific literacy, and promotes pro-social and pro-environmental behaviours [5,6].

In the air quality context, CS has shown potential in extending monitoring coverage and capturing localized pollution dynamics. However, its integration into scientific and policy frameworks remains constrained by several technical and operational challenges [7]. A key technical limitation lies in the quality and reliability of the data obtained through LCSs, as varying participant expertise may lead to inconsistencies in data collection, which may be affected by factors such as sensor drift, cross-sensitivity to interfering compounds, and improper usage conditions [3,8]. The degree of digitalization, sensor handling, and data recording practices across diverse participant groups can introduce inconsistencies, which raises concerns about the representativeness and accuracy of the collected data. In addition, social challenges such as ensuring sustained user participation over time and translating citizen-generated data into actionable evidence for public policy remain substantial. LCSs should be designed as intuitive sensor systems that can be incorporated into robust digital platforms with transparent feedback mechanisms that ensure engagement without imposing cognitive or technological barriers. LCSs must be designed to be physically unobtrusive, low-power, and user-friendly, while core features should be embedded for data validation, such as built-in diagnostics, remote logging, or adaptive sampling strategies [9–11]. Within this framework, the SOCIO-BEE (Wearables and Drones City Socio-Environmental Observations and BEhavioral ChangE) project addresses the challenges of citizen science-based air quality monitoring by introducing a methodology for conducting CS campaigns that is supported by a wearable wireless sensor node (WSN) capable of measuring NO_2, O_3, and $PM_{2.5}$. This device is integrated into a structured engagement framework to actuate measurement campaigns and to ensure that the data collected by citizens are of sufficient quality to support both scientific analysis and policy development. The WSN is employed to obtain data through campaigns implemented in Zaragoza (Spain), Maroussi (Greece), and Ancona (Italy) with different societal groups.

Data collected through the WSN by citizens during campaigns are analyzed and validated with a two-fold approach: (i) a direct comparison with official reference station data and (ii) a clustering-based validation method that employs K-Means algorithms to group daily measurements and assess intra-cluster coherence and variability, particularly in locations where reference data are not available. This research aims to present the technical innovation of SOCIO-BEE, which is supported by a multi-domain integration of citizens across all stages of the scientific process, including data collection, analysis, and interpretation; thus, this study enhances public awareness, fosters behavioral engagement, and reinforces the role of bottom-up contributions in evidence-based policymaking.

The structure of this paper is organized as follows. Section 2 reviews related work on air quality sensors adopted in CS initiatives with the aim of democratizing CS and improving the validity of crowdsourced air quality data. Section 3 provides an overview of the SOCIO-BEE methodology, the technical details of the WSN, the design of the measurement campaigns across pilot cities, and the data analysis to assess the data validity. Section 4 presents the results of the SOCIO-BEE project. Finally, Section 5 summarizes the conclusions of this paper and discusses the lessons learned throughout the project.

2. State of the Art

In line with the growing necessity of expanding air quality measurements in cities and with the adoption of CS, the past two decades have seen a notable increase in air quality monitoring sensors and related initiatives. Various projects have employed a wide range of sensor technologies and methodological frameworks to assess the feasibility, reliability, and scalability of low-cost sensing solutions [12,13]. As an example, the Luftdaten project [14] has notably empowered citizens to participate in air pollution data collection using accessible LCSs. This initiative has highlighted the importance of accessible technology in enabling widespread participation and has provided a framework for community-driven environmental monitoring. The OpenSense project [15] utilized mobile sensors and machine learning to analyze urban air quality. Similarly, a study conducted in Nairobi [16] analyzed the impact of the construction of a new highway. It involved distributing optical sensors (AlphaSense OPC-N2) among 15 households within 150 m of each other that were grouped into three clusters. Data collected over 48 h indicated air quality levels substantially above WHO-recommended thresholds, with a recorded average concentration of 28.7 $\mu g/m^3$, in contrast to the WHO limits of 5 $\mu g/m^3$ (annual) and 15 $\mu g/m^3$ (daily). Another study [17] aimed to demonstrate the effectiveness of high-resolution sensor networks, compare collected data with a reference station, analyze air pollutants, and assess personal exposure. Using Sensirion SPS30 sensors integrated into the BMD-340 module with Bluetooth connectivity, approximately 40 sensors were distributed to citizens with the aim of capturing particulate matter (PM), thermo-hygrometric parameters, and gases such as CO_2 and NO_2. Each sensors costs approximately \$250 at low production volumes, and less than \$100 at large scale production. The measurements from the low-cost sensors closely matched those of the reference station, with $PM_{2.5}$ averages of 3 $\mu g/m^3$ (sensors) and 4.5 $\mu g/m^3$ (reference). Bousiotis et al. proposed a hyperlocal, mobile source apportionment methodology in Birmingham that used a backpack-based sensor kit carried by citizen volunteers, integrating low-cost instruments such as the Alphasense OPC-N3. The study not only demonstrated the technical feasibility of attributing PM sources using positive matrix factorization (PMF) but also emphasized how mobile sensor-based CS can enhance the spatial resolution in urban pollution studies, especially when it is combined with structured calibration efforts against reference stations [18,19]. The literature also showed how the integration of low-cost CO_2 and PM sensors into a public building (via the

Guildford Living Lab) enabled real-time air quality visualization and public engagement. The study highlighted how short indoor events could elevate the CO_2 concentration above 1000 ppm, and how human presence increased the PM10 via resuspension effects.

The presented literature raises an important question regarding the quality of data collected by LCSs, as these devices typically exhibit greater uncertainty compared to official monitoring stations. Indeed, due to their susceptibility to anomalies and external influences, especially the relative humidity and temperature, it remains challenging to perform thorough and accurate calibration of these sensors. Nevertheless, the ambition of SOCIO-BEE has been, precisely, to provide a LCS (the price tag for the device designed in the project was EUR 800) that is capable of delivering measurements that are sufficiently reliable, as demonstrated through direct comparisons against ground truth provided by reference stations, and to go further by testing these devices in real test-beds.

The SOCIO-BEE initiative and experimental measurement campaigns aimed to go further beyond traditional CS frameworks by integrating advanced wearable technologies and real-time data acquisition. This research applies clustering techniques to analyze the datasets collected through CS campaigns, comparing the clustered results with measurements obtained from reference stations. This approach allows researchers to evaluate whether data provided by citizen science initiatives are sufficient to accurately characterize air quality conditions, or if further data collection and methodological refinements are necessary. Several studies have explored the application of K-means clustering to spatially distributed air quality data, with a particular focus on particulate matter concentrations. For instance, some works [14,15] have applied K-means to geo-referenced PM measurements collected across urban environments to identify localized pollution patterns. Studies have also used spatial clustering to define air quality zones and assess the spatial behavior of $PM_{2.5}$ and PM_{10} in relation to official monitoring infrastructure. These works incorporate geographic coordinates as clustering features, which allows for the derivation of spatial centroids and distance-based assessments, a methodology that is closely aligned with the approach followed in this research work.

3. SOCIO-BEE Methodology

This section presents the methodology developed by the SOCIO-BEE project to conduct CS experiments across different case studies and test the validity and quality of the data obtained thought the WSN, as well as the methodology used to measure the impact of the experiments.

3.1. The Bee Metaphor: The Social Structure of the Community of Practices

Drawing inspiration from nature, SOCIO-BEE, in the design and execution of its measurement campaigns, uses the metaphor of bee colonies to develop effective behavioral and engagement strategies for a diverse range of stakeholders. Figure 1 illustrates the roles that hive members, representing a community of practice, can assume within SOCIO-BEE's approach, and also highlights the core features that the SOCIO-BEE technical platform must exhibit to enable members to oversee and actively participate in CS campaigns. It is important to note that the entire process undertaken by hive members is the result of a collaborative co-creation effort involving multiple stakeholders. The roles are inspired by the organization of a bee colony, and are intended to facilitate collaborative citizen engagement. The Beekeeper is responsible for initiating and coordinating the campaign, forming the team (or "hive"), and ensuring smooth collaboration among participants. The Queen Bee designs the campaign by defining key parameters, such as the location, measurement frequency, and duration, using the SOCIO-BEE web platform and templates. The Queen Bee also oversees the distribution of

WSNs and manages the overall execution, including data collection, monitoring, and dissemination of the results. Worker Bees, the main data collectors, are notified via the SOCIO-BEE mobile app and guided to measurement locations by the MVE. They contribute air quality data, track campaign progress, and access feedback tools and heat maps during and after the campaign. Finally, Drone Bees act as ambassadors, sharing insights generated by the campaign and helping to raise awareness among institutions and the public.

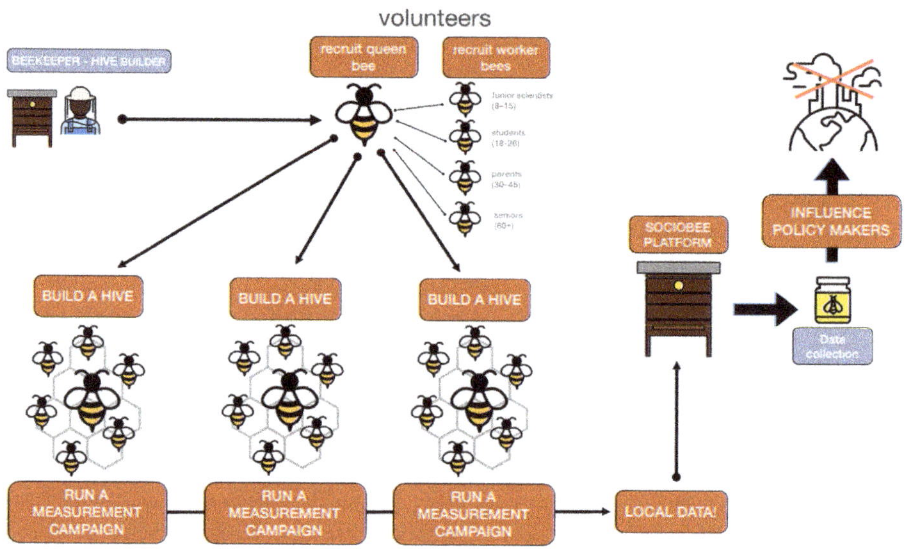

Figure 1. Bee metaphor employed by SOCIO-BEE and co-creation process carried out by a hive [16].

3.2. Wearable Sensor Node

The WSN (Figure 2) is a low-cost, wearable device designed by Bettair® Cities for making air quality measurements. The measurement device integrates two slots for toxic gas sensors, a PM sensor, and an environmental sensor for the internal temperature and outdoor relative humidity. Gas sensors are commercial electrochemical sensors (4 series form factor) designed to detect NO_2 and O_3. These sensors require precision analog electronics, including a stable potentiostat and a low-noise transimpedance amplifier, to convert nanoampere-level signals into millivolt readings. The PM sensor measures $PM_{2.5}$ concentrations. The WSN can operate autonomously on battery power, but it can also be connected to an external power source or power bank for continuous or prolonged monitoring. The WSN communicates with mobile phones via Bluetooth Low Energy (BLE), which enables the real-time transmission of data to the related cloud platform along with the phone's GPS coordinates. These sensors can also connect directly to the cloud via Wi-Fi, but GPS data are not included in this mode. For pre-deployment calibration, Bettair® Cities performed factory calibration using proprietary, high-performance technology validated in field conditions. The initial accuracy of the toxic gas sensors aligned with the Class 2 specifications under the TS17660 standard and was expected to remain within at least Class 3 throughout the measurement period. The WSNs are housed in compact enclosures, as illustrated in Figure 2. The internal sampling rate is configurable, with a typical setting of one measurement every 5 s. However, the data are only considered reliable after a minimum of one minute of static acquisition at a fixed location. Regarding the potential influence of human metabolism on the sensors' readings, it is important to note that, while indoor environments are more susceptible to localized effects caused by human occupancy, limited dispersion, and the resuspension of particles, outdoor environment and air pollution is predominantly driven by industrial activities and vehicular emissions. Consequently, indoor measurements are more vulnerable to human-induced interference and require careful consideration during calibration and sensor deployment. However, in the context of the present study, the SOCIO-BEE pilot measurement campaigns were conducted primarily in outdoor environments, where human emissions are significantly diluted and considered negligible relative to ambient pollutant sources [17–19]. Moreover, the WSNs were predominantly used in a stationary configuration, which allowed the sensors sufficient time to stabilize before each measurement. This strategy helps to reduce transient fluctuations and minimizes uncertainties associated with body heat, respiration, and movement. The supporting literature consistently indicates that indoor environments exhibit elevated particulate concentrations that are linked to human presence, while outdoor measurements are more influenced by environmental variability and are less affected by direct human metabolic outputs.

(**a**) (**b**) (**c**)

Figure 2. Wearable sensors designed for SOCIO-BEE, ready for use in a CS campaign. (**a**) Example of how to wear the WSN. (**b**) Fleet of WSNs developed for the project. (**c**) WSN design and case.

The WSN was developed together with the SOCIO-BEE infrastructure to facilitate CS initiatives that are focused on improving air quality and comprises the following additional components:

- AcadeMe platform: This is the main entry point to the SOCIO-BEE system. It manages participants' data, creates analytics, and connects with other SOCIO-BEE components (e.g., WSN). The AcadeMe platform is further divided into two core subcomponents:
  Mobile App: This application (see Figure 3) enables the participant to pair the WSN and consequently upload data collected by the WSN;
  Web app: Complementing the mobile app, the web app (see Figure 4) provides users with interactive visualizations of the WSN data, project updates, and opportunities to participate in discussions and forums with fellow citizen scientists and project stakeholders. The web app is important because a key feature is the possibility to define key areas to enable CS measurement campaigns through the use of WSNs;
- Micro-Volunteering Engine (MVE) [20]: This microservice plays a key role in organizing citizen engagement and ensuring the collection of valid data within SOCIO-BEE's CS campaigns. It provides users with tailored recommendations for specific locations where air quality measurements are needed [21]. The MVE optimizes volunteer efforts by guiding users to take measurements in the most critical areas, ensuring comprehensive and effective data coverage. Powered by diverse spatial computing strategies, it seeks to encounter the best balance between a given campaign's measurement objectives, e.g., five measurements per hour per cell in the area covered, and the users'

behavior patterns and needs, e.g., some users do not mind walking for longer periods whilst others want to stick to taking measurements in their immediate surroundings.

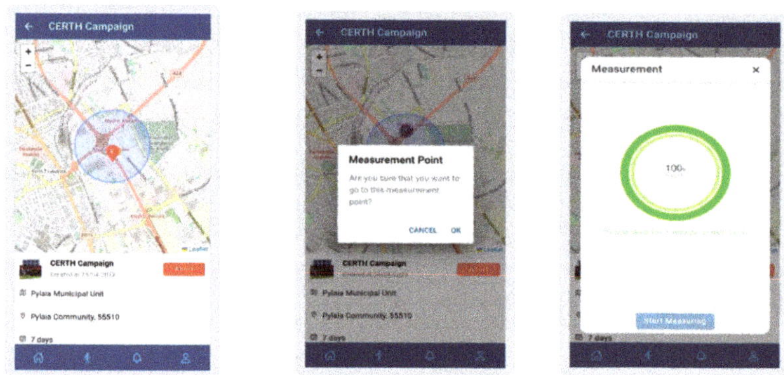

(**a**) (**b**) (**c**)

Figure 3. Mobile app showing how to (**a**) select a campaign, (**b**) choose an air quality measurement recommendation, and (**c**) take an actual measurement.

Figure 4. Web app used for designing a citizen science campaign.

3.3. *Measurement Campaign*

Prior to describing how the WSN were distributed to involved citizens, it is important to underline the fundamental concept pinpointed by SOCIO-BEE, which is based on the involvement of citizens in the measurement process. The experiments were designed to generate crowdsourced air quality data that can support actions to reduce pollution which are based on insights from data analysis. To evaluate the real-world applicability and the operational performance of the WSNs and the methodology, the system was deployed across three European cities, Ancona, Maroussi, and Zaragoza, through a structured execution. The implementation spanned three phases: (1) a pre-pilot phase focused on device testing, user training, and system refinement; (2) a pilot phase involving continuous field deployment of the wireless sensor nodes (WSNs), real-time data collection, and user engagement under varying urban conditions; and (3) a post-pilot phase aimed at evaluating sensor performance, collecting user feedback, and validating the data quality. This paper is focused on phase (2) and (3) of the pilot application.

In each pilot site, recruited participants were provided with dedicated training sessions to ensure proper use of the WSNs. These sessions included both technical and practical guidance on how to operate the devices, collect data reliably, and interact with the supporting digital tools (e.g., mobile app and web platform).

As part of the training, the users were informed about specific operational constraints identified during the preliminary testing and calibration phases to prevent the WSNs from negatively affecting data quality and ensuring proper device performance in real-world conditions. One of the first issues encountered was the tendency of the devices to overheat when charged in warm environments. The participants were advised to avoid placing the WSNs in direct sunlight or high-temperature conditions during charging, as excessive internal heat could affect the sensor's stability or even damage its components. In addition, it was noted that sensor readings taken immediately after powering on the devices tended to be unstable. To address this, users were instructed to turn on the WSNs well in advance, ideally a few hours before initiating a measurement session, to allow the sensors to reach thermal and operational stability. It is important to note that, when worn on the human body, wearable particulate matter sensors may be influenced by proximity effects such as thermal plumes, resuspension due to movement, and emissions from respiration, as highlighted in the recent literature [19,20]. These factors were considered in user training and data interpretation to minimize measurement artifacts. To minimize such interferences, the participants were instructed to remain still during the one-minute measurement intervals and to wear the device attached either to a belt or a backpack to ensure consistent positioning and airflow exposure.

By communicating these technical constraints clearly, the SOCIO-BEE team aimed to enhance participant awareness of potential measurement artifacts and ensure more consistent and meaningful data. Data used for the analysis were collected from citizens during the pilot phase, which involved continuous field deployment of the WSNs.

3.4. Pilot Description

Ancona, a coastal city in central Italy with notable air quality challenges due to traffic congestion and environmental conditions, was selected as a SOCIO-BEE pilot site. The campaign focused on assessing the exposure to $PM_{2.5}$ among vulnerable populations, particularly seniors, who represent 26% of the local population, to evaluate the sensors' performance under real-world conditions and encourage outdoor activity in low-exposure zones. The Ancona pilot lasted approximately three months. The participants were voluntarily recruited through local associations, and each user was given the device for about two weeks before returning it.

The second city was Maroussi, a densely trafficked district in the Athens metropolitan area that was selected as a SOCIO-BEE pilot site due to its high commuter flow and associated air pollution. The campaign focused on commuters as both contributors to and victims of poor air quality. WSNs were deployed among a diverse group of participants, including residents, students, employees, and public transport users, to monitor their exposure to air pollutants in transit and workplace environments. The pilot enabled evaluation of the sensors' performance across varied urban mobility patterns and supported inclusive engagement in real-world measurement scenarios. Similarly, the Maroussi pilot recruited citizens for a period of three months.

Zaragoza, a major urban center in northeastern Spain, hosted a SOCIO-BEE pilot that targeted students aged from 11 to 16 years old. The campaign aimed to assess the usability and educational impact of wearable low-cost air quality sensors within school-based activities. The students collected, analyzed, and interpreted environmental WSN data, contributing to both personal exposure assessment and broader urban air quality mapping. The pilot also served to test and refine the SOCIO-BEE platform's hardware and software tools in an educational context, leveraging Etopia's digital innovation hub and its school outreach network. Also in this case, measurement campaigns were conducted in school during lessons.

3.5. Data Analysis

First of all, while the system includes electrochemical sensors for NO_2 and O_3, its use was deliberately limited in the analytical phase of the study and the analysis was focused on $PM_{2.5}$ data. This decision was based on several operational constraints and scientific limitations identified during preliminary testing. To assess the reliability of the air quality data collected through the citizen-operated WSNs, two complementary data analysis approaches were applied, each adapted to the specific data availability of the pilot cities, which provided different data according to the population involved in the measurement process. Notably, all measurements during the pilot campaigns were conducted in uncontrolled real-world environments which differ significantly from the controlled settings used during initial calibration. This raised the need to evaluate whether such conditions negatively affect the accuracy and consistency of the data. As the literature currently lacks systematic studies assessing the reliability of wearable LCS data in real-world citizen science contexts, this work applies two distinct methods to explore this issue from multiple perspectives.

The first method involved a direct comparison between air quality measurements obtained from WSN devices and those from reference monitoring stations, and this analysis was conducted for the city of Ancona. This involved aligning temporal data collected from WSNs with those from the nearest official air quality reference stations. The reference instrument used for measuring $PM_{2.5}$ was a BAM 1020 Beta Attenuation Mass Monitor (Figure 5). Key statistical metrics were computed, including the mean absolute error (MAE), root mean square error (MSE), mean bias error, and coefficient of determination (R^2), to quantify the accuracy and consistency of the citizen-collected $PM_{2.5}$ measurements.

Figure 5. Pictures of the reference station in the city of Ancona.

The second analysis involved spatial clustering, and was applied to citizen-collected data points using K-means:

- The approach was different for the city of Ancona: for each identified cluster, the mean (μ_{ci}) and standard deviation (σ_{ci}) of the $PM_{2.5}$ values were computed. These values were then compared to the daily average values (μ_d) from the reference stations for the city of Ancona, since reference data were available. To evaluate the spatial distribution, the centroid in the GIS coordinates of each cluster was derived and its geographic distance to the reference station was measured. The relationship between this distance in meters from the reference station and the associated relative error (difference from the reference value in percentage) were analyzed to understand the influence of proximity on measurement reliability;

- For Maroussi and Zaragoza, where no official station data were available for comparison, the analysis focused on intra-cluster variability in the citizen-collected data. K-means clustering was applied to group measurements spatially, and for each cluster, the coefficient of variation (CV) was computed to evaluate the dispersion of values relative to the average value. This method allowed us to assess the reliability and homogeneity of the measurements within spatial clusters, and thus to obtain conclusions on data quality even in the absence of a formal reference system.

The variation within each cluster serves as an indicator of internal consistency: a low coefficient of variation implies homogeneity of the measurements and suggests that sensors in the same area captured similar exposure levels. In contrast, a high variation may indicate environmental heterogeneity, sensor discrepancies, or localized pollution events. Therefore, internal consistency metrics should become a proxy for data quality, as they allow researchers to assess whether citizen-collected measurements are sufficiently coherent to support meaningful interpretation.

3.6. Impact Assessment Instruments and KPIs for the Democratization Evaluation

The SOCIO-BEE project has developed and employed a comprehensive set of impact assessment instruments for the participants who are recruited and the effectiveness of its citizen science campaigns from a social perspective. Below are the key instruments used to measure SOCIO-BEE impact through the execution of its CS campaigns:

- Data Model Queries and Logging Mechanism: This tool tracks the usage of SOCIO-BEE's resources by executing SQL queries over its data model to gather quantitative key performance indicators (KPIs). For example, in its data model, it records the number of participants collecting air quality data via wearable sensors, it records every measure performed by each user with each concrete device, and so on;
- Demographic and Activity Satisfaction Questionnaires: These questionnaires collect demographic information (age range, gender, educational level, and so on) and assess participant satisfaction with various pilot activities;
- Evaluation Questionnaires: Designed for hive members involved in CS experiments or receiving results—including worker bees and drone bees—this evaluation process consists of the PRE and POST SOCIO-BEE assessment questionnaires:
  - The PRE SOCIO-BEE Citizen Science Activists Evaluation Questionnaire is completed before the campaign begins to establish a baseline, i.e., a departure point, regarding participants' awareness of the citizen science approach and its potential;
  - The POST SOCIO-BEE Citizen Science Activists Evaluation Questionnaire is completed after the campaign concludes to measure changes and learning experienced by participants across the CS campaign.

The above two questionnaire types (the sociodemographic and evaluation ones) assess critical factors such as socio-demographics, acceptance, interests, attitudes, satisfaction, accessibility, inclusiveness, awareness of air quality issues, usability, and behavioral change. Following an *ex-ante* (i.e., pre) and *ex-post* (i.e., post) evaluation approach [22], they enable the measurement of changes in those taking part in CS campaigns before and after they participate in CS campaigns.

4. Results

A summary analysis of the main characteristics of the campaigns conducted in the three pilot cities is displayed in Table 1 together with a visual representation of the positions in the three cities where CS campaigns were conducted (Figure 6). In the city of Ancona (Table 2), between 27 May and 27 June 2024, one campaign was conducted, with six different sub-campaigns, representing six different areas of the city (Figure 7); 10 recruited citizens were involved who collectively performed a total of 225 measurements. Over the course of just over a year, from 6 June 2023 to 28 June 2024 (measurements of 30 days in that period were collected), a total of 34 measurement campaigns were conducted in Maroussi (Table 3), with the active participation of 59 citizens. These campaigns involved 61 participants who collectively performed 1206 measurements, contributing valuable data on the air quality in the city (Figure 8).

Table 1. Characteristics of the measurement campaigns conducted in the pilot cities.

	Ancona	Maroussi	Zaragoza
Campaigns	6	34	13
Citizens	10	59	79
Overall duration [measurements' days]	31	30	29
Average daily number of measurements [$\mu \pm \sigma$]	12 ± 7	57 ± 24	24 ± 22

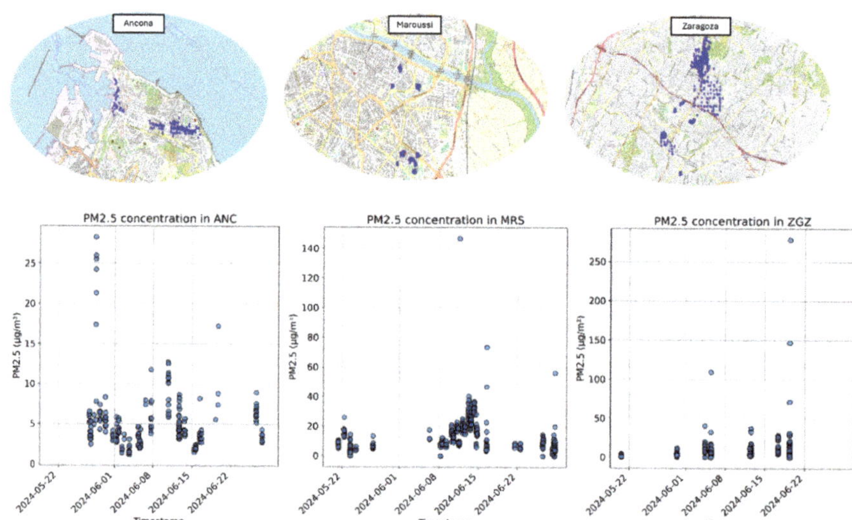

Figure 6. Upper panel: spatial visualization of the area of the city monitored through SOCIO-BEE initiatives. Lower panel: raw $PM_{2.5}$ data collected from the CS experimental campaigns.

Table 2. $PM_{2.5}$ data from Ancona (IT), collected by elderly people.

Campaign	Duration (Days)	Measurement Points	Citizens	$PM_{2.5}$ ($\mu \pm \sigma$), $\mu g/m^3$		
1	32	86	6	5	±	2.2
2	32	55	7	4.2	±	2.1
3	18	44	5	5.1	±	2.7
4	23	23	5	5.4	±	4.1
5	9	7	3	15.3	±	12.1
6	22	10	5	6.3	±	7.0

Figure 7. Example of where the measurements were conducted for some campaigns in the city of Ancona.

Table 3. $PM_{2.5}$ data from Maroussi (GR) collected by commuters.

Campaign	Duration (Days)	Measurement Point	Citizens	PM2.5 (μ ± σ), μg/m^3		
1	2	45	8	11.5	±	4.2
2	5	97	10	3.5	±	2.9
3	4	42	3	9.5	±	5.3
4	5	139	1	21.3	±	13.3
6	1	26	8	11.3	±	15.5
7	1	10	2	6.6	±	1.5
8	3	39	6	4.8	±	2.7
9	3	37	12	7.0	±	4.2
10	1	34	4	6.9	±	9.3

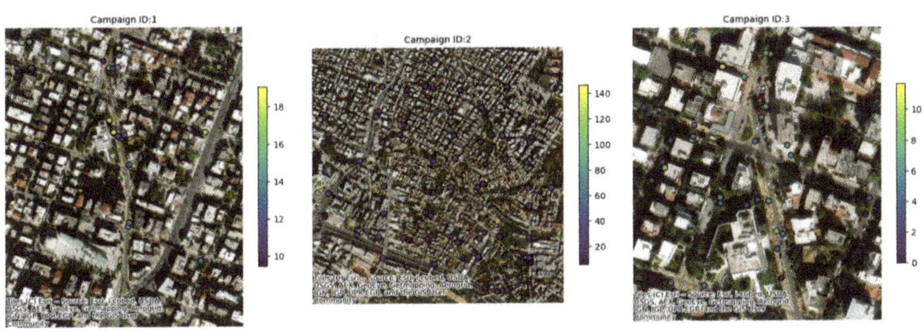

Figure 8. Example of where the measurements were conducted for some campaigns in the city of Maroussi.

In Zaragoza (Table 4), over the course of 29 days, from mid-May to the end of June 2024, a total of 13 measurement campaigns were conducted. These campaigns involved 79 participants who collectively performed 1206 measurements, contributing valuable data on the air quality in the city (Figure 9).

First, a summary analysis of the measurement campaigns that were conducted is presented. This includes relevant summary statistics that describe the overall trend of each campaign, including the number of users taking measurements, number of measurement points, and average $PM_{2.5}$ concentration over the monitored period.

Table 4. Single-day $PM_{2.5}$ measurements in Zaragoza, Spain, involving schoolchildren.

Campaign	Duration (Days)	Measurement Point	Citizens	$PM_{2.5}$ ($\mu \pm \sigma$), $\mu g/m^3$
1	1	27	7	1.4 ± 1.1
2	1	8	4	3.7 ± 1.1
3	1	25	10	3.8 ± 2.8
4	1	46	13	10.0 ± 6.2
5	1	46	14	7.9 ± 16.2
6	1	56	15	4.3 ± 3.2
7	1	19	6	6.3 ± 3.7
8	1	23	9	10.5 ± 9.6
9	1	29	12	4.9 ± 3.5
10	1	29	8	10.0 ± 7.1
11	1	40	7	8.7 ± 3.8
12	1	28	8	7.9 ± 6.2
13	1	29	7	25.0 ± 56.7

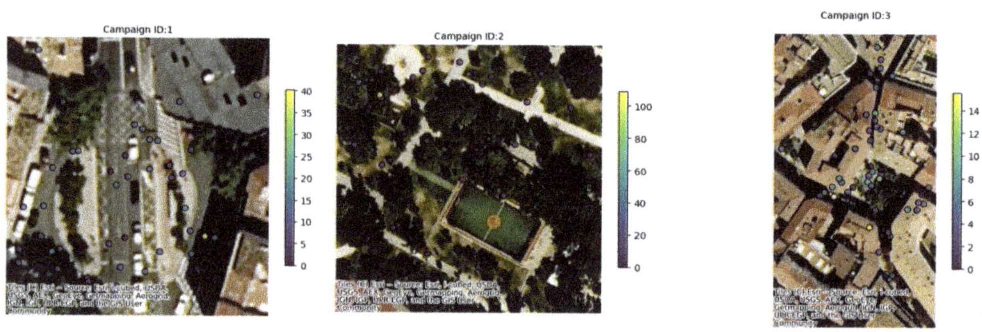

Figure 9. Example of where the measurements were conducted for some campaigns in the city of Zaragoza.

The Ancona seniors recorded lower, stable levels (residential areas), the Maroussi commuters showed traffic-linked peaks, and Zaragoza's schoolchildren had extreme short-term spikes (e.g., Campaign 13). The differences reflect the measurement duration, localized sources, and participant behavior, underscoring the potential of citizen-collected data and methodological considerations for reliability. The single-day $PM_{2.5}$ measurements in Zaragoza recorded an extreme outlier, likely due to transient pollution events near schools. The outlier in the data underscores the influence of localized, short-term environmental factors on air quality data. The campaigns highlight traffic-linked variability, with extremes such as Campaign 4 in Maroussi (21.3 ±13.3 $\mu g/m^3$). The short-term campaigns (1–5 days) and diverse measurement points (10–139) captured transient exposure patterns near transit routes.

4.1. Comparison Between Measurements Obtained from WSN Devices Reference Stations

In this subsection, the data examined in the previous section for Ancona, where access to reference station data was possible, are compared to the reference monitoring stations to verify the validity of the data obtained through the CS campaigns.

Figure 10 presents the temporal evolution of the $PM_{2.5}$ concentrations recorded by the official reference station in Ancona alongside the measurements collected during the CS campaigns, as well as the result of the regression analysis.

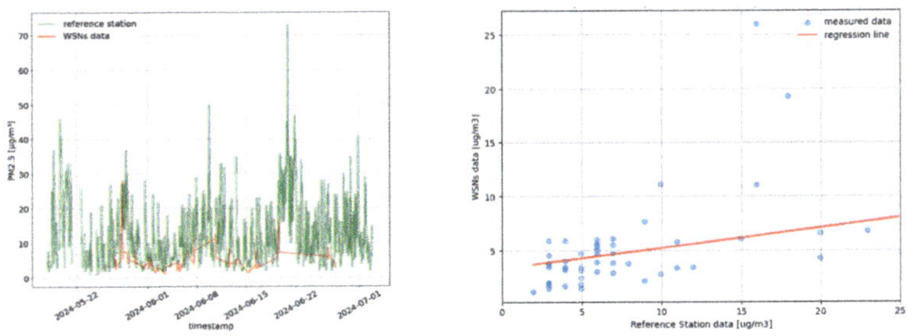

(a) (b)

Figure 10. (**a**) Temporal series of the reference station in Ancona and the measurement conducted during the CS campaigns. (**b**) Regression analysis.

To quantify this relationship, a linear correlation analysis was conducted, the results of which are summarized in Table 5. The accuracy, expressed as the average of the residuals, is 0.54 μg/m^3, and the precision (standard deviation of the residuals) is 2.2 μg/m^3, suggesting that most CS measurements fall within a reasonable range of the reference data. Overall, the statistical confidence, with a coverage factor (K) of two, is 3.3 μg/m^3 and the bias is 3.9 μg/m^3.

Table 5. Results of the correlation analysis between the reference station and the CS campaigns.

Accuracy (μg/m^3)	Bias (μg/m^3)	Precision (μg/m^3)	Statistical Confidence (K = 2) (μg/m^3)
0.54	3.9	2.2	3.3

4.2. Clustering Data Collected from the WSN

A cluster analysis was designed to identify discrepancies between the reference station and the citizen-collected campaign measurements. This analysis aims to categorize the daily data into distinct clusters using clustering methods. It serves a dual purpose: first, to determine whether each cluster collectively aligns with the trend observed at the reference station; and second, to evaluate the impact of the distances between the cluster centroids and the reference station's location. Additionally, the variability in the daily collected data within each cluster is examined to further characterize the consistency and distribution of measurements.

Ancona

For the city of Ancona, the data were pre-processed and subsequently aggregated into hourly and daily datasets. The daily data were then input into the k-means algorithm, considering two configurations: cluster numbers of K = 2 and K = 3. An example of the clustering results is provided in Figure 11.

For each cluster, we computed the daily mean of the measurements and compared it with the daily mean from the reference station, calculating the associated relative error. The comparison between the cluster measurements and reference station data revealed a correlation coefficient (r) of 0.75 and 0.63 for clusters 0 and 1, respectively, for the case of K = 2. For K = 3, the r was 0.74, 0.63, and 0.68 for K = 0, 1, and 2, respectively. Despite these differences, the high correlation between cluster measurements and the reference station suggests that the data can be considered qualitatively reliable (Figure 12).

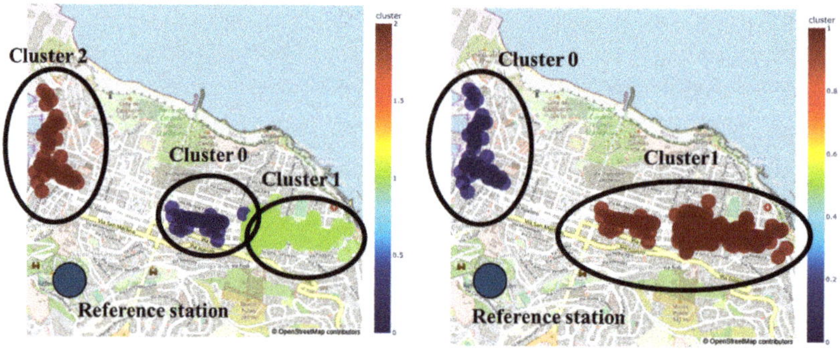

Figure 11. Example of how the clustering algorithm groups the $PM_{2.5}$ data (K = 3, **left** and K = 2, **right**).

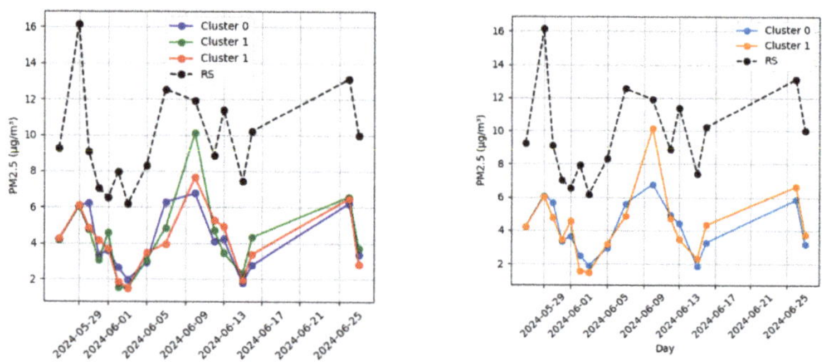

Figure 12. Average daily concentration of $PM_{2.5}$ measured using WSN during experimental campaign, according to the cluster grouping (K = 2, **left** and K = 3, **right**) and measured through reference stations (RSs).

From a measurement discrepancy perspective, we analyzed the mean percentage errors for K = 2 and K = 3, computed against the data collected from the RSs (Table 6).

Table 6. Summary of the comparison between clustered citizen science (CS) data and the reference station (Ancona). The table reports the mean percentage error, standard deviation, and Pearson correlation coefficient for each cluster under two clustering configurations (K = 2 and K = 3), providing insight into the accuracy and consistency of the CS measurements.

Cluster	Mean Relative Error [%]	Standard Deviation [%]	Pearson Correlation with the RS
		K = 2	
0	55.05	16.6	0.75
1	56.92	13.37	0.63
		K = 3	
0	55.59	18.2	0.74
1	57.88	11.5	0.63
2	56.07	13.3	0.68

For the Ancona pilot, CS measurements taken farther from the reference station were compared to understand whether they remained meaningful. For this purpose, the GIS coordinates of each cluster centroid were considered and, for each measurement day, the distance between the centroid and the reference station was computed. Table 7 reports the correlation coefficient r between the relative error for each day, and the average distance from the RS.

Table 7. Pearson correlation between the distance of cluster centroids from the reference station (Ancona) and the corresponding percentage error in $PM_{2.5}$ measurements for different clustering configurations (K = 2 and K = 3).

Cluster	Pearson
2	−0.02
3	+0.03

Maroussi

As previously mentioned, in this analysis, a comparison among data collected from the WSNs was conducted, in order to analyze the variability in these data. As a first step, the daily clustering approach was applied using K = 2 and K = 3. For each clustering configuration, the mean $PM_{2.5}$ concentration and the data dispersion (expressed as the standard deviation) were computed. The coefficient of variation (CV) was then calculated for each cluster, and the results were averaged across all days. Moreover, the analysis aimed to assess the internal consistency of the measurements and to determine whether closely spaced sensors tend to report similar values. A low CV would suggest homogeneous readings within a cluster, supporting the reliability of the data. Conversely, a high CV may indicate local anomalies, sensor instability, or contextual environmental differences, all of which may require further investigation or validation.

Figure 13 illustrates the distribution of the average $PM_{2.5}$ concentrations for each cluster on each day, along with the associated dispersion, highlighting both the temporal and spatial variability. Table 8 summarizes the overall results, showing that, for K = 2, the standard deviation averaged 5.9 ± 6.9 µg/m^3 with a mean CV of 0.50 ± 0.40, while for K = 3, slightly lower values were observed (5.8 ± 7.5 µg/m^3 and CV = 0.5 ± 0.4). These results suggest that increasing the number of clusters may slightly reduce the internal dispersion, potentially improving the spatial resolution of the analysis.

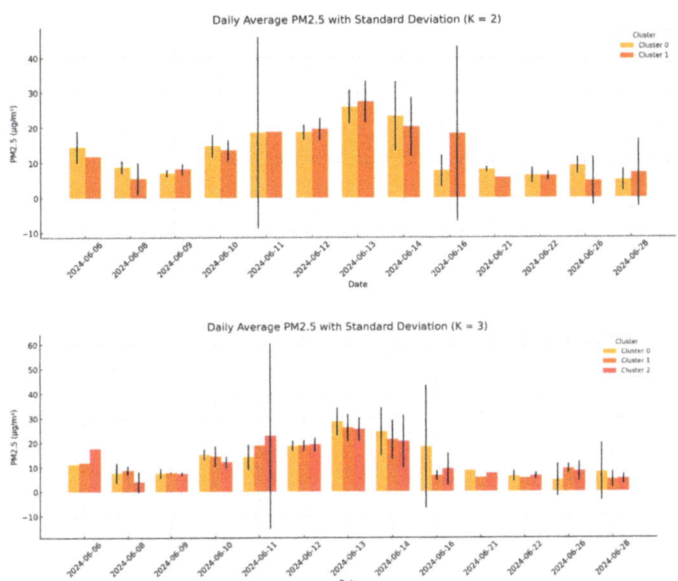

Figure 13. Distribution for each day of monitoring of the average $PM_{2.5}$ for each identified cluster and the related dispersion (Maroussi).

Table 8. Standard deviation of the $PM_{2.5}$ for each identified cluster and the related coefficient of variation (Maroussi).

	Standard Deviation [μg/m³]	Coeff of Variation
K = 2	5.9 ± 6.9	0.5 ± 0.4
K = 3	5.8 ± 7.5	0.5 ± 0.4

Zaragoza

Similarly, the analysis was conducted in Zaragoza. Table 9 summarizes the overall results, showing that, for K = 2, the standard deviation averaged 2.3 ± 1.6 μg/m^3 with a mean CV of 0.30 ± 0.30, while for K = 3, slightly lower values were observed (2.1 ± 1.7 μg/m^3 and CV = 0.28 ± 0.30). Also, in this case, increasing the number of clusters reduced the internal dispersion, potentially improving the spatial resolution of the analysis. Figure 14 illustrates the distribution of the average $PM_{2.5}$ concentrations for each cluster on each day, along with the associated dispersion, highlighting both the temporal and spatial variability.

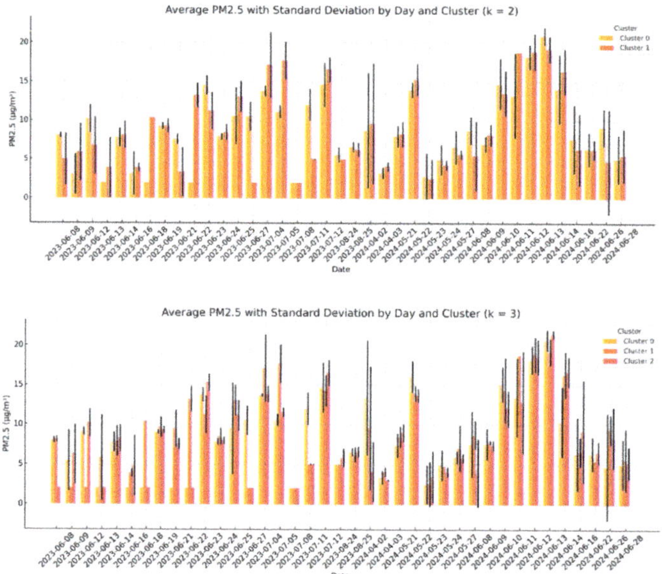

Figure 14. Distribution for each day of monitoring of the average $PM_{2.5}$ for each identified cluster and the related dispersion (Zaragoza).

Table 9. Standard deviation of the $PM_{2.5}$ for each identified cluster and the related coefficient of variation.

	Standard Deviation [μg/m³]	Coeff of Variation
K = 2	2.3 ± 1.6	0.3 ± 0.3
K = 3	2.1 ± 1.7	0.28 ± 0.3

4.3. Impact Assessment and KPIs

This section examines the impact of the SOCIO-BEE solution and measurement campaigns from multiple perspectives. It analyses age, gender, role, education level, digital proficiency, and employment status across the pilots. The data presented include all citizens who responded to PRE- or POST-campaign surveys, or both, in each city. Figure 8 illustrates the gender distribution of the participants. Most users chose to specify their gender, identifying as either male or female. Zaragoza and Maroussi showed a relatively balanced distribution between men and women, whereas Ancona had a higher proportion of male participants. Notice that in the case of Zaragoza, although measurements in the name of only 79 citizens were collected, those measurements were often performed by couples or trios of volunteers. This explains the total number of volunteers specified for Zaragoza, i.e., 233. Figure 9 illustrates the age distribution of the participants, showing that younger individuals were predominant in the Zaragoza pilot, middle-aged users were predominant in Maroussi, and older participants in Ancona. Additionally, due to the fact that complete classes of secondary education students were involved in Zaragoza, this pilot had a significantly higher number of participants compared to the others (Figure 15a,b).

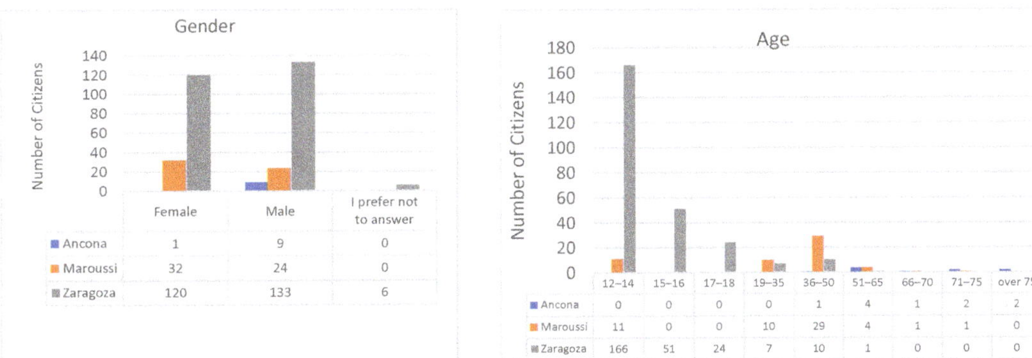

(a) (b)

Figure 15. (**a**) Gender analysis and (**b**) age distribution of citizens in pilots.

Regarding the role of citizens (Figure 16a), the "working bee" is the predominant role, as intended, since it represents the main type of members of the community of practice who collect and reflect upon the collected data, while the other roles are less represented. In terms of education level (Figure 16b), Maroussi had a higher proportion of citizens with a bachelor's or master's degree. In Zaragoza, most participants had a secondary education or a high school diploma. In Ancona, the majority held either a master's degree or a high school diploma.

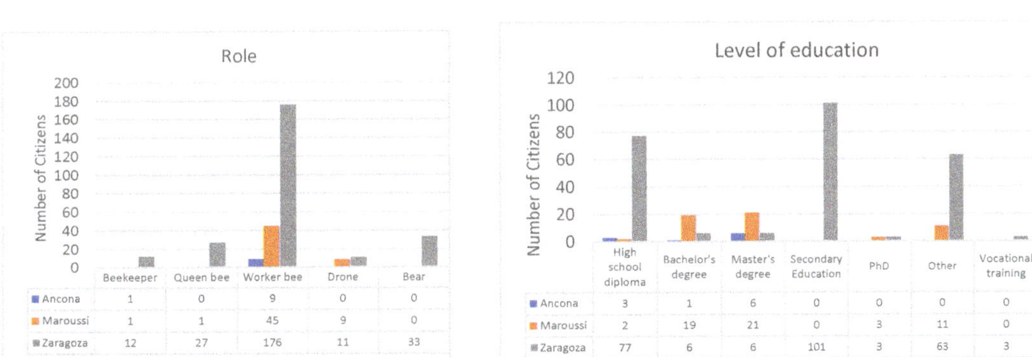

(a) (b)

Figure 16. (**a**) Role of citizens and (**b**) level of education.

Figure 17a illustrates the citizens' proficiency levels with digital tools. In Zaragoza, most participants had an intermediate or advanced level of proficiency, since they were digital native youngsters. In Maroussi, the majority possessed an advanced level, as they were middle aged people, while in Ancona, the users were evenly divided between advanced and intermediate proficiency levels. They were elderly people with a middle–high educational level.

Regarding work status (see Figure 17b), most of the volunteers in Zaragoza were students, while in Maroussi, the majority were employed people. In Ancona, most participants were retired, which aligns with the demographic analysis of the age distribution of the users.

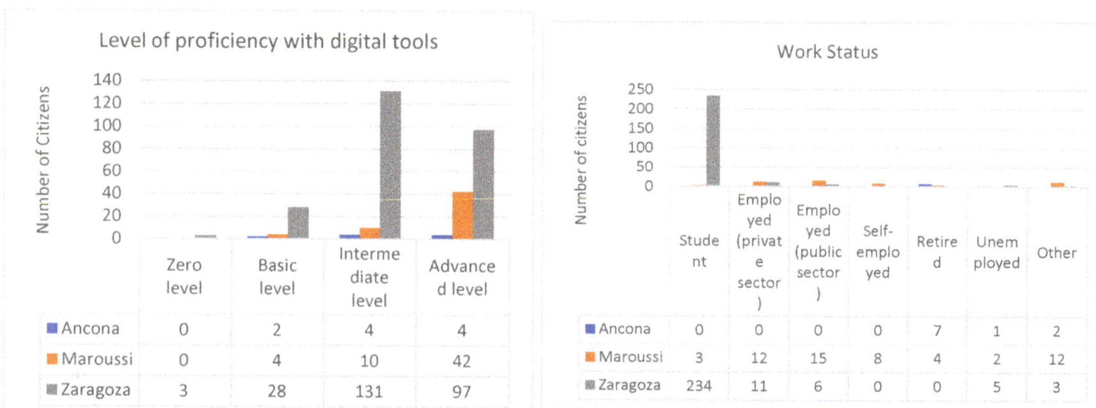

(**a**) (**b**)

Figure 17. (**a**) Digital proficiency level and (**b**) work status of citizens in pilots.

This section discusses the KPIs that were designed to evaluate the extent to which SOCIO-BEE democratizes CS experiments, particularly those related to air pollution. To assess whether SOCIO-BEE enhances the co-design and co-delivery of evidence-gathering CS experiments that support decision-making in air quality, data were collected from tool logs, databases, and responses from a set of Google Forms surveys created for the project. Below, the most significant KPIs used to inform evaluation dimensions are analyzed.

Only responses from users who completed both the pre- and post-surveys are considered; therefore, 7 citizens in Ancona, 50 in Maroussi, and 103 in Zaragoza were included in this analysis.

The KPIs can be divided into two types: (1) those that measure an increase in favorable user perception towards Citizen Science by comparing pre- and post-intervention dimensions, and (2) those evaluated solely based on post-intervention data. The first category includes metrics such as the increase in technology acceptance, heightened interest or engagement in science, the percentage of EU citizens who feel more aware of air pollution issues after participating in a SOCIO-BEE project, and user behavior changes driven by SOCIO-BEE. The second category includes KPIs that assess aspects such as user satisfaction with SOCIO-BEE (AcadeMe), tool accessibility, inclusiveness, and perceived usability. All values are calculated using the mean and standard deviation (SD) of user responses to the survey questions. Responses were measured on a Likert scale from 1 to 5, where 1 represents 'strongly disagree' and 5 indicates 'strongly agree'.

In the case of the first category of KPIs, the means and SD for both the pre- and post-intervention questionnaire answers, as well as the increase in mean values, were calculated. These values are shown in Tables 10–12, for the Ancona, Maroussi, and Zaragoza pilots, respectively.

Table 10. Ancona KPIs.

	Ancona ($\mu \pm \sigma$)		
KPIs	**Pre**	**Post**	**Increase (Average)**
Technology acceptance rate	4.06 ± 0.80	3.69 ± 1.00	-0.37 ± 0.21
Higher interest or engagement in science	3.66 ± 0.92	3.81 ± 0.98	0.13 ± 0.05
Percentage of EU citizens who feel more aware of air pollution issues after participating	4.36 ± 0.51	4.14 ± 0.69	-0.21 ± 0.18
User behaviour changes driven by the project.	3.67 ± 0.93	4.05 ± 0.78	$0.38 \pm (-0.15)$

Table 11. Maroussi KPIs.

	Maroussi ($\mu \pm \sigma$)		
KPIs	**Pre**	**Post**	**Increase (Average)**
Technology acceptance rate	4.14 ± 0.76	4.42 ± 0.76	0.28 ± 0.00
Higher interest or engagement in science	3.67 ± 0.92	3.83 ± 0.88	$0.17 \pm (-0.04)$
Percentage of EU citizens who feel more aware of air pollution issues after participating	4.06 ± 0.79	4.40 ± 0.81	0.34 ± 0.02
User behaviour changes driven by the project.	3.38 ± 1.02	4.04 ± 0.78	$0.66 \pm (-0.25)$

Table 12. Zaragoza KPIs.

	Zaragoza ($\mu \pm \sigma$)		
KPIs	**Pre**	**Post**	**Increase (Average)**
Technology acceptance rate	4.20 ± 0.80	4.20 ± 0.80	-0.32 ± 0.23
Higher interest or engagement in science	3.72 ± 1.04	3.58 ± 1.00	$-017 \pm (-0.04)$
Percentage of EU citizens who feel more aware of air pollution issues after participating	4.19 ± 0.71	3.86 ± 0.99	-0.33 ± 0.27
User behaviour changes driven by the project.	3.75 ± 1.02	3.66 ± 0.98	$-0.09 \pm (-0.04)$

In the Ancona pilot, some results (acceptance and air pollution awareness) showed a decline in the post-intervention values. This can be attributed to the average age of participants who had limited technological literacy and found the technology that they were offered not entirely usable. Consequently, the decrease in the technology acceptance rate KPI is understandable. Similarly, the fact that understanding the importance of air quality and other factors is not entirely straightforward may also explain the reduction in the percentage of citizens who felt an increased awareness and knowledge about air pollution. The respondents learned about air pollution, but they also realized that there is still a lot more to learn on this matter. Notably, all the average values obtained for the indicators considered in the case of Ancona were higher than 3.69 over 5.0, i.e., over 73% positive perception, which can be regarded as positive, since volunteers felt that their acceptance rate for SOCIO-BEE's technology and approach, their engagement in CS, and their knowledge about air pollution had increased and their behavior had changed positively.

The Maroussi pilot demonstrated a positive increase in all measured KPIs. Since this pilot involved middle-aged participants with a high level of technological acquaintance and high educational level, they were able to comprehend concepts about air quality and use the technology efficiently. Therefore, the values obtained indicate their positive opinion about SOCIO-BEE in terms of its technology and approach, how engagement is promoted by the project, their awareness about air quality, and their positive influence regarding practices that limit air pollution.

In the Zaragoza pilot, discrepancies were observed in the post-intervention results, with all examined KPIs showing a decrease, exhibiting the biggest decrease of 0.33 out of 5 points. This could be attributed to the involvement of younger participants who are experienced with modern mobile apps and are very sensitive to the usability and user experience of the apps they use on a daily basis. Other factors such as the short duration of the campaigns and technical issues encountered during the interventions could have also affected these results. Remember that the campaigns in Zaragoza were conducted in schools, and each session lasted no more than an hour. Therefore, comparing pre- and post-intervention evaluations in this case may not be highly significant, given the limited interaction period of the citizens with the system. For the second category of KPIs, comprising satisfaction, accessibility, inclusiveness, and usability, the mean and standard deviation were also calculated, but in this case, they were calculated exclusively based on the post-survey responses. The results for each measured value are presented in Table 13. All these values confirm that the measured factors—acceptability, satisfaction, inclusion, and usability—were positively rated across the three population groups corresponding to the three pilots. These findings support the core hypothesis of this paper, demonstrating that the SOCIO-BEE technology and its approach to organizing, executing, and evaluating citizen science (CS) campaigns effectively meet key criteria that are necessary for widespread adoption. Specifically, the solution achieved an average satisfaction agreement of over 70%, accessibility of over 76%, inclusiveness of over 73%, and usability above 60%. These strong results across fundamental evaluation dimensions indicate that SOCIO-BEE is making significant progress toward the democratization of CS campaigns. The consistently high levels of agreement across these multiple evaluation dimensions affirm the platform's potential to engage a broad and diverse range of participants in meaningful scientific contribution.

Table 13. Cross-pilot satisfaction, accessibility, inclusiveness, and usability KPIs.

Mean ± SD			
KPIs	**Ancona**	**Maroussi**	**Zaragoza**
User appreciation (satisfaction) with SOCIO-BEE (AcadeMe),	3.36 ± 1.43	4.26 ± 0.87	3.97 ± 1.00
The level of tool accessibility	4.29 ± 0.74	3.81 ± 1.08	3.92 ± 0.97
The level of inclusiveness of the tool	3.86 ± 0.9	3.65 ± 0.99	3.73 ± 0.96
Perceived usability score	3.09 ± 1.24	3.14 ± 0.87	3.51 ± 1.09

4.4. Citizen Science and Policy Integration: Lessons from SOCIO-BEE

One of the key challenges in citizen science (CS) initiatives is ensuring that citizen-generated data are not only collected but also meaningfully integrated into policymaking and urban planning. SOCIO-BEE has addressed this challenge by implementing a structured participatory framework across three European pilot cities—Ancona, Maroussi, and

Zaragoza. The project demonstrated that CS could bridge the gap between community engagement and policy impact by aligning citizen-collected air quality data with municipal decision-making processes.

The Ancona pilot exemplified how CS can inform targeted public health and urban mobility policies. The project engaged senior citizens in air quality monitoring, revealing pollution hotspots in high-traffic areas like the city center and port. The findings underscored the need for pedestrian-friendly zones and alternative traffic routes to minimize exposure to harmful pollutants. The Municipality of Ancona recognized the value of these data and proposed the integration of CS-driven insights into future urban planning and social assistance strategies.

In Maroussi, the project demonstrated the feasibility of integrating citizen-generated air quality data into municipal monitoring frameworks. By engaging commuters and deploying wearable sensors on public transport, SOCIO-BEE provided real-time pollution data that complemented traditional monitoring stations. The municipal government actively participated in data validation efforts and acknowledged the potential for CS to enhance evidence-based traffic management policies. Additionally, high-profile local policymakers supported the initiative, strengthening its legitimacy and potential for policy uptake.

The Zaragoza pilot showcased the long-term potential of CS by embedding air quality monitoring into local education systems. Engaging students in scientific data collection not only raised environmental awareness but also positioned CS as a continuous, scalable policy tool. The Municipality of Zaragoza integrated the project's findings into its broader climate neutrality strategy and encouraged further citizen participation in air quality governance.

The results of the SOCIO-BEE projects highlight that the successful policy integration of CS data depends on three critical factors:

1. Validation and standardization—ensuring that citizen-collected data meet scientific and regulatory standards to be considered in official decision-making;
2. Institutional collaboration—establishing formal partnerships between CS initiatives and municipal authorities to create structured pathways for policy adoption;
3. Community engagement—building long-term citizen involvement through education, awareness campaigns, and participatory urban planning.

By addressing these factors, SOCIO-BEE has established a replicable model for embedding citizen-generated environmental data into local and national policymaking. Future efforts should focus on scaling this approach, fostering cross-border collaborations, and further integrating CS methodologies into formal environmental governance structures.

5. Conclusions

The analysis of the validity of the data obtained through the WSNs developed in this paper is presented to ensure the air quality of the studied cities. This research, created in the context of SOCIO-BEE project, successfully demonstrates the potential of citizen science as an effective tool for air quality monitoring and community engagement. By integrating advanced technological solutions with a participatory framework, SOCIO-BEE overcomes key challenges in CS, including data quality assurance, sustained citizen involvement, and policy integration. The project's campaigns across Ancona, Maroussi, and Zaragoza validate that CS initiatives can produce reliable environmental data while fostering public awareness and action.

Future work will focus on scaling up SOCIO-BEE's approach to additional cities and refining its methodology to enhance its long-term engagement and impact. Further research will also explore the integration of AI-driven analytics to improve data validation processes and extend the project's applicability to other environmental challenges beyond air quality monitoring.

Author Contributions: Conceptualization, N.M., M.P.-B. and D.L.-d.-I.; methodology, N.M., M.P.-B. and D.L.-d.-I.; software, N.M., M.P.-B., S.U. and D.L.-d.-I.; validation, N.M., M.P.-B. and D.L.-d.-I.; formal analysis, N.M., M.P.-B. and D.L.-d.-I.; investigation, N.M., M.P.-B. and D.L.-d.-I.; resources, G.M.R. and D.L.-d.-I.; data curation, N.M., M.P.-B. and D.L.-d.-I.; writing—original draft preparation, N.M., M.P.-B. and D.L.-d.-I.; writing—review and editing, N.M., M.P.-B., O.G.-C., D.C.-M., M.K. and D.L.-d.-I.; visualization, N.M., M.P.-B. and D.L.-d.-I.; supervision, A.B., D.L.-d.-I., C.L.-G., S.C. and G.M.R.; project administration, D.L.-d.-I. and G.M.R.; funding acquisition, D.L.-d.-I. and G.M.R. All authors have read and agreed to the published version of the manuscript.

Funding: This work has been sponsored by SOCIO-BEE Horizon Europe project with Grant ID 101037648 and DEUSTEK5 (IT1582-22) Basque University system's A grade research team grant.

Institutional Review Board Statement: Not applicable.

Informed Consent Statement: Informed consent was obtained from all subjects involved in the study.

Data Availability Statement: Data is contained within the article.

Conflicts of Interest: Author Sergi Udina was employed by the company Bettair Cities SL. The remaining authors declare that the research was conducted in the absence of any commercial or financial relationships that could be construed as a potential conflict of interest.

Abbreviations

The following abbreviations are used in this manuscript:

CS	Citizen Science
PM	Particle Matter
CO_2	Carbon dioxide
SOCIO-BEE	Wearables and drones CIty Socio-Environmental Observations and BEhavioral ChangE
WSN	Wireless Sensor Nodes
MVE	Micro-Volunteering Engine
FSN	Fixed Sensor Nodeds
RS	Reference Station
KPIs	Key Performance Indicators
BLE	Bluetooth Low Energy

References

1. European Commission. Fitness Check of the EU Ambient Air Quality Directives. Available online: https://commission.europa.eu/publications/fitness-check-eu-ambient-air-quality-directives_en (accessed on 24 February 2025).
2. Hegger, D.L.T.; Mees, H.L.P.; Driessen, P.P.J.; Runhaar, H.A.C. The Roles of Residents in Climate Adaptation: A Systematic Review in the Case of the Netherlands. *Environ. Policy Gov.* **2017**, *27*, 336–350. [CrossRef]
3. Buček, P.; Maršolek, P.; Bílek, J. Low-Cost Sensors for Air Quality Monitoring—The Current State of the Technology and a Use Overview. *Chem.-Didact.-Ecol.-Metrol.* **2022**, *26*, 41–54. [CrossRef]
4. Kortoçi, P.; Motlagh, N.H.; Zaidan, M.A.; Fung, P.L.; Varjonen, S.; Rebeiro-Hargrave, A.; Niemi, J.V.; Nurmi, P.; Hussein, T.; Petäjä, T. Air Pollution Exposure Monitoring Using Portable Low-Cost Air Quality Sensors. *Smart Health* **2022**, *23*, 100241. [CrossRef]
5. Bonney, R.; Cooper, C.B.; Dickinson, J.; Kelling, S.; Phillips, T.; Rosenberg, K.V.; Shirk, J. Citizen Science: A Developing Tool for Expanding Science Knowledge and Scientific Literacy. *BioScience* **2009**, *59*, 977–984. [CrossRef]
6. Vergara, F.; Olivares-Rodríguez, C.; Guenaga, M.; López-De-Ipiña, D.; Puerta-Beldarrain, M.; Sánchez-Corcuera, R. Enhancing Citizen Science Engagement Through Gamification: A Case Study of the SOCIO-BEE Project. In Proceedings of the 2024 9th International Conference on Smart and Sustainable Technologies (SpliTech), Bol and Split, Croatia, 25–28 June 2024; 2024; pp. 1–7.

7. Juhola, S. Adaptation to Climate Change in the Private and the Third Sector: Case Study of Governance of the Helsinki Metropolitan Region. *Environ. Plan. C Gov. Policy* **2013**, *31*, 911–925. [CrossRef]
8. Lim, C.C.; Kim, H.; Vilcassim, M.J.R.; Thurston, G.D.; Gordon, T.; Chen, L.-C.; Lee, K.; Heimbinder, M.; Kim, S.-Y. Mapping Urban Air Quality Using Mobile Sampling with Low-Cost Sensors and Machine Learning in Seoul, South Korea. *Environ. Int.* **2019**, *131*, 105022. [CrossRef]
9. Gómez-Carmona, O.; Casado-Mansilla, D.; López-de-Ipiña, D.; García-Zubia, J. Human-in-the-Loop Machine Learning: Reconceptualizing the Role of the User in Interactive Approaches. *Internet Things* **2024**, *25*, 101048. [CrossRef]
10. Levontin, L.; Gilad, Z.; Shuster, B.; Chako, S.; Land-Zandstra, A.; Lavie-Alon, N.; Shwartz, A. Standardizing the Assessment of Citizen Scientists' Motivations: A Motivational Goal-Based Approach. *CSTP* **2022**, *7*, 25. [CrossRef]
11. Mousa, H.; Mokhtar, S.B.; Hasan, O.; Younes, O.; Hadhoud, M.; Brunie, L. Trust Management and Reputation Systems in Mobile Participatory Sensing Applications: A Survey. *Comput. Netw.* **2015**, *90*, 49–73. [CrossRef]
12. Ceccaroni, L.; Piera, J.; Wernand, M.R.; Zielinski, O.; Busch, J.A.; Woerd, H.J.V.D.; Bardaji, R.; Friedrichs, A.; Novoa, S.; Thijsse, P.; et al. Citclops: A next-Generation Sensor System for the Monitoring of Natural Waters and a Citizens' Observatory for the Assessment of Ecosystems' Status. *PLoS ONE* **2020**, *15*, e0230084. [CrossRef] [PubMed]
13. Campbell, C.J.; Barve, V.; Belitz, M.W.; Doby, J.R.; White, E.; Seltzer, C.; Di Cecco, G.; Hurlbert, A.H.; Guralnick, R. Identifying the Identifiers: How iNaturalist Facilitates Collaborative, Research-Relevant Data Generation and Why It Matters for Biodiversity Science. *BioScience* **2023**, *73*, 533–541. [CrossRef]
14. ScienceDirect. Assessment of the Madrid Region Air Quality Zoning Based on Mesoscale Modelling and K-Means Clustering. Available online: https://www.sciencedirect.com/science/article/pii/S1352231022003235 (accessed on 7 April 2025).
15. Meltus, Q.J.; Karanja, F.N. Mapping Air Quality Using Remote Sensing Technology: A Case Study of Nairobi County. *Open J. Air Pollut.* **2024**, *13*, 1–22. [CrossRef]
16. Choi, W.; Ho, C.-H.; Lee, Y. Temporal Pattern Classification of PM2.5 Chemical Compositions in Seoul, Korea Using K-Means Clustering Analysis. *Sci. Total Environ.* **2024**, *927*, 172157. [CrossRef] [PubMed]
17. López-de-Ipiña, D.; Puerta, M.; Barco, A.; Casado-Mansilla, D.; Gómez-Carmona, O.; Morresi, N.; Kotzagianni, M.; López-Gomollón, C.; Udina, S. Towards the Democratization of Citizen Science Campaigns for Air Quality Assessment. In *Proceedings of the International Conference on Ubiquitous Computing and Ambient Intelligence (UCAmI 2024)*; Bravo, J., Nugent, C., Cleland, I., Eds.; Springer Nature: Cham, Switzerland, 2024; pp. 729–740.
18. Wang, P.; Zhou, W.; Niu, Z.; Huo, D.; Zhou, J.; Li, H.; Cheng, P.; Wu, S.; Xiong, X.; Chen, N. An Approach for Assessing Human Respiration CO_2 Emissions Using Radiocarbon Measurements and Bottom-Up Data Sets. *J. Geophys. Res. Atmos.* **2024**, *129*, e2023JD040578. [CrossRef]
19. Han, Q.; Huang, H.; Li, C.; Tang, H. An Experimental Study on the Characteristics of Human Respiratory Aerosol Emission in Different Activity Intensity States. *Build. Environ.* **2025**, *267*, 112314. [CrossRef]
20. Subramanian, P.S.G.; Puthussery, J.V.; Mao, Y.; Salana, S.; Nguyen, T.H.; Newell, T.; Verma, V. Influence of Human Activities and Occupancy on the Emission of Indoor Particles from Respiratory and Nonrespiratory Sources. *ACS EST Air* **2024**, *1*, 386–396. [CrossRef]
21. Puerta-Beldarrain, M.; Gómez-Carmona, O.; Chen, L.; López-de-Ipiña, D.; Casado-Mansilla, D.; Vergara-Borge, F. A Spatial Crowdsourcing Engine for Harmonizing Volunteers' Needs and Tasks' Completion Goals. *Sensors* **2024**, *24*, 8117. [CrossRef] [PubMed]
22. Gardner, C.; Marsden, J.R.; Pingry, D.E. DSS Evaluation: A Comparison of Ex-Ante and Ex-Post Evaluation Methods. In *Recent Developments in Decision Support Systems*; Holsapple, C.W., Whinston, A.B., Eds.; Springer: Berlin/Heidelberg, Germany, 1993; pp. 439–455.

Disclaimer/Publisher's Note: The statements, opinions and data contained in all publications are solely those of the individual author(s) and contributor(s) and not of MDPI and/or the editor(s). MDPI and/or the editor(s) disclaim responsibility for any injury to people or property resulting from any ideas, methods, instructions or products referred to in the content.

sensors

MDPI

Article

The Impact of Domain Shift on Predicting Perceived Sleep Quality from Wearables

Nouran Abdalazim *, Leonardo Alchieri , Lidia Alecci , Pietro Barbiero and Silvia Santini

Faculty of Informatics, Università della Svizzera italiana, 6900 Lugano, Switzerland;
leonardo.alchieri@usi.ch (L.A.); lidia.alecci@usi.ch (L.A.); pietro.barbiero@usi.ch (P.B.); silvia.santini@usi.ch (S.S.)
* Correspondence: nouran.abdalazim@usi.ch

Abstract: Machine learning models for personal informatics systems are typically trained offline on *records of a specific population of users*, resulting in *population models*. These models may suffer performance degradation in real-world settings due to *domain shift*, i.e., differences in data distributions across users and contexts. Domain adaptation techniques can address this *issue* by, *e.g.*, personalizing models with user-specific data. *In this paper, we quantify the impact of domain shift* on *the performance* of both population and personalized models *in a specific scenario:* sleep quality recognition. *To this end, we also collect and make available to the research community the new BiheartS dataset*. Our analysis shows *that* domain shift *causes the* accuracy of population models *to decrease* by up to 18.54 percentage points, when *used* on *new data*. Personalized models, *instead*, show robust performance across datasets. However, *crafting personalized models typically requires using new data or user-provided labels*, limiting their *applicability in real settings*. To *mitigate* the limitations *of both population and personalized models*, we propose a novel unsupervised domain adaptation approach: the cluster-based population model (CBPM). CBPM achieves accuracy improvements of up to 13.45 percentage points *w.r.t. population model* without requiring *the use of* user-specific records or *labels*.

Keywords: personal informatics systems; sleep behaviour monitoring; sleep quality recognition; dataset bias; domain shift; covariate shift; online learning; wearable devices; machine learning models

1. Introduction

Personal informatics (PI) systems are ubiquitous computing technologies designed to help users collect, analyze, and reflect on personal information to enhance self-awareness and achieve personal goals [1]. These systems support self-reflection and self-knowledge by providing the users with insights about their daily behaviour and practices, e.g., health, productivity, emotions, and well-being [2]. PI systems often rely on the use of wearable devices, e.g., smartwatches, digital rings, earbuds, to collect raw data about their users' actions, physiological responses, emotional states, context, and more.

Machine learning (ML) models are often used in PI systems to infer high-level constructs from sensor data, for instance mental health [3–6], sleep behaviour [7], students' emotional engagement in classrooms [8,9], or behavior in the workplace [10–12]. To train ML models used in PI systems, researchers and practitioners typically rely on existing datasets containing the data of a population of users. The obtained model is referred to as a *population model* or *one-size-fits-all model* [4] and can be applied unchanged to perform inference from the data collected about the (previously unseen) target user of a PI system.

However, the distribution of both input data and labels may shift significantly across users within the population used to train the model and the target unseen user(s). This issue, known as *domain shift* in the ML literature, is caused by *interpersonal variability* across users and may cause population models to perform poorly in real settings [4,13]. To cope with this problem, *domain adaptation* techniques can be used to bridge the distribution mismatch between the training data and the data of the target user(s). For example, a pre-trained model can be adapted to the specific characteristics or data of the target user, leading to what in the PI systems literature is typically referred to as a *personalized model* [7,14].

In this work, we provide a comparative analysis of the performance of personalized models in a specific use case of PI systems: sleep monitoring from human physiological signals. Specifically, we first assess the impact of personalization on model performance using data records from a single dataset, as has been carried out in similar works in the literature [4,7,15,16]. We thereby use a subset of the data for model training and the remaining data for model personalization and evaluation. Yet, since the records pertain to the same dataset, they tend to have lesser interpersonal variability than if the data was picked from different datasets—a phenomenon also known as *dataset bias* [3,17–19].

Thus, we repeat our analysis using entirely different datasets for model training, personalization and evaluation. To this end, we rely on both the existing M2sleep dataset by Gashi et al. [7], and on the novel BiheartS dataset which we collected in the context of this work and make publicly available to the research community. BiheartS contains data records of 10 participants collected in a real-world setting over 30 days. The records include physiological signals gathered using three wearable devices—an Oura ring (third-generation) and two Empatica E4 wristbands (worn one on each wrist)—as well as self-reports about participants' sleep and daily behaviour.

Lastly, we describe a novel domain adaptation technique that does not require labelled data from the target user for personalization. The method matches each sleep session of the target user to one of previously determined *clusters* of training data, thereby capturing both interpersonal variability between users as well as intra-personal variability of the target user. Since the method generates a cluster-based population model, we refer to it as CBPM.

Our results indicate that personalized models outperform population models by up to 6.81 percentage points. When different datasets are used for evaluation, the performance of personalized models attains similar levels with respect to the single-dataset scenario. However, the performance of population models drops by up to 18.54 percentage points. This highlights the risk of overestimating the performance of population models when dataset bias is not taken into account. If our CBPM technique is used instead, the mean accuracy of the population model increases by up to 13.45 percentage points even when different datasets are used for training and evaluation.

The remainder of this paper is organized as follows: Section 3 summarizes relevant background information and related work. Section 3 describes the BiheartS dataset, contributed as part of this study, whereas Section 4 describes the M2Sleep dataset and the data pre-processing procedures. Section 5 provides evidence of the existence of dataset bias between the BiheartS and M2sleep datasets, proving that they are adequate for our analysis. In Section 6, we show how dataset bias causes significant performance degradation of population models, whereas personalized models can adapt effectively to the data of target users irrespective of the setting. Lastly, in Section 7, we introduce our CBPM technique and analyse its performance. Section 8 discusses the obtained results and their implications, whereas Section 10 concludes the paper.

The BiheartS dataset is available upon signing a data sharing agreement. All software artifacts developed as part of this work are open-source and available at the following GitHub (vesion 1) repository: https://github.com/nouran-abdalazim/sleep_quality_recognition_models.git (accessed on 17 June 2025). The link contains an excerpt of the repository for reviewing purposes. The full code base is going to be released upon acceptance of this paper).

2. Background and Related Work

Since our research centers on domain adaptation for sleep quality recognition, we examine the pertinent literature on domain adaptation broadly, as well as specifically for sleep recognition.

2.1. Domain Adaptation

In the ML literature [17,20–22], a *domain* D consists of the input feature vectors X, the output labels Y, and the joint probability $P(X,Y)$, i.e., $D = (X, Y, P(X,Y))$. The domain over which a model is trained is referred to as the *source domain* $D_s = (X_s, Y_s, P_s(X_s, Y_s))$, whereas the domain over which the model is tested (or used in real settings) is instead referred to as the *target domain* $D_t = (X_t, Y_t, P_t(X_t, Y_t))$. In the PI systems literature, such a model is typically referred to as a *population model* [4] because it captures the characteristics of the population of users over which it has been trained.

The differences between the distribution $P_s(X_s, Y_s)$ in the source domain, which is used for model training, and the distribution $P(X_t, Y_t)$ in the target domain, where the model is deployed, leads to the so-called *domain shift* problem [13,23]. This mismatch between the source and target distribution makes the ML model unable to perform reliably when confronted with the new data in the target domain, which was unseen at the time of training. In other terms, the model does not *generalize* well to the target domain(s). There are different types of domain shift: covariate shift, label shift, and concept shift [13].

Our work focuses in particular on *covariate shift*, which occurs when the distribution of the marginal probability of the feature vectors in the source domain $P(X_s)$ differs from the same distribution in the target domain $P(X_t)$. In PI systems, this occurs for instance when the distribution of the features computed from physiological data in the source domain differ systematically from the distribution of features in the target domain due to, e.g., significant age differences in the population of users whose data was included in the source domain and the age of the target user. In general, variability in users' demographic characteristics, behaviour, preferences, contexts, and perceptions can lead to covariate shift.

Domain adaptation (DA) is a paradigm that has been widely used in the computer vision and natural language processing literature to deal with the covariate shift problem [24–27]. DA techniques typically exploit the feature vectors and/or labels from the target domain to adapt the model that is previously trained on the source domain. In the PI systems literature, the model resulting from this adaption is usually referred to as a *personalized model* [4,7].

There exist several categories of DA techniques depending on the specific differences between the feature vectors and/or the corresponding labels in the source and target domains.

Wang and Deng [25], for instance, specify that DA can be applied either in *homogeneous* settings, where the input feature space and the label space are the same in the source and target domains, or in *heterogeneous* settings, where the feature space and the label space differ across domains. Also, Farahani et al. [24] refer to the setting where the label space and the feature space in the source and target domains coincide as a *closed-set* setting [24].

Further, Patel et al. [26] highlight that DA techniques can be categorized based on the availability of labelled feature vectors from the source and the target domains. In particular, they differentiate between semi-supervised and unsupervised approaches. In semi-supervised DA approaches, the ML model exploits labelled data from the target domain to adapt itself. In PI systems, this implies that the target user(s) can provide labels for the data collected by the system after the deployment of the ML model. In unsupervised DA approaches, the model leverages only unlabelled data from the target domain. In PI systems, this implies that new user data collected by the system after the deployment of the ML model is used for adaptation, but that the corresponding labels are not available.

Lastly, Wang and Deng [25] also show that DA can occur either in one step or in multiple steps, depending on whether transferring knowledge from the source to the target domain can be accomplished in one or multiple steps.

In this paper, we address the covariate shift problem using DA in homogeneous closed-set settings. We explore the use of both a personalized model generated using a one-step, semi-supervised approach and a cluster-based model derived by using a multi-step, unsupervised DA technique.

2.2. Domain Adaption in Sleep Quality Recognition

In the PI systems literature in general, and in sleep quality recognition in particular, several authors have explored DA techniques to cope with the poor generalizability of population models [7,15,16,28].

Gashi et al. [7] collect and use physiological signals along with self-reports to build and evaluate ML models for sleep quality recognition and sleep/wake recognition tasks. They compare the results of two models—population models and personalized models—using the M2sleep dataset. Their experimental results confirm that personalized models achieve higher performance than population models in both tasks. In particular, the personalized model achieves a maximum balanced accuracy of 61.51, with an improvement of 14 percentage points in the sleep quality recognition task. Also, Moebus and Holz [28] use the M2sleep dataset proposed by Gashi et al. [7]. They add features that represent cardiovascular activity instead of the features extracted from electrodermal activity as Gashi et al. [7]. Their results reveal the personalized model achieves a maximum accuracy of 70%, with an improvement of 11 percentage points in the model performance compared to Gashi et al. [7]'s approach. Van et al. [16] use multiple datasets to estimate sleep efficiency. They rely on activities, emotions, environmental, and sleep data collected using the Fitbit wearable device in their ML task. They train an ML model on one dataset (population model), and then they adapt this model based on the data of the target test user (personalized model). They derive the final estimation from the results of both population model and personalized model. Their results show that the integration of the population model and the personalized model achieves an improvement of 0.5 in root mean square error compared to the baselines. Khademi et al. [15] use actigraphy data to build a personalized model to distinguish sleeping/waking events and estimate sleep parameters from golden standard polysomnography (PSG). They compare a personalized model, built using the data of the target user, with a population model. Their experimental results show that the performance of the personalized model is higher than that of the population model in estimating sleep parameters. The results show no statistical difference between the sleep parameters estimated by the personalized model and the PSG. Gan et al. [29] propose a framework that integrates objective and subjective data from the Fitbit wristband, a sports logging app, and questionnaires, incorporating daily activity, affective states, sleep information, and demographic data. Specifically, they train a deep learning model using objective and subjective data to estimate sleep quality. Moreover, they use pattern mining to discover different relations between the data from different modalities and sleep quality. Then, they rely on the model output and pattern mining to provide the personalized feedback about sleep habits to the users.

Grandner et al. [30] compare two ML models adopted on a new wearable device. The two models are a population model and a personalized model in sleep stage recognition and sleeping/waking recognition tasks. They use the parameters from the golden standard PSG as the ground truth. Their results show that the personalized model surpasses the population model.

2.3. Novelty of This Work with Respect to the Existing Literature

The use of DA techniques to cope with the domain shift problem has been widely explored in the computer vision and natural language processing literature [17,20–22]. However, less attention has been paid to this problem in PI systems [3,13]. In particular, in sleep quality recognition, several studies in the literature attempt to investigate the impact of DA i.e., personalized model, e.g., [7,15,16,28]. However, the majority of the existing studies employ only one single dataset in their experimental setup. Such an experimental setup is less prone to the effect of domain shift, since users in the same dataset tend to share common characteristics. This can lead to misleading over-estimated results for the performance of the personalized model.

In this study, we evaluate the performance of both the personalized and population models in a multiple-dataset experimental setup. To this end, we introduce the new BiheartS dataset, collected using a data collection protocol similar to that of the existing M2Sleep dataset. These two datasets enable us to investigate the presence of covariate shift between datasets with comparable characteristics, e.g., wearable devices, geographical regions, and demographic characteristics of participants. Furthermore, we propose a novel cluster-based approach that leverages unsupervised DA to enhance the performance of the population model across diverse users in different datasets, eliminating the need for user feedback in order to adapt the model.

3. BiheartS Dataset

To analyze the impact of distribution mismatch across datasets, we collect a new dataset named BiHeartS. With the BiheartS dataset, we provide the research community with a new dataset that includes extensive self-reports about the daily behaviour, physical activity, daily fatigue, daily stress and the sleep behaviour of participants. The design of the data collection protocol aligns with existing studies in the literature [7,31,32]. However, some key novel elements in the BiheartS dataset are the use of two wearable devices, which collect sensors' data from both sides of the body simultaneously, and the extensive daily self-reports. To the best of our knowledge, the BiheartS dataset is the first dataset that is collected in the wild and includes sensors' data from both sides of the body. The BiheartS dataset is available after signing a data sharing agreement. Please contact the corresponding author of the paper to make a request.

Given the extensive self-reports and sensor signals collected in the BiheartS dataset, researchers can use this dataset in several research studies. They can use it to investigate personalized sleep quality recognition, similar to [7,28]. The BiheartS dataset also includes raw sensor data from multiple wearable devices, enabling direct comparisons between different wearables, similar to [31,33]. Additionally, since BiheartS contains raw sensor data from both the left and right sides of the body, researchers can explore lateralization analysis, as in [34,35]. In this paper, we focus on the sleep quality recognition task. The data collection campaign has been reviewed and approved by our faculty's delegate for ethics.

Participants and Devices We recruit 10 participants (1 female and 9 males) aged between 20 and 25 years old. All the participants are students. We use the term user and participant interchangeably in this work. Participants are asked to wear one finger-worn wearable device and two wrist-worn wearable devices for 30 consecutive nights: the Oura ring (Gen. 3) (https://ouraring.com/product/rings), (accessed on 17 June 2025) which measures sleep parameters with a Photoplethysmography (PPG) sensor, and two Empatica E4 (https://www.empatica.com/en-gb/research/e4/) (accessed on 17 June 2025) wristbands, one on each wrist. The Empatica E4 is a research-grade device equipped with four sensors: electrodermal activity (EDA), an accelerometer (ACC), skin temperature (ST), and PPG sensors [36]. The E4 wristband and the Oura ring, especially the second generation, are adopted in several studies in the literature, e.g., [7,31,33,37–41].

Self-reports To collect the information about the participants, their sleep behaviour and their daily routine, we rely on self-reports, similar to existing studies in the literature, e.g., in [7,31,32]. We divide the self-reports into a *pre-study questionnaire*, *daily self-reports*, and a *post-study questionnaire*. The pre-study questionnaire integrates demographics questions and several validated questionnaires: the Pittsburgh Sleep Quality Index (PSQI) [42], to assess sleep behaviour; the Morningness–Eveningness (Chronotype) questionnaire, to determine the chronotype [43]; and the Big 5 inventory questionnaire to estimate personality traits [44].

We provide the participants with two options to log their daily self-reports: a pen-and-paper diary and an application-based diary, the RealLife Exp mobile application (https://www.lifedatacorp.com) (accessed on 17 June 2025) . We split the daily self-reports into *morning self-reports* and *evening self-reports*. The morning self-reports include information about the sleep behaviour of the previous night: bedtime, wake-up time, latency, number of awakenings, perceived sleep quality score, recovery after sleep score, and reasons for sleep disturbance. All the questions in the morning self-reports are formulated based on the validated PSQI questionnaire [42] and the Daily User Caregiver Sleep Survey (DUCSS) questionnaire [45]. We perform adaptations to some of the questions from the validated questionnaire to match our study design, e.g., we adapt the question *"During the past month, How long (in minutes) does it usually take you to fall asleep each night?"* from the PSQI questionnaire to be *"How long (in minutes) did it take you to fall asleep last night?"*. We present more details about the questions used in the morning self-reports and the corresponding adaptions in the Supplementary materials (A). The evening self-reports include information about the routine and the behaviour during the current day: stress score, sleepiness score, fatigue score, physical health, and physical activity. For the evening self-reports, we adapt questions from the following questionnaires: the Perceived Stress Scale Questionnaire [46], PSQI questionnaire [42], the Quality of Life Enjoyment and Satisfaction Questionnaire [47], and the International Physical Activity Questionnaire [48]. Finally, the post-study questionnaire includes the PSQI questionnaire to assess the overall sleep behaviour of the participants over the whole study duration. We designed the data collection campaign and the questions in the self-reports in collaboration with an expert in psychology and organizational behaviour. We illustrate the details of the morning and the evening self-reports used in the BiheartS data collection campaign in the Supplementary materials (A).

Data Collection Procedure

We design the data collection procedure following similar studies in the literature, e.g., [7,31–33]. In the beginning, we provide the participants with information about the objective of the study and the study protocol, then the participants sign an informed consent form. Finally, we give the wearable devices to the participants alongside instructions for setting up the designated synchronization applications for obtaining the raw data from the wearable devices.

Before the study, the participants respond to a pre-study questionnaire. Furthermore, we request that the participants put the pen-paper diary beside their beds so they can easily remember to log their daily self-reports. We use the RealLife Exp mobile application to send two notifications in the morning and two notifications in the evening to the participants as a reminder to compile the daily self-reports. To minimize the effect of the recall bias, the participants are allowed to log only self-reports of the current day.

We ask the participants to wear the Oura ring on the index finger of their left hand. They wear one Empatica E4 on their left wrist and one on the right wrist, every night. We call the sleep period of a participant during the night a *sleep session*. The participants wear the devices one hour before sleep and log the evening self-reports. The next day, participants complete the morning self-report and then take off the devices one hour after waking up. During the day, participants synchronize the data of each device. To ensure that the participants synchronize the data from the wearable devices on a daily basis, we run daily checks. After the study, the participants respond to the post-study questionnaire.

4. Datasets and Data Preparation

To investigate the effect of domain shift in PI systems, we use two datasets: the BiheartS dataset, which we collect for the purpose of this study, and the M2sleep dataset [7]. Both the BiheartS and the M2sleep datasets are collected for sleep behaviour analysis in the wild with similar data collection protocol. They are collected using similar wearable devices, in the same geographical zone, from participants with a comparable age range and occupation. The similarities between the two datasets make them suitable for studying covariate shift in the same task: sleep quality recognition. Table 1 shows the summary of the two datasets.

We describe the BiHeartS dataset in the dedicated Section 3. Moreover, Figure 1 presents a 30-second sample of raw signals collected from one participant in one sleep session from the BiHeartS dataset. Moreover, we briefly present the M2sleep dataset in this section, as well as the pre-precessing procedure for both datasets.

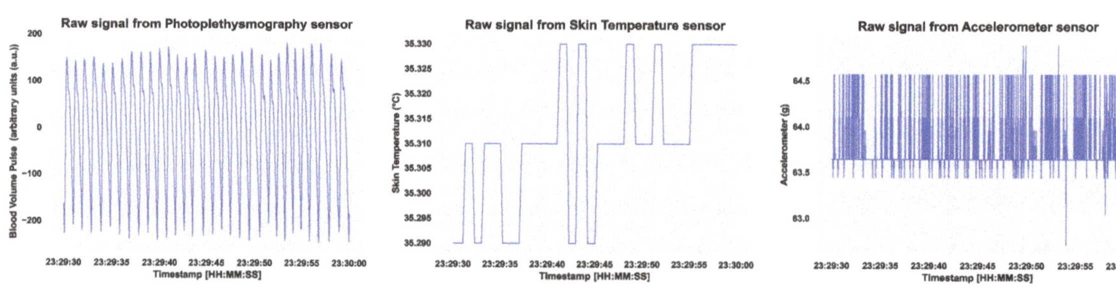

(a) (b) (c)

Figure 1. Raw signals from photoplethysmography, skin temperature, and accelerometer sensors collected using the Empatica E4 wristband wearable device. (**a**) Photoplethysmography sensor. (**b**) skin temperature sensor. (**c**) accelerometer sensor.

Table 1. Summary of the two datasets used in this paper. Symbol (✓) indicates that the dataset is collected in the wild.

Dataset	In the Wild	Users	Duration	Devices
M2sleep	✓	16	30	E4 wristband
BiheartS	✓	10	30	Two E4 wristbands and Oura ring Gen.3

4.1. The M2sleep Dataset

The M2sleep dataset, collected by Gashi et al. [7], is widely used in studies involving sleep behaviour in the literature, e.g., [7,28,49]. It is publicly available upon signing a data sharing agreement with the authors. It contains data collected from 16 participants for 30 nights in the wild. Some 69% of the participants are male, and 56% of the participants are students in the age range of 21–35 years. The dataset includes sensor data collected using the Empatica E4 wristband, along with self-reports about the previous sleep session including bedtime, wake-up time, awakening, and perceived sleep quality.

4.2. Pre-Processing, Feature Extraction, and Labelling

In our experiments, we rely on sensor signals collected from the Empatica E4 wristband: heart rate (HR), inter-beat interval (IBI), ST, and ACC signals. We also extract the respiration patterns (RP) from the photoplethysmogram (PPG) signal using a similar methodology as described in [50]. We refer to each of the IBI, ST, ACC and RB signals as *modality*). These modalities are widely used in sleep quality recognition studies, e.g., [7,28,31,51,52]. From each sleep session—i.e., the sleep period from one participant—we extract hand-crafted features from each modality individually. We refer to the set of features extracted from the same sensor signal as a *single modality*. For HR, ACC, ST and RP signals, we extract eight time-domain features for the whole signal as in [7] and four frequency-domain features by transforming each signal using the fast Fourier transform (FFT), similar to [7]. We use the `Scikit-Learn` [53] Python library to extract these features. To extract features related to heart rate variability (HRV) from the IBI signal, we "clean" the signal by removing IBI values outside the 300–2000 ms range [54]. Then, we remove the ectopic beats based on Malik's rule, as suggested in [54]. Finally, we extract 12 time-domain features and 7 frequency-domain features using the `FLIRT` [54] and the `hrv-analysis` (https://aura-healthcare.github.io/hrv-analysis/readme.html accessed on 12 June 2025) Python libraries.

In addition to features extracted from single sensor modalities, we employ contextual features from self-reports. These features are objective measures for each participant's previous sleep session. Several studies in the literature show that contextual features are correlated with perceived sleep quality [14,28]. Consequently, they can be used as an indicator of the participants' perception of their sleep quality.

Moreover, we group features from the single sensor modalities to form sets of features, which we refer to as *multi-modality*. The first multi-modality set consists of features from all sensor modalities together, which we define for simplicity as the *all-sensor-features modality*. The second set combines features from all the sensor modalities as well as the contextual modality, which we refer to as *all-features modality*. Several studies in the literature rely on the multi-modality in their experiments on sleep quality recognition [7,28]. In Table 2, we present a summary of all the modalities used in our experiments. We provide more details for the feature extraction from each modality in the Supplementary materials (B).

We formulate the sleep quality recognition task as a binary ML task, similar to Gashi et al. [7], Moebus and Holz [28], Abdalazim et al. [31], Moebus et al. [49]. Participants in both datasets report their sleep quality score using Likert scales; for the M2Sleep dataset, the scale is from 1 (lowest quality) to 5 (highest quality), while for the BiHeartS, the scale is from 1 (lowest) to 10 (highest). In order to obtain binary labels, following similar works in the literature [7,28,31], we consider the following. For the M2Sleep dataset, we select as "high sleep quality" all scores of 4 or above (top 20% of the 5-point Likert scale), and as "low sleep quality" all scores of 3 or less, similar to [7,31]. For the BiHeartS dataset, scores of 7 or above (top 20% of the 10-point Likert scale) are labelled as "high sleep quality", while those of 6 or less are "low sleep quality".

By applying the same feature extraction procedure to both datasets and the same threshold to define positive and negative labels, we ensure the homogeneity of the feature and the label spaces. This allows the investigation of the covariate shift and the cross-dataset generalization capability of ML models, as recommended by Xu et al. [3], Meegahapola et al. [13].

Table 2. Summary of the seven modalities used in our experiments. The (||) symbol represents the concatenation of two vectors.

Modality	Feature Vector (fv)
HR	$fv_{HR} \in \mathbb{R}^{12}$
HRV	$fv_{HRV} \in \mathbb{R}^{19}$
ST	$fv_{ST} \in \mathbb{R}^{12}$
ACC	$fv_{ACC} \in \mathbb{R}^{12}$
RP	$fv_{RP} \in \mathbb{R}^{12}$
All-sensor-features	$fv_{all-sensor-features} = [fv_{HR} \parallel fv_{HRV} \parallel fv_{ST} \parallel fv_{ACC} \parallel fv_{RP}] \in \mathbb{R}^{67}$
Contextual	$fv_{Contextual} \in \mathbb{R}^{4}$
All-features	$fv_{all-features} = [fv_{all-sensor-features} \parallel fv_{contextual}] \in \mathbb{R}^{71}$

5. Domain Shift Analysis

In this section, we describe how we assess the domain shift between the two datasets, BiHeartS and the M2sleep. To this end, we first conduct a statistical analysis across the datasets, as is common practice in the literature, e.g., [3,19], using correlation and statistical tests. Then, we perform a name-the-dataset ML classification task to estimate the degree of dataset shift between the feature vectors in the M2sleep and the BiheartS datasets [3,17].

5.1. Statistical Comparison Between Datasets

First, we compute the Spearman's rank correlation coefficient between the perceived sleep quality as reported on the Likert scale in the morning self-reports and each extracted feature. With this experiment, we aim to investigate the variability of the correlation patterns across the features from both datasets. This variability is an indication of the difference in the features' distribution across the M2sleep and the BiheartS dataset. We present in Figure 2 a scatter plot that shows the relation between the correlation coefficient of each feature in the two datasets. We observe that the correlation coefficient of the majority of the features is different in both datasets. These results hint at differences in the distribution of the features across the datasets, and, accordingly, these differences can lead to covariate shift. Consequently, we compare the distribution of each feature statistically in the two datasets. We provide more results for the correlation analysis for each feature in the Supplementary materials (C).

We perform the two-sample Kolmogorov–Smirnov non-parametric statistical test ($\alpha = 0.05$) [55] between the distributions of each feature when extracted from the BiHeartS and the M2Sleep datasets. Figure 3 presents the distribution of the p-values computed for each feature, divided by modality, between the two datasets when using the Kolmogorov–Smirnov statistical test. These results show that for all the modalities, except the ACC modality, the majority of the features are statistically different in the two datasets. In particular, we observe that 61.11% of the features are statistically different between the two datasets. Also, we perform Bonferroni corrections to the α, such that $\alpha_n = \frac{\alpha}{n}$, where n is the number of features. When performing the correction, we find that 27.78% of the features are statistically different, with $\alpha_n \approx 0.00069$. These results confirm the existence of a distribution mismatch between most of the features and imply the existence of covariate shift between the two datasets.

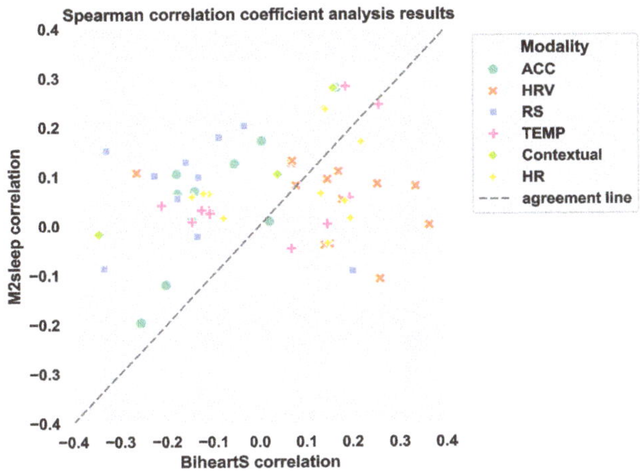

Figure 2. Spearman's rank correlation coefficient between extracted features and the reported sleep quality. The diagonal line represents the agreement between correlation coefficients in both datasets.

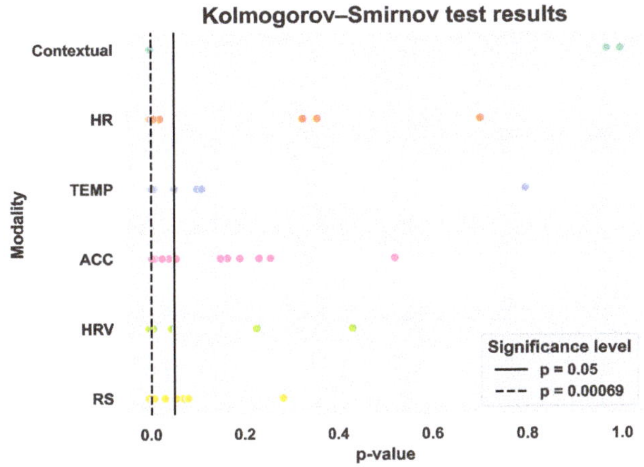

Figure 3. Visualization of the distribution of the p-value computed for each feature in the two datasets using the Kolmogorov–Smirnov non–parametric statistical test.

5.2. Name-the-Dataset Machine Learning Task

In the computer vision literature, the presence of domain shift is often evaluated using the name-the-dataset (NED) ML task [17,18]. Xu et al. [3] used the same task to quantify the difference among datasets in the human behaviour modelling domain. Accordingly, we employ the NED task on our two datasets, in order to evaluate the ability of ML models to distinguish between the feature vectors of the two datasets. We formulate the task as a binary classification task, and we use features from the *all-features modality* described in Section 4.2. We use a support vector machine (SVM) classifier, as suggested in [17,18]. We also sample varying proportions of data from each dataset, similar to [17,18], while ensuring that users present in the training set are not in the test set, as in [3]. As an evaluation metric, we use the balanced accuracy.

Figure 4 shows the performance of the SVM classifier and the RG baseline. We conduct this experiment multiple times with a different number of random samples in each round to investigate the behaviour of the SVM classifier, as in [18]. With this different number of samples, we can observe the change in the performance of the SVM classifier as more data from each dataset is added to the training set. From Figure 4, we find that the balanced accuracy of the SVM classifier increases as more samples from both datasets are included in the experiment. Moreover, the difference in the balanced accuracy between the SVM and the baseline is statistically significant when measured using the Wilcoxon statistical test ($\alpha = 0.05$).

The results of the NED classification task support our observation about the distribution mismatch between the two datasets, since the SVM classifier can distinguish between the feature vectors from the two datasets with a balanced accuracy of 90.72% when all the samples from the two datasets are used. This implies the existence of covariate shift between the two datasets. Accordingly, we explore the impact of covariate shift on the performance of population models and personalized models in a sleep quality recognition task.

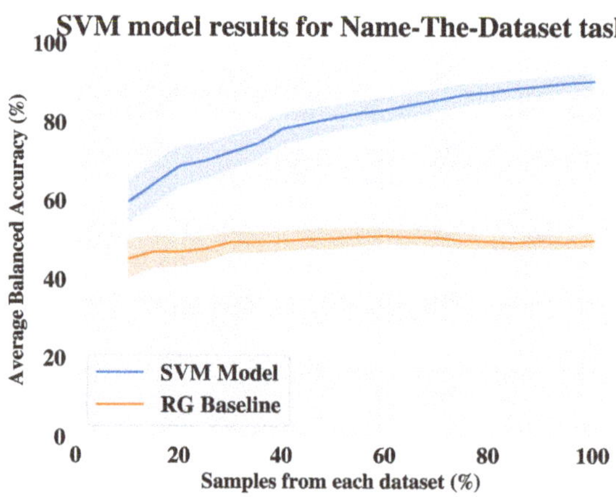

Figure 4. Results of the support vector machine (SVM) model and the random guess baseline in name-the-dataset ML task. The experiment is conducted multiple times with a different number of samples in each round to investigate the behaviour of the SVM model. The shadows represent the standard error for each iteration.

6. Impact of Domain Shift on Sleep Quality Recognition

In the previous Section 5, our results demonstrate the existence of covariate shift between the BiheartS and the M2sleep datasets. In this section, we investigate the impact of covariate shift on the performance of ML models in a sleep quality recognition task. To this end, we rely on different evaluation scenarios, two training settings, and two different ML classifiers. We rely on the feature sets from multiple modalities —all-sensor-features and all-features— as well as the contextual modality, given that existing studies demonstrate that they achieve higher results than single-modality feature sets [7,28]. Also, the multi-modality feature set allows us to assess whether our findings are consistent across different types of input features.

6.1. Evaluation Scenario

We implement three distinct setups (hereafter *evaluation scenarios*) to train and evaluate the sleep quality recognition models. The three evaluation scenarios are *single-dataset*, *mixed-dataset*, and *multiple-dataset*. With these scenarios, we have three different setups for the datasets used for model training and evaluation.

Single-dataset scenario: This evaluation scenario relies on the same dataset for model training and evaluation. The implementation of this scenario is similar to the leave-one-participant-out evaluation method, where in each training/evaluation iteration, one user from the designated dataset is left out as the target test user for model evaluation.

Multiple-dataset scenario: It is implemented as the leave-one-dataset-out evaluation method, where one dataset is left out at each training/evaluation iteration. The model is then evaluated on each user in the left-out dataset. In other words, this scenario is a composite of the leave-one-participant-out evaluation method and the leave-one-dataset-out evaluation method.

Mixed-dataset scenario. In this evaluation scenario, the two datasets are combined together as one dataset. Then, the leave-one-participant-out evaluation method is used to evaluate the ML model per user, i.e., the left-out target user. This scenario acts as a middle ground between the single-dataset scenario and the multiple-dataset scenario, as it ensures that the training set of the model includes feature vectors from the same dataset as the target left-out user. In this scenario, the model is trained on the data from users within the same dataset as the target user as well as the data from all the other datasets.

In all the previous scenarios, the data from each user is sorted in chronological order to avoid any temporal data leakage. We provide more details and mathematical formulations for the three scenarios in the Supplementary materials (D.1).

6.2. Training Settings and Baselines

To investigate the impact of covariate shift on sleep quality recognition models, we implement two possible real-world settings for PI systems. The first setting relies on ***population models*** that are trained once using the available training data then deployed without adjustments to the target unseen users. In our experiments, the population model is trained on the available training set depending on the scenario (Section 6.1). Then, it is evaluated on target unseen users in the test set. This model is an offline model because the model is used by the target users without any adjustments.

The second setting relies on the *online learning* approach [56] to adapt the model over time for the target test user. This setting allows the users to provide their feedback about the insights provided by the PI systems. Then, this feedback is used to adjust the model, which we call, in this case, *personalized models*. In our experiments, the personalized model is trained using the available training set based on the scenario. Then, the model is updated incrementally for the target unseen user when a new labelled data point from this designated user is available. This model relies on the discrepancy-based domain adaptation approach with a class criterion [25]. The personalized model is considered online model because the model's weights are updated with the arrival of a new data point.

To implement both models, we use two different ML classifiers in our experiments. The first one is the Passive Aggressive Classifier (PAC) (https://scikit-learn.org/1.5/modules/generated/sklearn.linear_model.PassiveAggressiveClassifier.html) (accessed on 17 June 2025) [57]. The PAC is widely used in online learning settings [57], in particular for personalization purpose as demonstrated by existing studies, e.g., Winger et al. [58,59]. The second one is the Multilayer Perceptron (MLP) (https://scikit-learn.org/1.5/modules/generated/sklearn.neural_network.MLPClassifier.html#sklearn.neural_network.MLPClassifier) (accessed on 17 June 2025) [60,61], also known as a feed-forward neural network and one of the simplest deep learning techniques [62]. The MLP classifier is employed in several studies focused on monitoring sleep behavior and classifying sleep quality using data from wearables [62–64]. For instance, Sathyanarayana et al. [62] report that the MLP achieves performance comparable to deep learning models such as Convolutional Neural Networks (CNNs) and Long Short-Term Memory (LSTM) networks in recognizing sleep quality from actigraphy signals. This finding is further supported by [64]. Based on this evidence, we include the MLP classifier in our experiments, particularly given our context, which does not involve large datasets that typically benefit complex deep learning models. We implement both models using the `Scikit-Learn` [53] Python library (version 1.6.1).

We employ three baselines: the random guess (RG) baseline, the biased random guess (BRG) baseline, and the personalized biased random guess (PBRG) baseline. The *random guess baseline* makes sleep quality predictions by randomly extracting positive and negative labels from a uniform distribution. The *biased random guess baseline* makes predictions by randomly selecting the positive and negative labels from the labels' distribution in the training set. Both the RG and the BRG baselines are offline models, since the distribution of the labels in the training set does not change. Finally, the *personalized biased random guess baseline* is a variation of the BRG baseline. It extracts the predictions by considering both the distribution of the labels in the training set and the new available data point from the target test user. This baseline is an online model, since it considers data from the target user to update the label distribution in the training set.

6.3. Evaluation Metrics

Offline models are typically evaluated using the leave-out method, which splits the dataset into training and test sets [56]. Online models, however, are often assessed with the predictive sequential (prequential) method [56,65,66], also known as interleaved test-then-train. This approach evaluates the model on each test data point before using it for training, enabling performance assessment over time.

However, to train and evaluate the performance of online models and offline models simultaneously, Fekri et al. [56] propose the prequential–leave-out evaluation method. This method uses the available training data for training the offline models and the online models. However for model evaluation, the online models are evaluated using the prequential evaluation method and the test data, while the offline models are evaluated using the traditional hold-out evaluation method using the same test data. This approach guarantees that regardless of the available training data for both models, the two models are evaluated using the same test data.

In our experimental setup, we rely on the prequential–leave-out evaluation method to evaluate the offline models—the population model, biased random guess, and random guess—and the online models, i.e., the personalized model and personalized biased random guess.

For evaluation, we use accuracy as the metric for both online and offline models. Prequential accuracy is commonly used to assess online models [66]; however, it is mathematically equivalent to accuracy when computed on the entire test set, assuming that the target test user provides feedback for each inference result.

Moreover, to statistically compare the performance of the population model and the personalized model in the different scenarios, we use Cliff's δ effect size [67]. Cliff's δ is non-parametric effect size measure, with a range of $[-1, +1]$, that quantifies the overlap between two distributions, where values near the two extremes indicate minimal overlap and values close to 0 indicate high overlap. We use the thresholds proposed by Vargha and Delaney [68] to interpret the effect size values as negligible, small, medium, and large.

6.4. Experimental Results for Sleep Quality Recognition

We aim to recognize the binary sleep quality labels in three evaluation scenarios (a single-dataset scenario, a multiple-dataset scenario, and a mixed-dataset scenario) and two model training settings (a personalized model and a population model). In each scenario, we evaluate the population model and the personalized model using seven different modalities. To mitigate the effect of random initialization of the models, we run each experiment using 50 different random seeds, and we compute the average accuracy over all iterations for each model per each modality. In this section, we present the results for the PAC and the MLP classifiers in the single-dataset scenario and the multiple-dataset scenario with the all-sensor-features, the contextual, and the all-features modalities in Table 3. Moreover, we report the difference in the models' performance across the different scenarios in Table 3. We present more results on the mixed-dataset scenario and the other modalities in Supplementary materials (D.2).

Table 3. Results of the sleep quality recognition binary task for single- and multiple-dataset scenarios. We report the mean accuracy (%)±standard error. Symbols indicate statistical significance of the models: (*) vs. random guess, (**) vs. random and biased random guess, (†) vs. biased random guess. Bold values show significant differences between population and personalized models within the same modality and scenario. All statistical comparisons are carried out using the Wilcoxon paired non-parametric test ($\alpha = 0.05$).

Passive Aggressive Classifier Performance				
Modality	**Single-dataset Scenario**		**Multiple-dataset Scenario**	
	Population Model	**Personalized Model**	**Population Model**	**Personalized Model**
	Test on the M2sleep dataset			
All-sensor-features	51.25 ± 0.01 *	**63.40 ± 0.00** **	53.06 ± 0.01 **	**62.97 ± 0.00** **
Contextual	56.57 ± 0.01 **	**64.48 ± 0.01** **	48.50 ± 0.01	**64.12 ± 0.00** **
All-features	50.59 ± 0.01 *	**63.75 ± 0.00** **	52.41 ± 0.01 **	**63.55 ± 0.00** **
BRG baseline	53.48 ± 0.01	51.67 ± 0.00	46.93 ± 0.01	49.24 ± 0.00
RG baseline		47.44 ± 0.01		
	Test on the BiheartS dataset			
All-sensor-features	64.41 ± 0.01 **	**69.85 ± 0.01** **	49.63 ± 0.01 †	**62.77 ± 0.01** **
Contextual	48.27 ± 0.01	**67.12 ± 0.01** **	48.62 ± 0.01	**65.66 ± 0.01** **
All-features	62.02 ± 0.01 **	**70.39 ± 0.01** **	49.02 ± 0.01 †	**64.52 ± 0.01** **
BRG baseline	51.60 ± 0.01	51.12 ± 0.01	46.06 ± 0.01	47.51 ± 0.01
RG baseline		52.81 ± 0.01		
Multilayer Perceptron Classifier Performance				
	Test on the M2sleep dataset			
Modality	**Single-dataset Scenario**		**Multiple-dataset Scenario**	
	Population Model	Personalized Model	Population Model	Personalized Model
All-sensor-features	55.53 ± 0.01 *	**63.74 ± 0.01** **	57.40 ± 0.01 **	**62.37 ± 0.01** **
Contextual	60.69 ± 0.01 **	60.42 ± 0.01 **	43.99 ± 0.01	**59.96 ± 0.01** **
All-features	54.83 ± 0.01 *	**64.95 ± 0.01** **	56.55 ± 0.01 **	**62.97 ± 0.01** **
BRG Baseline	53.48 ± 0.01	51.67 ± 0.00	46.93 ± 0.01	49.24 ± 0.0
RG Baseline		47.44 ± 0.01		
	Test on the BiheartS dataset			
All-sensor-features	**63.73 ± 0.01** **	59.03 ± 0.01 †	45.19 ± 0.01	**56.48 ± 0.01** †
Contextual	51.17 ± 0.01	52.26 ± 0.01	48.89 ± 0.01 †	51.32 ± 0.01 †
All-features	59.85 ± 0.01 **	**61.28 ± 0.01** **	43.21 ± 0.01	**55.39 ± 0.01** †
BRG Baseline	51.60 ± 0.01	51.12 ± 0.01	46.08 ± 0.01	47.51 ± 0.01
RG Baseline		52.81 ± 0.01		

Sleep Quality Recognition in the Single-Dataset Evaluation Scenario. The personalized models generally achieve higher performance than the population models across most modalities in both datasets. Additionally, both models outperform the baseline performance in most cases. Specifically, for the M2Sleep dataset, the personalized model achieves the highest accuracy of 64.95% using the all-features modality with the MLP classifier. In the BiheartS dataset, it reaches a maximum accuracy of 70.39% with the same modality and the PAC classifier. On the other hand, the population model attains its best performance on the M2sleep dataset with 60.69% accuracy using the contextual modality and MLP classifier, and 64.41% accuracy on the BiheartS dataset using the all-sensor-features modality and PAC classifier. These results show that personalized models outperform population models in most cases. However in some cases, both models have comparable performance—for instance, using the all-sensors-features modality and the MLP classifier for the BiheartS dataset and using the contextual modality and the MLP classifier for the M2sleep dataset. This implies that when the target test user belongs to the same dataset used for model training, both the personalized model and the population model can achieve comparable performance. Accordingly, the impact of the covariate shift is not evident in the single-dataset scenario.

Sleep Quality Recognition in the Multiple-Dataset Evaluation Scenario. We find that the personalized models consistently outperform the population model across all modalities in both datasets, achieving the highest accuracies of 64.12% and 65.66% using the contextual modality with the PAC classifier for the M2sleep and the BiheartS datasets, respectively. Moreover, they outperform the RG and BRG baselines in most cases. On the other hand, for the population model, there is a drop in the performance of the population model in this scenario compared to the single-dataset scenario. For instance, the population model achieves a maximum accuracy of 57.40% with the M2sleep dataset using the MLP classifier and the all-sensor-features modality. Also for the same modality and the BiheartS dataset, it reaches a maximum performance of 49.63% using the PAC classifier.

These results imply that the leading performance of the personalized model is consistent across the different scenarios. This indicates that personalization consistently improves model's performance regardless of modality. However, there is a drop in the performance of the population model when using one dataset for model training and evaluating the model using data from an unseen new user that belongs to a different dataset. Accordingly, the impact of the covariate shift is evident in this multiple-dataset scenario with the population model. These results encourage us to explore and quantify the difference in the performance of the population models and the personalized models across the different scenarios.

Comparison Between Models' Performance in the Evaluation Scenarios

To assess the impact of the covariate shift on the performance of the population model and the personalized model, we first compare the performance of each individual model across the evaluation scenarios—i.e., with a single-dataset or multiple-dataset—before we then assess the performance gap between the personalized model and the population model in each scenario using Cliff's δ effect size. In both experiments, we use Cliff's δ effect size to quantify the difference in performance.

In Figure 5, we show Cliff's δ effect size between the performance of the models in the different evaluation scenarios. From Figure 5, we find that the effect size for the performance of the personalized model in the two evaluation scenarios ranges between negligible and small. This means that the performance of the personalized model is consistent across the two evaluation scenarios. On the other hand, the effect size for the performance of the population model across the two evaluation scenarios ranges between medium and high for most of the modalities. These results imply that the performance of the population model changes based on the evaluation scenario.

Moreover, we compare the performance gap between the personalized model and the population model in the two evaluation scenarios. In Figure 6, we report the effect size between the performance of the two models in the single-dataset and the multiple-dataset scenario. From Figure 6a,b, we observe that the effect size between the performance of the personalized model and the population model in the multiple-dataset scenario is always larger than the effect size for the single-dataset scenario, regardless of the modality and the classifier. This implies that the difference in the performance of the population model and the personalized model is bigger in the multiple-dataset scenario than in the single-dataset scenario.

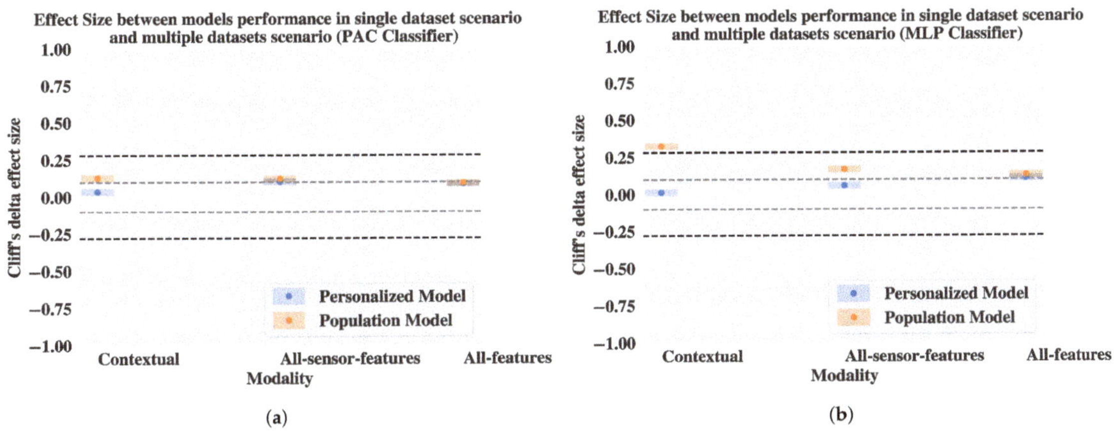

Figure 5. Cliff's δ effect size between the performance of the models in two evaluation scenarios. The dashed grey and back horizontal lines represent the thresholds for the small and the medium effect size as indicated by Vargha and Delaney [68]. (**a**) Passive Aggressive (PAC) classifier. (**b**) Multilayer Perceptron (MLP) classifier.

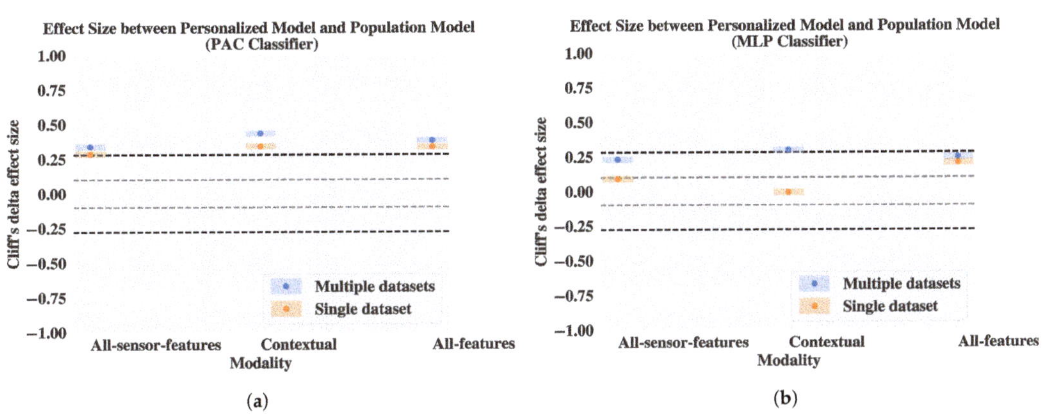

Figure 6. Cliff's δ effect size between the performance of the personalized model and the population model in the single-dataset and the multiple-dataset scenarios. The dashed grey and back horizontal lines represent the thresholds for the small and the medium effect size as indicated by Vargha and Delaney [68]. (**a**) Passive Aggressive (PAC) classifier. (**b**) Multilayer Perceptron (MLP) classifier.

In other words, these results indicate that the performance gap between the personalized model and the population model becomes larger when different datasets are used for training and evaluation. This confirms that a mismatch in the features' distribution between the training and evaluation sets, known as covariate shift, negatively impacts the performance of the population model on unseen users. We provide more results for the performance of the population and the personalized models across the three scenarios in the Supplementary materials (D.2).

7. Unsupervised Domain Adaptation with Population Model

In the preceding Section 6, we show that covariate shift hinders the performance of population models when using data from users in different datasets for model training and evaluation. On the other hand, we also show that personalized models are less affected by the presence of covariate shift than population models. However, researchers have shown that personalized models are difficult to implement in the real-world, since they heavily rely on user feedback to continuously adapt [13].

In this section, we propose a novel approach which bridges the gap between population and personalized models. We call this approach the *cluster-based population model* (CBPM). We use a multi-step instance-based unsupervised DA method [25] to improve the performance of the population model in the presence of covariate shift. The proposed approach leverages an unsupervised clustering algorithm with a population model, hypothesizing that narrowing the feature space for a population model by including only data from similar sleep sessions in the training set of the model can improve its performance. Moreover, the CBPM model focuses on clustering similar sleep sessions together instead of similar users, acknowledging that an individual's behaviour can vary over time due to factors such as context, health conditions, emotional states, mental health [69], and *session-level* clustering rather than *user-level* clustering. For our novel approach, we perform experiments using the M2sleep and the BiheartS datasets in the *multiple-dataset scenario* only, since we aim to simulate real-world settings where the ML model is trained on a set of users from publicly available dataset(s) and then used by new unseen users after deployment. Figure 7 shows a demonstration of the proposed approach.

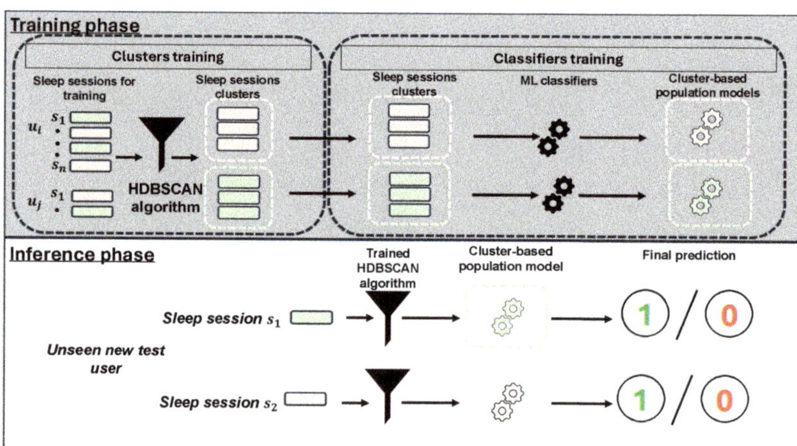

Figure 7. Illustration of the proposed cluster-based population model (CBPM) that integrates the Hierarchical Density-Based Spatial Clustering of Applications with Noise (HDBSCAN) clustering algorithm with the population model. In the training phase, the feature vectors of the training dataset are clustered using the HDBSCAN algorithm, and the feature vectors labeled as noise are discarded. Then, a population model is trained for each cluster independently. During the inference phase, each feature vector from the target test user is assigned to one of the existing clusters. The population model of the corresponding cluster is used to predict the binary sleep quality.

7.1. Configuration of the Clusters

We employ the Hierarchical Density-Based Spatial Clustering of Applications with Noise (HDBSCAN) algorithm [70] in our cluster-based population model. An ablation study is conducted to tune two key hyperparameters: minimum cluster size and minimum samples. Using the Silhouette Coefficient [71] to evaluate clustering quality, we select the configuration yielding the highest coherence based on training set feature vectors in a cross-dataset setup. Hyperparameter tuning is performed separately for each modality. Feature vectors labeled as noise by HDBSCAN are excluded. Details on the selected hyperparameters are provided in Supplementary materials (E). To implement the HDBSCAN clustering algorithm, we use the `HDBSCAN` Python library (https://hdbscan.readthedocs.io/en/latest/api.html) (accessed on 17 June 2025).

7.2. Training of Cluster-Based Population Model

After the selection of the optimal hyperparameters for each modality, using the feature vectors from the training set only, we train a population model for each cluster individually. To this end, we use both the PAC and the MLP classifiers, as in Section 6.2. In case the feature vectors in the same cluster have only a single class label, we randomly select other feature vectors with a different class label from the other clusters.

7.3. Evaluation of Cluster-Based Population Model

We evaluate our CBPM model using the left-out dataset. During the evaluation phase, for each feature vector in the left-out dataset, we first assign the test feature vector using the HDBSCAN clustering algorithm to one of the existing clusters. Then, we retrieve the population model that corresponds to the cluster of the feature vector. We use this population to predict the label for the test feature vector, then we evaluate the prediction using the accuracy evaluation metric, as carried out in Section 6.3. In case the HDBSCAN algorithm clusters the test feature vector as noise, we assign this test feature vector to the nearest cluster by computing the distance between the test feature vector and the centroids of all the existing clusters.

7.4. Experimental Results for the Cluster-Based Population Model

In this set of experiments, we compare the performance of the CBPM model with respect to the performance of the population model and the personalized model in a multiple-dataset scenario.

In Figure 8, we present the performance of the three models using the PAC and the MLP classifiers. Using Figure 8a,b, we can compare the performance of the population model, the personalized model, and the CBPM model using the PAC classifier. The results show that the performance of the CBPM model outperforms the population model using the contextual modality in both datasets. In particular, the CBPM model achieves a maximum improvement of 9.03 and 4.39 percentage points using the contextual modality for the M2sleep and the BiheartS dataset, respectively. These improvements are statistically significant compared to the population model, based on the Wilcoxon statistical test ($\alpha = 0.05$).

Moreover, in Figure 8c,d we compare the performance of three models using the MLP classifier. The results show that the CBPM model outperforms the population model for most of the modalities in the two datasets. In particular, we observe maximum improvement of 13.45 and 9.67 percentage points using the contextual modality in the M2sleep and the BiheartS datasets, respectively. Moreover, the CBPM model outperforms the population model in the all-sensor-features and the all-features modalities in the BiheartS dataset, with an improvement of 4.22 and 7.39 percentage points, respectively. These improvements are statistically significant when using the Wilcoxon statistical test ($\alpha = 0.05$). The CBPM model also outperforms the personalized model when using the contextual modality and when testing on the BiheartS dataset. Finally, we observe that the CBPM model outperforms the two baselines in most of the modalities.

These results show that performance of the CBPM model is positioned between that of the personalized and population models.

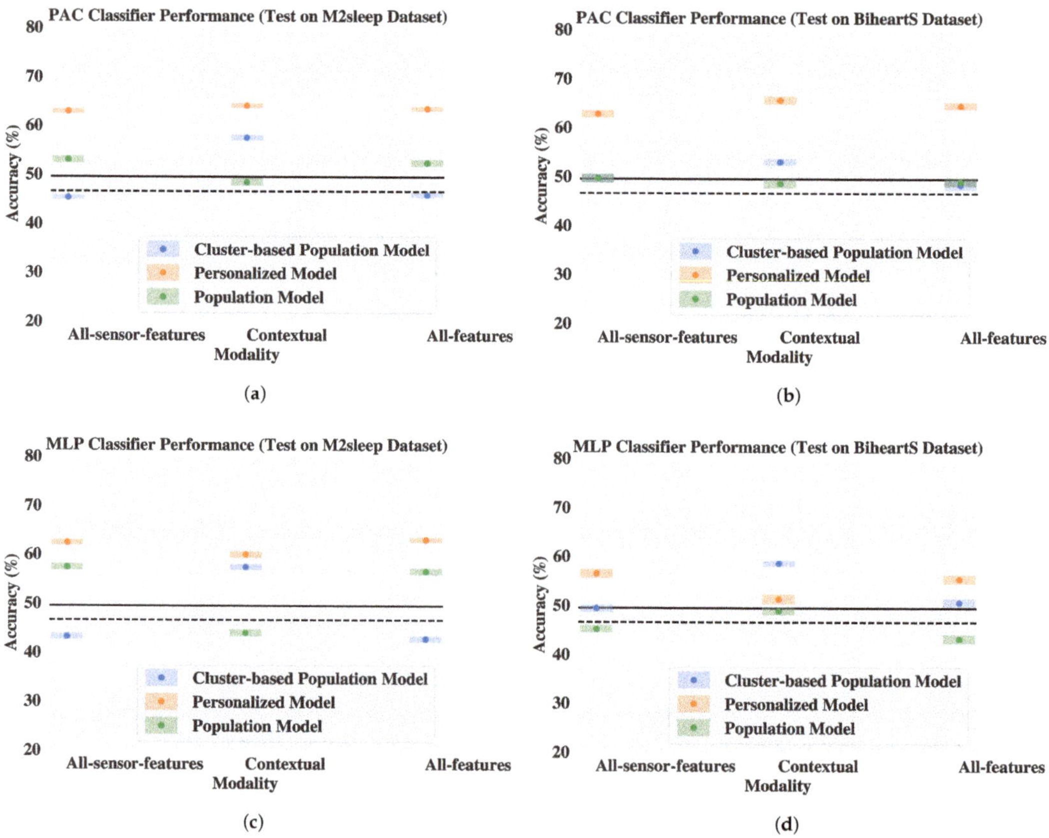

Figure 8. The performance of the cluster-based population model using Passive Aggressive Classifier (PAC) and Multilayer Perceptron (MLP) on the M2sleep and the BiheartS datasets. The dashed and the solid horizontal lines represent the random guess and the biased random guess baselines, respectively. (**a**) PAC classifier trained on the BiheartS dataset and tested on the M2sleep dataset. (**b**) PAC classifier trained on the M2sleep dataset and tested on the BiheartS dataset. (**c**) MLP classifier trained on the BiheartS dataset and tested on the M2sleep dataset. (**d**) MLP classifier trained on the M2sleep dataset and tested on the BiheartS dataset.

8. Discussion

In this section, we present the discussion and the implications of the obtained experimental results.

The first set of experiments investigate the existence of the distribution shift in personal informatics datasets collected for the same task. In this experiment, we rely on statistical analysis, similar to Meegahapola et al. [19] and Xu et al. [3], to investigate the covariate shift between the BiheartS and the M2sleep datasets. The results of the statistical analysis indicate that the majority of the features have statistically different distributions in the two datasets. Moreover, the correlation analysis shows that the associations between the features and the sleep quality labels are different in the two datasets. These results suggest the existence of covariate shift between the two datasets.

Moreover, we implement the *"Name-The-Dataset"* task, since it is suggested by several studies in the computer vision literature as an indicator of the existence of distribution shift across different domains [17,18]. The goal of the task is to analyze the capability of a classifier to distinguish between feature vectors from the two datasets. This task was also recently adopted in a human behavioural analysis study [3]. Our results show that an ML model differentiates between the feature vectors from the M2sleep dataset and the BiheartS dataset with a balanced accuracy up to 90.72%. This implies that the features in each dataset have distinguishable distributions and that each dataset has a unique identifiable signature. The results of the first set of experiments reveal the presence of a distribution shift, specifically, a covariate shift between the BiheartS and M2sleep datasets. These findings underscore that covariate shift can arise even between datasets that share similar characteristics (such as the use of wearable devices, self-reported labels, geographic region, and user demographics) and that are collected for the same task. In real-world applications, such shifts can significantly affect the performance of machine learning models trained on one dataset when deployed to new and unseen users. While the impact of covariate shift has been extensively studied in fields such as computer vision and natural language processing, it has received relatively little attention in the context of PI systems.

In the second set of experiments, we investigate the impact of the covariate shift on the performance of personalized models and population models, in a sleep quality recognition task. We extract features from the ST and the ACC modalities as in [7]. Moreover, we use features from the HR, the HRV, and the contextual modalities as in [28]. The HR and the HRV signals correlate with sleep stages and sleep quality as highlighted by several studies, e.g., [52,72,73]. We rely on self-reports to collect the ground truth for the sleep quality recognition task, similar to [7,28]. Finally, we formulate the sleep quality recognition task as a binary classification problem as in [7,28]. In addition, we use two datasets, BiHeartS and M2sleep, to estimate the impact of the covariate shift on the performance of personalized and population models in three distinct evaluation scenarios. Our experimental results indicate a decline in the performance of the population models (up to 18.54 percentage points) when trained and evaluated on data from different datasets, compared to when both sets originate from the same dataset. These findings demonstrate that the existence of covariate shift between the users in the different datasets hinders the performance of population models. In other words, training a machine learning model using data from a cohort of users with similar characteristics as the expected target user is insufficient for ML models in the context of PI systems.

On the other hand, for the personalized models, our experimental results indicate that personalized models achieve comparable performance regardless of whether the training and testing datasets originate from the same source. This suggests that personalized models effectively address both interpersonal and intra-personal variability by incorporating user-specific feedback to continuously adapt over time. These findings align with those of Gashi et al. [7], which underscore the importance of personalization in sleep quality recognition, given the subjective nature of self-reported ground truth and the inherent variability in input features across and within individuals. However, the personalized model requires feedback from unseen target users about the model's predictions to adjust the model incrementally. This continuous user feedback is cognitively demanding and requires high engagement from the user, as highlighted by Meegahapola et al. [13].

Accordingly, in this paper, we propose a novel approach that adopts unsupervised DA to improve the performance of the population model for unseen users, without the need for the user feedback. This novel approach relies on the multi-step instance-based domain adaptation method. It integrates the HDBSCAN clustering algorithm with the population model. This cluster-based population model does not require labelled data from the target user. Instead, it relies on creating clusters out of the available training data and trains a population model for each cluster individually. Then, during the inference phase, it assigns the sleep sessions of the target user to the nearest cluster and uses the population model of the corresponding cluster to infer the sleep quality. The key novelty of the cluster-based population model (CBPM) lies in its clustering approach; instead of grouping the similar users together in the same cluster, the CBPM model uses feature vectors extracted from sleep sessions to group the same sessions together. This strategy addresses intra-personal variability by allowing sessions from the same user to be assigned to different clusters, acknowledging that an individual's behaviour can vary over time due to factors such as context, health conditions, emotional states, and mental health [69]. These factors can influence a user's preferences, perceptions, and physiological responses. By focusing on session-level clustering rather than user-level grouping, the CBPM model adapts more effectively to such temporal changes. Additionally, by training population models only on data from similar sleep sessions, the CBPM narrows the feature space, improving model performance. These characteristics contribute to improved performance in sleep quality recognition tasks, which inherently require personalization due to the subjectivity of self-reported ground truth and the interpersonal variability among users, as emphasized by Gashi et al. [7].

Our results show that this approach improves the performance of population models in the presence of covariate shift, with a maximum improvement of 13.45 and 9.67 percentage points using the contextual features for the M2sleep and the BiheartS datasets, respectively. Our findings highlight the potential of the unsupervised DA approach to accommodate for the interpersonal variability among users and the intra-personal variability within users. This approach eliminates the need for labelled data from the target user, instead leveraging unlabelled data. Accordingly, it reduces the burden of continuous feedback from the users that can impact the overall user experience.

9. Limitations and Future Work

A limitation of this study is that we focus our analysis on covariate shift. As part of our future research, we plan to investigate the existence of different types of domain shift, e.g., label shift and concept shift.

For the personalized model, we also assume that new unseen users will consistently provide feedback. The personalized model uses this feedback to adapt incrementally to the target user. In practice, however, users may not always offer consistent feedback, which could hinder the model's ability to personalize its predictions effectively. Future work can address this by using generative artificial intelligence for data augmentation, given its potential in the image domain [74].

As part of our future research, we aim to extend this work by exploring more objective tasks, such as activity recognition, to evaluate whether the need for personalization—as emphasized in the current study—remains important in other contexts. Additionally, we plan to collect a larger and more diverse dataset, as the current BiheartS dataset is limited to only 10 users and exhibits a significant gender imbalance. With a larger dataset, we will also be able to investigate the use of deep learning classifiers, which typically require substantial data to train effectively and may offer further improvements in modelling complex patterns in wearable sensor data.

10. Conclusions

PI systems help users track their daily habits. These systems include ML models trained using existing datasets which are then deployed for use by unseen arbitrary users, i.e., *population models*. However, distribution differences between the training data and the data of new arbitrary users lead to a decline in these models' performance.

In this paper, we use two datasets, the new BiHeartS and the existing M2sleep, collected from participants with comparable demographics and geographical areas and using similar wearable devices. Our experimental results confirm the existence of covariate shift between the two datasets that share similar characteristics.

Moreover, we explore the impact of covariate shift on the performance of population and personalized models. Our results highlight that personalized models effectively address covariate shift by continuously adapting to the target user, but they require ongoing user feedback, thus limiting practical applicability. On the other hand, population models are affected by covariate shift when trained and tested on different datasets, compared to when using the same dataset for training and testing.

To overcome the limitations of population models, which are affected by domain shift, and personalized models, which require ongoing user feedback, we propose the novel cluster-based population model (CBPM), which employs an unsupervised domain adaptation approach. Our results show that our CBPM approach improves population model performance by up to 13.45 percentage points in terms of accuracy when using different datasets for model training and testing, while also not requiring continuous user feedback. This method provides a practical and efficient solution, narrowing the performance gap between personalized and population models in the presence of domain shift.

Supplementary Materials: The following supporting information can be downloaded at: https://www.mdpi.com/article/10.3390/s25134012/s1.

Author Contributions: Conceptualization, N.A., P.B. and S.S.; Data curation , N.A., L.A. (Lidia Alecci), L.A. (Leonardo Alchieri) and S.S.; investigation, N.A., L.A. (Lidia Alecci), L.A. (Leonardo Alchieri), P.B. and S.S.; methodology, N.A., P.B. and S.S.; software, N.A.; Writing—original draft, N.A.; Writing—review and editing, N.A., L.A. (Lidia Alecci), L.A. (Leonardo Alchieri), P.B. and S.S.; Supervision, S.S.; project administration, S.S.; funding acquisition, S.S. All authors have read and agreed to the published version of the manuscript.

Funding: This work is partially funded by the Swiss National Science Foundation (SNSF) through the grant 205121_197242 for the project "PROSELF: Semi-automated Self-Tracking Systems to Improve Personal Productivity".

Institutional Review Board Statement: The study was conducted in accordance with the Declaration of Helsinki, and approved by the delegate for ethics of Università della Svizzera Italiana.

Informed Consent Statement: Informed consent was obtained from all subjects involved in the study. Also, the data collection study is approved by our Faculty's delegate for ethics.

Data Availability Statement: Both datasets used in this work are available for the research community upon signing a data sharing agreement request. In particular, for the BiheartS dataset, please contact the corresponding author. Also, for the M2sleep dataset, more details can be found in [7].

Acknowledgments: The authors express their sincere gratitude to Maike Debus for her invaluable contributions to the design of the data collection protocol. Generative AI technologies were utilized to assist the optimization or editing of selected parts of this work, in particular text paragraphs. The generated materials have always been post-processed by the authors.

Conflicts of Interest: The funders had no role in the design of the study; in the collection, analyses, or interpretation of data; in the writing of the manuscript; or in the decision to publish the results. The authors declare no conflict of interest.

References

1. Li, I.; Dey, A.; Forlizzi, J. A stage-based model of personal informatics systems. In Proceedings of the SIGCHI Conference on Human Factors in Computing Systems, Pittsburgh , PA , USA, 10–15 April 2010; pp. 557–566.
2. Raskovic, D.; Martin, T.; Jovanov, E. Medical monitoring applications for wearable computing. *Comput. J.* **2004**, *47*, 495–504. [CrossRef]
3. Xu, X.; Liu, X.; Zhang, H.; Wang, W.; Nepal, S.; Sefidgar, Y.; Seo, W.; Kuehn, K.S.; Huckins, J.F.; Morris, M.E.; et al. GLOBEM: Cross-dataset generalization of longitudinal human behavior modeling. *Proc. Acm Interact. Mob. Wearable Ubiquitous Technol.* **2023**, *6*, 1–34. [CrossRef]
4. Xu, X.; Chikersal, P.; Dutcher, J.M.; Sefidgar, Y.S.; Seo, W.; Tumminia, M.J.; Villalba, D.K.; Cohen, S.; Creswell, K.G.; Creswell, J.D.; et al. Leveraging collaborative-filtering for personalized behavior modeling: A case study of depression detection among college students. *Proc. ACM Interact. Mob. Wearable Ubiquitous Technol.* **2021**, *5*, 1–27. [CrossRef]
5. Adler, D.A.; Yang, Y.; Viranda, T.; Xu, X.; Mohr, D.C.; Van Meter, A.R.; Tartaglia, J.C.; Jacobson, N.C.; Wang, F.; Estrin, D.; et al. Beyond Detection: Towards Actionable Sensing Research in Clinical Mental Healthcare. *Proc. ACM Interact. Mob. Wearable Ubiquitous Technol.* **2024**, *8*, 1–33. [CrossRef]
6. Adler, D.A.; Choudhury, T. Ubiquitous Computing in Action: Infrastructure to Support Sensing and Mental Health Research in Practice. In Proceedings of the Companion of the 2024 on ACM International Joint Conference on Pervasive and Ubiquitous Computing, Melbourne, Australia, 5–9 October 2024; pp. 301–306.
7. Gashi, S.; Alecci, L.; Di Lascio, E.; Debus, M.E.; Gasparini, F.; Santini, S. The role of model personalization for sleep stage and sleep quality recognition using wearables. *IEEE Pervasive Comput.* **2022**, *21*, 69–77. [CrossRef]
8. Di Lascio, E.; Gashi, S.; Santini, S. Unobtrusive assessment of students' emotional engagement during lectures using electrodermal activity sensors. *Proc. ACM Interact. Mob. Wearable Ubiquitous Technol.* **2018**, *2*, 1–21. [CrossRef]
9. Gao, N.; Shao, W.; Rahaman, M.S.; Salim, F.D. n-gage: Predicting in-class emotional, behavioural and cognitive engagement in the wild. *Proc. ACM Interact. Mob. Wearable Ubiquitous Technol.* **2020**, *4*, 1–26. [CrossRef]
10. Di Lascio, E.; Gashi, S.; Hidalgo, J.S.; Nale, B.; Debus, M.E.; Santini, S. A multi-sensor approach to automatically recognize breaks and work activities of knowledge workers in academia. *Proc. ACM Interact. Mob. Wearable Ubiquitous Technol.* **2020**, *4*, 1–20. [CrossRef]
11. Di Lascio, E.; Gashi, S.; Debus, M.E.; Santini, S. Automatic recognition of flow during work activities using context and physiological signals. In Proceedings of the 2021 9th International Conference on Affective Computing and Intelligent Interaction (ACII), Nara, Japan, 28 September–1 October 2021; IEEE: Piscataway, NJ, USA, 2021; pp. 1–8.
12. Mathur, A.; Van den Broeck, M.; Vanderhulst, G.; Mashhadi, A.; Kawsar, F. Tiny habits in the giant enterprise: Understanding the dynamics of a quantified workplace. In Proceedings of the 2015 ACM International Joint Conference on Pervasive and Ubiquitous Computing, Osaka, Japan, 9–11 September 2015; pp. 577–588.
13. Meegahapola, L.; Hassoune, H.; Gatica-Perez, D. M3BAT: Unsupervised Domain Adaptation for Multimodal Mobile Sensing with Multi-Branch Adversarial Training. *Proc. ACM Interact. Mob. Wearable Ubiquitous Technol.* **2024**, *8*, 1–30. [CrossRef]
14. Alecci, L.; Abdalazim, N.; Alchieri, L.; Gashi, S.; Santini, S. On the mismatch between measured and perceived sleep quality. In Proceedings of the Adjunct 2022 ACM International Joint Conference on Pervasive and Ubiquitous Computing and the 2022 ACM International Symposium on Wearable Computers, Cambridge, UK, 11–15 September 2022; pp. 148–152.
15. Khademi, A.; El-Manzalawy, Y.; Master, L.; Buxton, O.M.; Honavar, V.G. Personalized sleep parameters estimation from actigraphy: A machine learning approach. *Nat. Sci. Sleep* **2019**, *11*, 387–399. [CrossRef]
16. Van, N.T.P.; Son, D.M.; Zettsu, K. AE-Sleep: An Adaptive Enhancement Sleep Quality System Utilizing Data Mining and Adaptive Model. In *Sensing Technology: Proceedings of ICST'15*; Springer: Berlin/Heidelberg, Germany, 2023; pp. 31–47.
17. Tommasi, T.; Patricia, N.; Caputo, B.; Tuytelaars, T. A deeper look at dataset bias. *Domain Adapt. Comput. Vis. Appl.* **2017** , 37–55.
18. Torralba, A.; Efros, A.A. Unbiased look at dataset bias. In Proceedings of the IEEE Conference on Computer Vision and Pattern Recognition CVPR 2011, Colorado Springs, CO, USA, 20–25 June 2011; IEEE: Piscataway, NJ, USA, 2011; pp. 1521–1528.
19. Meegahapola, L.; Droz, W.; Kun, P.; De Götzen, A.; Nutakki, C.; Diwakar, S.; Correa, S.R.; Song, D.; Xu, H.; Bidoglia, M.; et al. Generalization and personalization of mobile sensing-based mood inference models: An analysis of college students in eight countries. *Proc. ACM Interact. Mob. Wearable Ubiquitous Technol.* **2023**, *6*, 1–32. [CrossRef]

20. Jaipuria, N.; Zhang, X.; Bhasin, R.; Arafa, M.; Chakravarty, P.; Shrivastava, S.; Manglani, S.; Murali, V.N. Deflating dataset bias using synthetic data augmentation. In Proceedings of the IEEE/CVF Conference on Computer Vision and Pattern Recognition Workshops, Seattle, WA, USA, 14–19 June 2020; pp. 772–773.
21. Clark, C.; Yatskar, M.; Zettlemoyer, L. Don't take the easy way out: Ensemble based methods for avoiding known dataset biases. *arXiv* **2019**, arXiv:1909.03683.
22. He, H.; Zha, S.; Wang, H. Unlearn dataset bias in natural language inference by fitting the residual. *arXiv* **2019**, arXiv:1908.10763.
23. Ganin, Y.; Lempitsky, V. Unsupervised domain adaptation by backpropagation. In Proceedings of the International Conference on Machine Learning, PMLR, Lille, France, 6–11 July 2015; pp. 1180–1189.
24. Farahani, A.; Voghoei, S.; Rasheed, K.; Arabnia, H.R. A brief review of domain adaptation. In *Advances in Data Science and Information Engineering: Proceedings from ICDATA 2020 and IKE 2020*; Springer: Berlin/Heidelberg, Germany, 2021; pp. 877–894.
25. Wang, M.; Deng, W. Deep visual domain adaptation: A survey. *Neurocomputing* **2018**, *312*, 135–153. [CrossRef]
26. Patel, V.M.; Gopalan, R.; Li, R.; Chellappa, R. Visual domain adaptation: A survey of recent advances. *IEEE Signal Process. Mag.* **2015**, *32*, 53–69. [CrossRef]
27. Wang, J.; Lan, C.; Liu, C.; Ouyang, Y.; Qin, T.; Lu, W.; Chen, Y.; Zeng, W.; Philip, S.Y. Generalizing to unseen domains: A survey on domain generalization. *IEEE Trans. Knowl. Data Eng.* **2022**, *35*, 8052–8072. [CrossRef]
28. Moebus, M.; Holz, C. Personalized interpretable prediction of perceived sleep quality: Models with meaningful cardiovascular and behavioral features. *PLoS ONE* **2024**, *19*, e0305258. [CrossRef]
29. Gan, W.; Dao, M.S.; Zettsu, K. Monitoring and Improving Personalized Sleep Quality from Long-Term Lifelogs. In Proceedings of the 2022 IEEE International Conference on Big Data (Big Data), Osaka, Japan, 17–20 December 2021; IEEE: Piscataway, NJ, USA, 2022; pp. 4356–4364.
30. Grandner, M.A.; Bromberg, Z.; Hadley, A.; Morrell, Z.; Graf, A.; Hutchison, S.; Freckleton, D. Performance of a multisensor smart ring to evaluate sleep: In-lab and home-based evaluation of generalized and personalized algorithms. *Sleep* **2023**, *46*, zsac152. In Proceedings of the 2015 IEEE 12th International Conference on Wearable and Implantable Body Sensor Networks (BSN), Cambridge, MA, USA, 9–12 June 2015; IEEE: Piscataway, NJ, USA, 2015; pp. 1–6. [CrossRef]
31. Abdalazim, N.; Alchieri, L.; Alecci, L.; Santini, S.; Gashi, S. Heart Rate During Sleep Measured Using Finger-, Wrist- and Chest-Worn Devices: A Comparison Study. In Proceedings of the 2022 Pervasive Computing Technologies for Healthcare, Thessaloniki, Greece, 12–14 December 2022.
32. Sano, A.; Phillips, A.J.; Amy, Z.Y.; McHill, A.W.; Taylor, S.; Jaques, N.; Czeisler, C.A.; Klerman, E.B.; Picard, R.W. Recognizing Academic Performance, Sleep Quality, Stress Level, and Mental Health Using Personality Traits, Wearable Sensors and Mobile Phones. In Proceedings of the IEEE 12th International Conference on Wearable and Implantable Body Sensor Networks (BSN 2015), Cambridge, MA, USA, 9–12 June 2015; IEEE: Piscataway, NJ, USA, 2015.
33. Stone, J.D.; Rentz, L.E.; Forsey, J.; Ramadan, J.; Markwald, R.R.; Finomore, V.S.; Galster, S.M.; Rezai, A.; Hagen, J.A. Evaluations of commercial sleep technologies for objective monitoring during routine sleeping conditions. *Nat. Sci. Sleep* **2020**, *12*, 821–842. [CrossRef]
34. Alchieri, L.; Abdalazim, N.; Alecci, L.; Gashi, S.; Gjoreski, M.; Santini, S. Lateralization Effects in Electrodermal Activity Data Collected Using Wearable Devices. *Proc. ACM Interact. Mob. Wearable Ubiquitous Technol.* **2024**, *8*, 1–30. [CrossRef]
35. Alchieri, L.; Abdalazim, N.; Alecci, L.; Gashi, S.; Di Lascio, E.; Santini, S. On the Impact of Lateralization in Physiological Signals from Wearable Sensors. In Proceedings of the Adjunct 2022 ACM International Joint Conference on Pervasive and Ubiquitous Computing and the 2022 ACM International Symposium on Wearable Computers, Cambridge, UK, 11–15 September 2022; pp. 472–477.
36. Siirtola, P.; Koskimäki, H.; Mönttinen, H.; Röning, J. Using sleep time data from wearable sensors for early detection of migraine attacks. *Sensors* **2018**, *18*, 1374. [CrossRef] [PubMed]
37. Chee, N.I.; Ghorbani, S.; Golkashani, H.A.; Leong, R.L.; Ong, J.L.; Chee, M.W. Multi-night validation of a sleep tracking ring in adolescents compared with a research actigraph and polysomnography. *Nat. Sci. Sleep* **2021**, *13*, 177–190. [CrossRef] [PubMed]
38. Ghorbani, S.; Golkashani, H.A.; Chee, N.I.; Teo, T.B.; Dicom, A.R.; Yilmaz, G.; Leong, R.L.F.; Ong, J.L.; Chee, M.W.L. Multi-Night at-Home Evaluation of Improved Sleep Detection and Classification with a Memory-Enhanced Consumer Sleep Tracker. *Nat. Sci. Sleep* **2022**, *14*, 645–660. [CrossRef]
39. Altini, M.; Kinnunen, H. The promise of sleep: A multi-sensor approach for accurate sleep stage detection using the oura ring. *Sensors* **2021**, *21*, 4302. [CrossRef]
40. Cakmak, A.S.; Poian, G.D.; Willats, A.; Haffar, A.; Abdulbaki, R.; Ko, Y.A.; Shah, A.J.; Vaccarino, V.; Bliwise, D.L.; Rozell, C.; et al. An unbiased, efficient sleep–wake detection algorithm for a population with sleep disorders: Change point decoder. *Sleep* **2020**, *43*, zsaa011. [CrossRef]

41. Assaf, M.; Rizzotti-Kaddouri, A.; Punceva, M. Sleep detection using physiological signals from a wearable device. In Proceedings of the EAI HealthyIoT 2018-5th EAI International Conference on IoT Technologies for HealthCare, Guimaraes, Portugal, 21–23 November 2018; Springer: Berlin/Heidelberg, Germany, 2018.
42. Buysse, D.J.; Reynolds III, C.F.; Monk, T.H.; Berman, S.R.; Kupfer, D.J. The Pittsburgh Sleep Quality Index: A new instrument for psychiatric practice and research. *Psychiatry Res.* **1989**, *28*, 193–213. [CrossRef]
43. Horne, J.A.; Östberg, O. A self-assessment questionnaire to determine morningness-eveningness in human circadian rhythms. *Int. J. Chronobiol.* **1976**, *4*, 97–110.
44. John, O.P.; Donahue, E.M.; Kentle, R.L. Big five inventory. *J. Personal. Soc. Psychol.* **1991**.
45. Sadeghi, R.; Banerjee, T.; Hughes, J.C.; Lawhorne, L.W. Sleep quality prediction in caregivers using physiological signals. *Comput. Biol. Med.* **2019**, *110*, 276–288. [CrossRef]
46. Cohen, S.; Kamarck, T.; Mermelstein, R. A global measure of perceived stress. *J. Health Soc. Behav.* **1983**, 385–396. [CrossRef]
47. Endicott, J.; Nee, J.; Harrison, W.; Blumenthal, R. Quality of Life Enjoyment and Satisfaction Questionnaire: A new measure. *Psychopharmacol. Bull.* **1993**, *29*, 321–326. [PubMed]
48. Craig, C.L.; Marshall, A.L.; Sjöström, M.; Bauman, A.E.; Booth, M.L.; Ainsworth, B.E.; Pratt, M.; Ekelund, U.; Yngve, A.; Sallis, J.F.; et al. International physical activity questionnaire: 12-country reliability and validity. *Med. Sci. Sport. Exerc.* **2003**, *35*, 1381–1395. [CrossRef]
49. Moebus, M.; Wolfensberger, J.; Holz, C. Predicting sleep quality via unsupervised learning of cardiac activity. In Proceedings of the 2024 46th Annual International Conference of the IEEE Engineering in Medicine and Biology Society (EMBC), Orlando, FL, USA, 15–19 July 2024; IEEE: Piscataway, NJ, USA, 2024; pp. 1–4.
50. Hinatsu, S.; Suzuki, D.; Ishizuka, H.; Ikeda, S.; Oshiro, O. Photoplethysmographic subject identification by considering feature values derived from heartbeat and respiration. In Proceedings of the 2020 42nd Annual International Conference of the IEEE Engineering in Medicine & Biology Society (EMBC), Montreal, QC, Canada, 20–24 July 2020; IEEE: Piscataway, NJ, USA, 2020; pp. 902–905.
51. Sadeghi, R.; Banerjee, T.; Hughes, J. Predicting sleep quality in osteoporosis patients using electronic health records and heart rate variability. In Proceedings of the 2020 42nd Annual International Conference of the IEEE Engineering in Medicine & Biology Society (EMBC), Montreal, QC, Canada, 20–24 July 2020; IEEE: Piscataway, NJ, USA, 2020; pp. 5571–5574.
52. Chalmers, T.; Hickey, B.A.; Newton, P.; Lin, C.T.; Sibbritt, D.; McLachlan, C.S.; Clifton-Bligh, R.; Morley, J.W.; Lal, S. Associations between Sleep Quality and Heart Rate Variability; Implications for a Biological Model of Stress Detection Using Wearable Technology. *Int. J. Environ. Res. Public Health* **2022**, *19*, 5770. [CrossRef]
53. Pedregosa, F.; Varoquaux, G.; Gramfort, A.; Michel, V.; Thirion, B.; Grisel, O.; Blondel, M.; Prettenhofer, P.; Weiss, R.; Dubourg, V.; et al. Scikit-learn: Machine Learning in Python. *J. Mach. Learn. Res.* **2011**, *12*, 2825–2830.
54. Föll, S.; Maritsch, M.; Spinola, F.; Mishra, V.; Barata, F.; Kowatsch, T.; Fleisch, E.; Wortmann, F. FLIRT: A feature generation toolkit for wearable data. *Comput. Methods Programs Biomed.* **2021**, *212*, 106461. [CrossRef]
55. Dodge, Y. *The Concise Encyclopedia of Statistics*; Springer Science & Business Media: Berlin/Heidelberg, Germany, 2008.
56. Fekri, M.N.; Patel, H.; Grolinger, K.; Sharma, V. Deep learning for load forecasting with smart meter data: Online Adaptive Recurrent Neural Network. *Appl. Energy* **2021**, *282*, 116177. [CrossRef]
57. Crammer, K.; Dekel, O.; Keshet, J.; Shalev-Shwartz, S.; Singer, Y.; Warmuth, M.K. Online passive-aggressive algorithms. *J. Mach. Learn. Res.* **2006**, *7*, 551–585.
58. Winger, T.; Ozdemir, C.; Narasimhan, S.L.; Srivastava, J. Time-Adaptive Machine Learning Models for Predicting the Severity of Heart Failure with Reduced Ejection Fraction. *Diagnostics* **2025**, *15*, 715. [CrossRef]
59. Winger, T.; Chellamuthu, V.; Guzenko, D.; Aloia, M.; Barr, S.; DeFranco, S.; Gorski, B.; Mushtaq, F.; Garcia-Molina, G. Fine tuned personalized machine learning models to detect insomnia risk based on data from a smart bed platform. *Front. Neurol.* **2024**, *15*, 1303978. [CrossRef]
60. Popescu, M.C.; Balas, V.E.; Perescu-Popescu, L.; Mastorakis, N. Multilayer perceptron and neural networks. *WSEAS Trans. Circuits Syst.* **2009**, *8*, 579–588.
61. Hornik, K.; Stinchcombe, M.; White, H. Multilayer feedforward networks are universal approximators. *Neural Netw.* **1989**, *2*, 359–366. [CrossRef]
62. Sathyanarayana, A.; Joty, S.; Fernandez-Luque, L.; Ofli, F.; Srivastava, J.; Elmagarmid, A.; Arora, T.; Taheri, S. Sleep quality prediction from wearable data using deep learning. *JMIR MHealth UHealth* **2016**, *4*, e6562.
63. Yeckle, J.; Manian, V. Automated Sleep Stage Classification in Home Environments: An Evaluation of Seven Deep Neural Network Architectures. *Sensors* **2023**, *23*, 8942. [CrossRef]
64. Arora, A.; Chakraborty, P.; Bhatia, M. Analysis of data from wearable sensors for sleep quality estimation and prediction using deep learning. *Arab. J. Sci. Eng.* **2020**, *45*, 10793–10812. [CrossRef]

65. Gama, J.; Sebastiao, R.; Rodrigues, P.P. On evaluating stream learning algorithms. *Mach. Learn.* **2013**, *90*, 317–346. [CrossRef]
66. Hidalgo, J.I.G.; Maciel, B.I.; Barros, R.S. Experimenting with prequential variations for data stream learning evaluation. *Comput. Intell.* **2019**, *35*, 670–692. [CrossRef]
67. Cohen, S. Perceived stress in a probability sample of the United States. *Soc. Psychol. Heal./Sage* **1988**.
68. Vargha, A.; Delaney, H.D. A critique and improvement of the CL common language effect size statistics of McGraw and Wong. *J. Educ. Behav. Stat.* **2000**, *25*, 101–132.
69. Van Geert, P.; Van Dijk, M. Focus on variability: New tools to study intra-individual variability in developmental data. *Infant Behav. Dev.* **2002**, *25*, 340–374. [CrossRef]
70. McInnes, L.; Healy, J.; Astels, S. hdbscan: Hierarchical density based clustering. *J. Open Source Softw.* **2017**, *2*, 205. [CrossRef]
71. Rousseeuw, P.J. Silhouettes: A graphical aid to the interpretation and validation of cluster analysis. *J. Comput. Appl. Math.* **1987**, *20*, 53–65. [CrossRef]
72. Sajjadieh, A.; Shahsavari, A.; Safaei, A.; Penzel, T.; Schoebel, C.; Fietze, I.; Mozafarian, N.; Amra, B.; Kelishadi, R. The association of sleep duration and quality with heart rate variability and blood pressure. *Tanaffos* **2020**, *19*, 135. [PubMed]
73. Fantozzi, M.T.; Artoni, F.; Faraguna, U. Heart rate variability at bedtime predicts subsequent sleep features. In Proceedings of the 2019 41st Annual International Conference of the IEEE Engineering in Medicine and Biology Society (EMBC), Berlin, Germany, 23–27 July 2019; IEEE: Piscataway, NJ, USA, 2019; pp. 6784–6788.
74. Ktena, I.; Wiles, O.; Albuquerque, I.; Rebuffi, S.A.; Tanno, R.; Roy, A.G.; Azizi, S.; Belgrave, D.; Kohli, P.; Cemgil, T.; et al. Generative models improve fairness of medical classifiers under distribution shifts. *Nat. Med.* **2024**, *30*, 1166–1173. [CrossRef] [PubMed]

Disclaimer/Publisher's Note: The statements, opinions and data contained in all publications are solely those of the individual author(s) and contributor(s) and not of MDPI and/or the editor(s). MDPI and/or the editor(s) disclaim responsibility for any injury to people or property resulting from any ideas, methods, instructions or products referred to in the content.

sensors

MDPI

Article

Experimental Analysis on the Effect of Contact Pressure and Activity Level as Influencing Factors in PPG Sensor Performance

Francesco Scardulla [1], Gloria Cosoli [2,*], Cosmina Gnoffo [1], Luca Antognoli [3], Francesco Bongiorno [1], Gianluca Diana [1], Lorenzo Scalise [3], Leonardo D'Acquisto [1] and Marco Arnesano [2]

1 Department of Engineering, University of Palermo, Viale delle Scienze, Ed. 8, 90128 Palermo, Italy; francesco.scardulla@unipa.it (F.S.); cosminagnoffo0@gmail.com (C.G.); francesco.bongiorno01@unipa.it (F.B.); gianluca.diana@unipa.it (G.D.); leonardo.dacquisto@unipa.it (L.D.)

2 Department of Theorethical and Applied Sciences, eCampus University, v. Isimbardi 10, 22060 Novedrate, Italy; marco.arnesano@uniecampus.it

3 Department of Industrial Engineering and Mathematical Sciences, Marche Polytechnic University, v. Brecce Bianche 12, 60131 Ancona, Italy; l.antognoli@staff.univpm.it (L.A.); l.scalise@staff.univpm.it (L.S.)

* Correspondence: gloria.cosoli@uniecampus.it

Abstract: Photoplethysmographic (PPG) sensors are small and cheap wearable sensors which open the possibility of monitoring physiological parameters such as heart rate during normal daily routines, ultimately providing valuable information on health status. Despite their potential and distribution within wearable devices, their accuracy is affected by several influencing parameters, such as contact pressure and physical activity. In this study, the effect of contact pressure (i.e., at 20, 60, and 75 mmHg) and intensity of physical activity (i.e., at 3, 6, and 8 km/h) were evaluated on a sample of 25 subjects using both a reference device (i.e., an electrocardiography-based device) and a PPG sensor applied to the skin with controlled contact pressure values. Results showed differing accuracy and precision when measuring the heart rate at different pressure levels, achieving the best performance at a contact pressure of 60 mmHg, with a mean absolute percentage error of between 3.36% and 6.83% depending on the physical activity levels, and a Pearson's correlation coefficient of between 0.81 and 0.95. Plus, considering the individual optimal contact pressure, measurement uncertainty significantly decreases at any contact pressure, for instance, decreasing from 15 bpm (at 60 mmHg) to 8 bpm when running at a speed of 6 km/h (coverage factor k = 2). These results may constitute useful information for both users and manufacturers to improve the metrological performance of PPG sensors and expand their use in a clinical context.

Keywords: photoplethysmography; PPG; PPG sensor; contact pressure; contact force; activity level; influencing parameters; wearables; heart rate; measurement accuracy

1. Introduction

During the last decades, the persistent improvement in clinical and pharmacological treatments has resulted in a global improvement in the quality of life and life expectancy. Indeed, life expectancy increased by more than six years between 2000 and 2019, as reported by the World Health Organization (WHO) [1]. While on the one hand this represents a great achievement, on the other hand it has led to an increasing number of elderly and frail people requiring healthcare, and people saved from serious acute pathological events but not from the consequent chronic degenerative pathologies [2]. Thus, healthcare providers nowadays operate under unsustainable conditions, which leads to decreased quality and sustainability in the healthcare assistance offered.

Chronic diseases and cognitive impairments pose a crucial burden that requires a change in the healthcare paradigm. Indeed, this is expected to result in new low-cost, high-efficacy interventions and technologies that facilitate new preventive treatments using monitoring systems which prevent rather than cure disease. It is well known that a large proportion of the population could potentially avoid hospitalization or the worsening of their physiological conditions if they were continuously monitored with wearable sensors, not only within a clinical setting but also during normal daily life [3]. Indeed, society wants to age well, pursuing a sustainable lifestyle which enables people to live independently without interference [4], and which also guarantees a personalized comfort level within indoor environments by seeking solutions to support buildings energy efficiency [5]. Furthermore, innovations in both sensing and digital technologies have reached unprecedented levels (partly thanks to artificial intelligence—AI), which can certainly contribute to non-traditional medical approaches which support personalized healthcare that relies (where possible) on wearable, digital, and compact tools that do not require specialized personnel [6].

In this context, wearable multidomain sensors (e.g., smartwatches and smartbands) can undoubtedly play a pivotal role as they are already adopted and used daily by people who wish to self-monitor their physical activities and their health status [7,8], assess their baseline parameters (e.g., heart rate—HR, respiratory rate, energy expenditure, sleep quality, SpO_2), and detect any alteration in their normal values. Wearable sensors include a wide variety of monitoring devices which differ in terms of the adopted technology (e.g., optical, piezoelectric, and piezoresistive), the body location (e.g., ears, chest, wrist, and forehead), and the detected parameters. Among the wearable devices, the most widely used technology for the detection of cardiovascular parameters is photoplethysmography (PPG), which is based on an optical sensor and often known simply as a PPG sensor. Figure 1 shows the growing trend of interest in PPG sensors during the last decades within the scientific community.

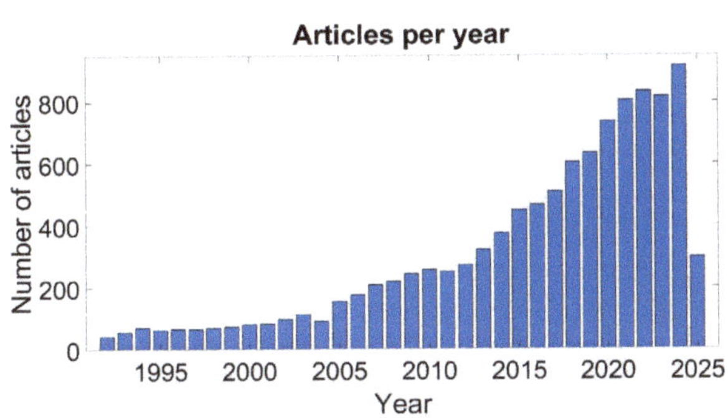

Figure 1. Report on the number of PPG-related articles since 1990. Source: PubMed, April 2025; search query: ((PPG) OR (photoplethysmography) OR (photoplethysmographic)).

Their implementation is supported by several aspects: small size, low cost, constructive simplicity, and the potential variety of cardiovascular parameters that can be detected [9,10]. These factors have driven the integration of this technology within various wearable devices of mass consumption, such as smartwatches, making this technology available to a significant portion of the population in developed countries [11].

The principal components of a PPG sensor are (i) a light source (LED), which is usually between 525 and 780 nm, and (ii) a photodetector (PD). These components can be arranged in two different configurations: transmission mode, in which the measuring site is in between the LED and the PD, and reflective mode, in which the LED and the PD are on the same plane with the measuring site underneath. This latter configuration, which is capable of providing a good signal [12], is suitable for integration into wearable devices such as smartwatches.

The PPG sensor can detect blood volume variations: every time the heart beats it pumps blood within the vessels, which in turn increases their volume to accommodate the larger amount of blood. This variation in blood volume is detected by the PPG sensor as a variation on the electrical output of the sensor, which is proportional to the light absorbed by blood. The PPG signal constitutes a direct current (DC) component, which corresponds to all the tissues and does not vary rapidly over time, and an alternating current (AC) component, which corresponds mainly to the blood volume variation that occurs at every heartbeat. Through the analysis of the PPG signal, it is possible to derive multiple parameters of the cardiorespiratory system, comprising HR [9], blood oxygenation [13,14], blood pressure (BP) [15], respiratory rate [16], and arterial stiffness [17], among others. Thus, the PPG sensor has huge potential for monitoring a subject's health status such as via patient monitoring [10] and daily activities monitoring [18], in the training of athletes [19], and for ensuring workers' safety [20].

Despite these strengths and its transversal applicability, the amplitude of the AC component represents just a fraction of the whole PPG signal, exposing it to different influencing parameters and limiting its accuracy and clinical usefulness [9], ultimately making it difficult to acquire a reliable PPG signal during daily routine activities and physical exercise. Among the different parameters of influence [9] the most significant are as follows: the contact pressure (CP) [21–23], which can be defined as the pressure between the skin and the PPG sensor, the wavelength of the light emitter [24], the skin tone [25], the LED-PD distance and configuration [26], motion artifacts [27], ambient light [28], and temperature [29]. Among these, the CP is reported to have a significant influence on the PPG amplitude and on minimizing the other influencing parameters such as the ambient light and the relative motion between the sensor and the skin during physical activities [9]. Specifically, a reduced CP may cause relative motion between the sensor and the skin which results in noise, and it allows ambient light to further degrade the signal quality. Conversely, an excessively high CP can alter the measurement site microcirculatory conditions because of the compression of the capillary bed at the point where the information is obtained. Thus, it appears even more evident that there is a need for sound testing and validation methodologies, delivering homogeneous and comparable results that robustly qualify a device from a metrological point of view [30].

From the perspective of the broader usability of PPG-based wearable sensors in clinical and (more in general) physiological monitoring settings, it is mandatory to improve their metrological performance; the development of sensor configurations and settings will also enhance the reproducibility of the measurement results generated in wildlife conditions. On another hand, it is fundamental to define rigorous test methodologies and measurement procedures to obtain reproducible results that can be compared with other studies in the literature [31]. This can undoubtedly contribute to a better understanding of the fundamentals of PPG functioning and its influencing parameters [9] and pave the way to a wider implementation of this sensor in clinical settings as well as other application fields. In particular, the present study has been carried out using the framework described in the WEPOP (WEarable PlatfOrm for OptImised Personal comfort) project, aiming at defining a wearable system (also relying on PPG technology) for the assessment of personalized comfort in living environments. The sensors employed allow us to gather data exploitable for the definition of Personal Comfort Models (PCMs) that can predict the actual status of the occupants' and to improve the conditions from a human-centric view, whilst also enabling an improvement from an energy consumption point of view. It is beyond doubt that the quality of the data collected directly impacts the accuracy of such models and hence influences the final contribution that such a methodology can have to the overall enhancement of the quality of the global ecosystem, considering both the living environment and its dwellers.

Based on these considerations, the future of medicine and personal physiological monitoring in everyday life lies in the uniformity and reliability of the testing procedures and on the metrological quality of measurements. Therefore, it is essential to improve the quality of the PPG sensors through a better understanding of the principal influencing parameters, such as CP. Assessing the optimal contact pressure, which may differ depending on each individual [9,32], is still challenging but may have significant benefits in improving the overall accuracy and usability of the PPG sensors. Hence, the aim of this paper is to perform a metrological characterization of a PPG-based wearable prototype in relation to different CP values (i.e., 20, 60, and 75 mmHg) and diverse walking/running speeds (i.e., 3, 6, and 8 km/h). The tests were carried out on 25 healthy volunteer subjects; it is worth underlining that this is not a clinical study and does not aim to certify any device as a medical product. A cardiac belt (based on electrocardiography, ECG) was employed as a reference sensor (gold standard); it is worth underlining that this metrological characterization does not aim to certify the accuracy of the sensors for clinical purposes nor for marketing, hence a medical-grade gold standard has not been employed in the test campaign.

2. Material and Methods

The tests were performed in two different laboratories: one at Università degli Studi di Palermo (Palermo, Italy) and one at Università Politecnica delle Marche (Ancona, Italy). The tests were conducted according to the same test protocol and in compliance with the ethical guidelines of the two universities. Before starting the tests, the aims and methodologies of the study were clearly explained to the participants; they signed an informed consent form highlighting the test protocol and conditions and stating that no risks were present. The participation was voluntary, and no compensation was provided. HR measurements acquired through a PPG sensor with controlled contact pressure and an ECG-based reference system were performed to evaluate how different contact pressure levels affect the PPG signal during different physical activity intensity levels. The general methodology complies with an adopted standard which regulates the procedures and aspects that should be observed during performance tests conducted on PPG sensors. Data analysis reflects the most adopted approaches for this type of study; all the data were managed according to the General Data Protection Regulation (GDPR). In the following subsections, the experimental setup, the experimental protocol, and the data analysis are described in detail.

2.1. Experimental Setup and Equipment

A PPG sensor equipped with a single LED with a wavelength of 520 nm (DFRobot, Beijing, China) was integrated in a 3D-printed polylactide frame, designed to be worn on the wrist via a bracelet and to host a bottom load cell (FX1901, Meas.Spec., Fremont, CA, USA) stack on the PPG sensor itself, as shown in Figure 2. Such a configuration allows the load cell to provide the desired contact pressure between the PPG sensor and the skin.

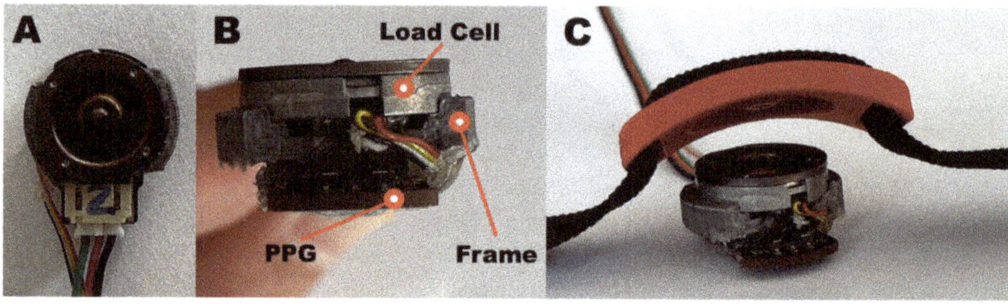

Figure 2. PPG-based device comprised of a PPG sensor and a compression load cell; (**A**) top view, (**B**) front view, and (**C**) integrated with a bracelet.

Specifically, the printed components are composed of a frame and a cell press (Figures 2B and 2C, respectively). The first component has two main functions: it is designed to act as an interface between the PPG sensor and the load cell and to keep the two surfaces parallel. The second component, which has two slots for the 2 cm nylon bracelet, is used to transmit the pressure exerted by the bracelet to the load cell, facilitated by the matching surface which is specular to that of the load cell. The combination of the two components allows us to normally transmit the force generated by the tightening of the strap directly on the load cell and on the underlying PPG sensor. A ratchet was used as a fastening system for the bracelet so that once the desired contact pressure is reached, the lock system prevents any loosening.

The calibration of the load cell, which directly provides the value of the pressure exerted by the sensor on the skin in mmHg (1 mmHg = 133,322 Pa), was conducted by stacking a series of calibrated masses of 110, 300, and 360 g on the whole load cell.

During the tests, to determine the higher contact pressure values without causing an uncomfortable feeling of constriction on the participants' wrists, the printed circuit board on the PPG sensor was milled into the lateral side with no electrical circuits (Figure 3).

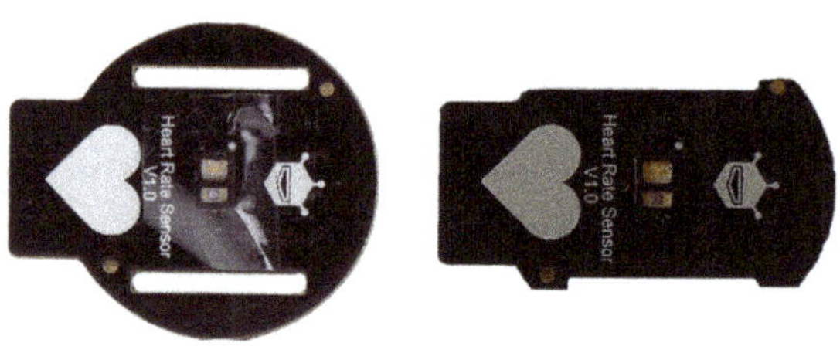

Figure 3. Original PPG sensor surface (on the left) and final (on the right) obtained after the milling process.

This allowed us to reduce the sensor area from approximately 460 mm^2 to an area of about 355 mm^2, thus a reduction of 25%.

The signal provided by the load cell was amplified through the integrated HX-711 (SparkFun Electronics, Niwot, CO, USA) and all the components were managed using an Arduino Uno microcontroller at a sampling frequency of 80 Hz.

The HR reference data were acquired via an ECG-based chest strap (H9, Polar Electro, Kempele, Finland) and a mobile app (Elite HRV Inc., Austin, TX, USA) which recorded a text file containing all the R-R intervals from which the HR was finally calculated and synchronized with the PPG-derived HR.

All the physical activities were performed on a treadmill (JK Fitness, Movifitness MF Top Slim, Padua, Italy) at different speeds (i.e., 3 km/h, 6 km/h, and 8 km/h; 1 km/h = 0.278 m/s) with the slope fixed at zero degrees. These values were chosen to reflect walking, jogging, and light running activities; additionally, these speed values are consistent with the ANSI/CTA-2065 recommendations [33].

Blood pressure was registered before every physical activity session through an automatic oscillometric device (Omron, M2, Kyoto, Japan) validated according to the Universal Protocol [34].

2.2. Experimental Protocol

The test protocol was designed in accordance with the standard ANSI/CTA-2065 [33], proposed and released in 2018 by the Consumer Technology Association; it aims to define the process necessary to test and validate the accuracy of a device for HR monitoring under different conditions, providing indications on different aspects including the test conditions and the protocol. It also suggests an ECG system as a reference device. Although chest straps are widely used and accepted in the literature [31,35], it is worth mentioning that studies for assessing the measurement accuracy of wrist-worn devices have also adopted 12-lead ECG systems as a reference [36].

Twenty-five subjects (10 males and 15 females; age: 23.9 ± 2.8 years, range 22–36; height: 171 ± 8 cm; weight: 67 ± 11 kg—values are reported as mean ± standard deviation) were enrolled in the study after signing informed and data consent forms. None of the enrolled subjects reported any known cardiovascular diseases nor reported taking any medications that can influence the main parameters of the cardiovascular system. No details on the general medical history or the subjects' anamnesis were requested from the participants; additionally, no hypertensive or hypotensive subjects were included in the study.

All the tests, which had an overall average duration of approximately 60 min each, were conducted in laboratories with room temperatures ranging from 22 °C to 24 °C.

Before starting the tests, the consensus and the anthropometric information for each subject were collected and managed in compliance with the General Data Protection Regulation in order to preserve the participants' privacy. The skin color of each participant was determined by the same single operator by comparing each participant's arm (inner part, where the PPG sensor would come into contact) with the Fitzpatrick chromatic scale displayed on a high-definition screen with a fixed brightness; three subjects presented a skin classification of type 1, nine subjects had a skin classification of type 2, 13 subjects had a skin classification of type 3.

At the very beginning of each test, a first blood pressure measurement (BP_1) was recorded. Then, the study coordinator asked each participant to wear the Polar H9 chest strap and then to sit and rest on a chair while placing the prototypal bracelet on the left wrist, ensuring it was tightened to the first contact pressure of 20 mmHg (CP_1). The wires from the bracelet were then fixed onto the arm using adhesive tape to prevent any unwanted displacement of the system during the physical activity execution. Thus, each subject began the planned activity as depicted in Figure 4.

Specifically, each participant was asked to perform three test sessions on a treadmill at different speeds (i.e., 3, 6, and 8 km/h for 90 s at each speed), with 60 s rest in between. After the third session at 8 km/h, each participant was asked to sit still at rest for 5 min while acquiring the PPG signal and then to have the last blood pressure measurement taken (BP_4).

At the end of each session, the coordinator, after verifying that the tightening pressure of the bracelet had not decreased during the physical exercise, tightened the bracelet to the next pressure level (i.e., 60 mmHg (CP_2) and 75 mmHg (CP_3)) and asked the participant to repeat the treadmill session (Figure 4).

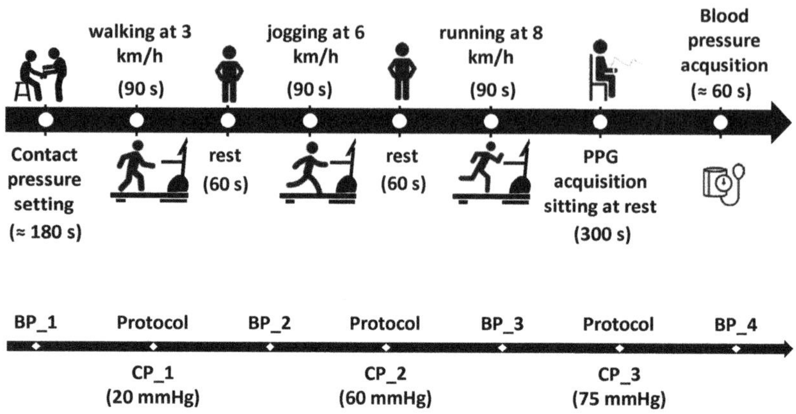

Figure 4. Description of the protocol (top), which is repeated for each contact pressure (i.e., 20, 60 and 75 mmHg) and a summary of the blood pressure acquisitions for each subject (bottom).

2.3. Data Analysis

The collected ECG and PPG data were post-processed in the MATLAB® (v. 2024b) environment. At first, the digitized PPG waveform was filtered with a combination of a fourth order Butterworth IIR band-pass filter (0.5–5.0 Hz) and a Hampel filter in order to remove outliers [37,38]. The Butterworth filter is extensively used in processing PPG signals as it guarantees the least signal distortion [39,40]. Nevertheless, it is worth mentioning that it may have introduced phase distortion to the pulse waveform [41].

HR values were computed from both the ECG (reference) and the PPG (test) signals; for the former, the RR values provided were converted into HR [bpm], whereas for the latter, the signal peaks were detected first, hence the PP intervals (i.e., the inter-beat intervals) were computed and the HR values were derived (using the same procedure adopted for the ECG sensor signals). Hence, the HR values as extracted from the ECG and PPG signals were synchronized by searching for the maximum correlation values (i.e., looking for the minimum difference in terms of beat per minute). Then, the test and reference HR values (the 50 central beats from each acquisition) were compared; the metrics employed to evaluate the metrological performance of the test PPG sensor were as follows: (i) the mean absolute percentage error (MAPE) and (ii) the confidence interval at 95% derived from the Bland–Altman plot. The strength of the linear correlation between the HR values derived from the test and reference devices was assessed using the Pearson's correlation coefficient (ρ). Also, the measurement differences (i.e., residuals) were analyzed; their mean and standard deviation values can be related to the measurement accuracy (or rather, trueness) and the precision, respectively.

3. Results

It is worth underlining that the CP set before starting physical activity changed during the test execution due to the arm movements and the contraction of the muscles. However, at the end of the test, the CP value returned to within ± 5 mmHg of the preset value. This means that the tightening system equipped with a ratchet was adequate and did not loosen during the activities execution.

In total, nine of the 225 acquisitions were corrupted or unavailable due to technical issues such as cable disconnection, hence 216 series were finally analyzed.

Table 1 reports the BP acquisitions during the tests, as described in Figure 4, together with the HR values provided by the oscillometric device. The average systolic and diastolic blood pressure measurements were equal to (113 ± 12) mmHg and (75 ± 7) mmHg, respectively (reported as mean ± standard deviation values), with a difference of (3 ± 7) mmHg for the systolic and (1 ± 8) mmHg for the diastolic blood pressure between the first and last measurement from each test. No significant correlation between the results of the tests and the blood pressure was found. Average HR during the BP acquisitions displayed a positive trend, increasing from 73 bpm to 87 bpm. The average HR provided by the ECG-based device was equal to (105 ± 5) bpm at 3 km/h, (120 ± 11) bpm at 6 km/h, and (136 ± 15) bpm at 8 km/h.

Table 1. Blood pressure values acquired during the protocol (Figure 4). HR values were provided by the oscillometric device together with the BP measurements and are reported as mean ± standard deviation.

	Systolic BP [mmHg]	Diastolic BP [mmHg]	HR [bpm]
BP_1	114 (12)	74 (7)	73 (12)
BP_2	115 (15)	75 (6)	82 (14)
BP_3	113 (12)	75 (7)	86 (17)
BP_4	111 (11)	74 (7)	87 (16)

An example of the synchronized test and reference signals is reported in Figure 5 for the test (PPG sensor) and the reference (ECG-based device) systems.

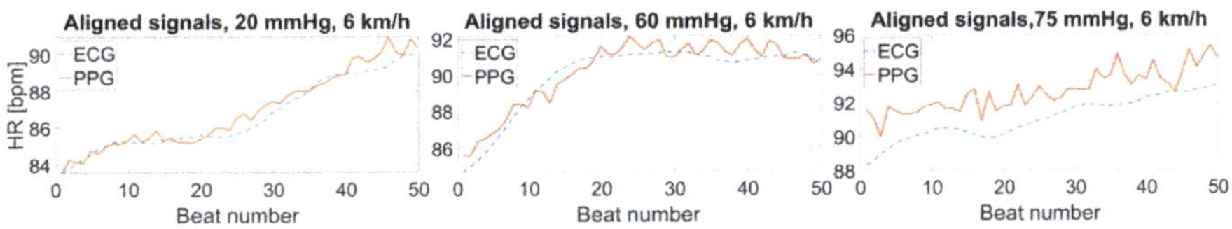

Figure 5. Examples of the test (PPG) and reference (ECG-based) acquired signals (speed: 6 km/h; contact pressure: 20, 60, 75 mmHg; skin color: 2; subject 23).

The accuracy and precision at each speed and for each contact pressure were assessed using the MAPE (Table 2) and the Pearson's correlation coefficient (Table 3), as well as through an analysis of the deviations between the HR values measured by the PPG-based and the ECG-based systems (Table 4).

Table 2. Performance metrics at different speeds at each contact pressure: MAPE at each physical activity level and for each contact pressure.

	MAPE [%] (σ)		
	CP [mmHg]		
Speed [km/h]	20	60	75
3	4.99 (6.38)	3.36 (3.49)	4.37 (3.73)
6	6.11 (6.54)	3.62 (4.59)	4.73 (5.67)
8	9.12 (9.44)	6.83 (6.38)	11.85 (8.56)

Table 3. Performance metrics at different speeds at each contact pressure: Pearson's correlation coefficient of PPG-HR.

	ρ [−]		
	CP [mmHg]		
Speed [km/h]	**20**	**60**	**75**
3	0.77	0.88	0.83
6	0.62	0.80	0.75
8	0.60	0.75	0.44

Table 4. Analysis of residuals: mean and standard deviation values, linked to measurement accuracy and precision, respectively.

	Residuals $\mu \pm \sigma$ [bpm]		
	CP [mmHg]		
Speed [km/h]	**20**	**60**	**75**
3	2 ± 7	0 ± 5	0 ± 6
6	2 ± 9	0 ± 7	0 ± 8
8	6 ± 15	0 ± 11	-4 ± 19

The comparison between the measurement systems was further tested using the Bland–Altman plot with confidence intervals at 95% (Figure 6) and through the distribution of deviations with respect to the reference device (Figure 7).

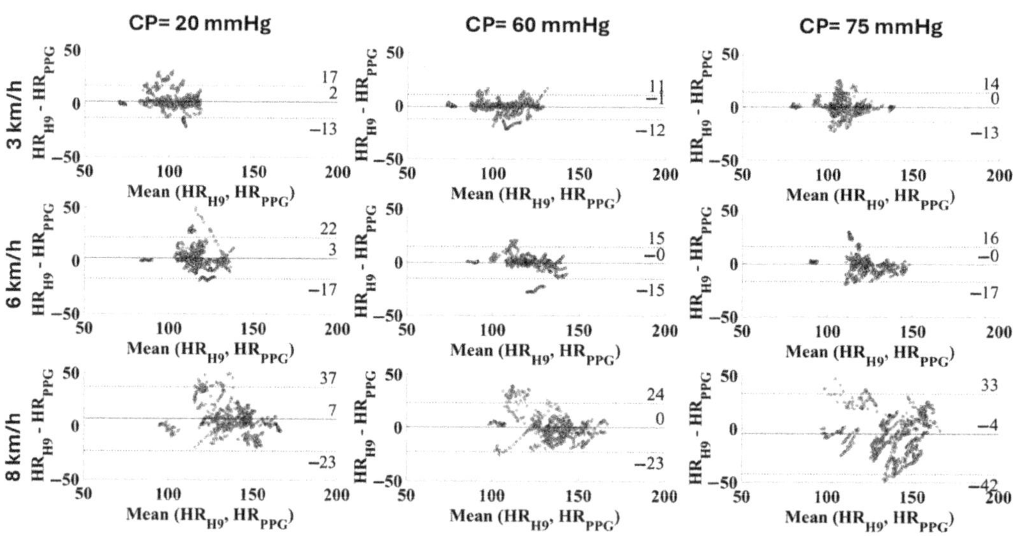

Figure 6. Bland–Altman plot related to each combination of CP and speed in the cohort of subjects (expressed as average difference ± 1.96 standard deviation of the measurement differences).

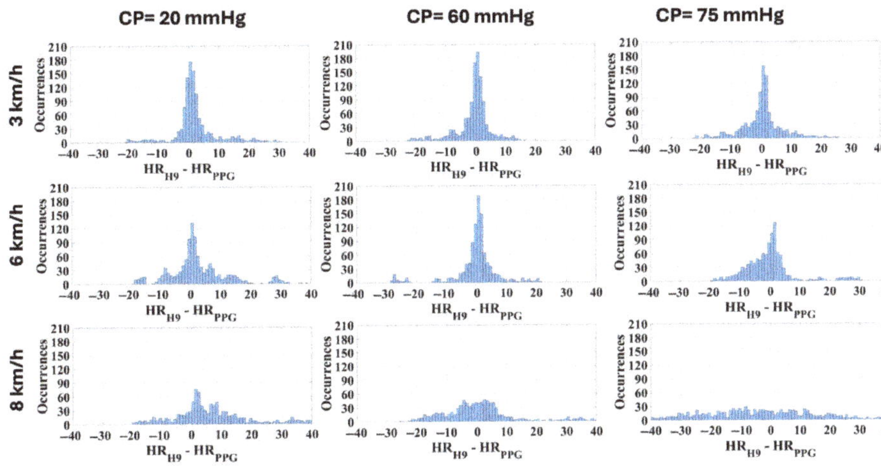

Figure 7. Distribution of deviations related to each combination of CP and speed in the cohort of subjects (width: 1 bpm).

Finally, the most reliable CP (i.e., the one which provides the lowest MAPE) is not a single value for all subjects. Beyond the physical activity level, it also depends on several factors including the specificities of each individual in terms of their vasculature properties; therefore, the optimal individual contact pressure was identified for each subject. Then, the same metrics previously adopted were recalculated; these are presented in Table 5 and in Figure 8.

Table 5. Performance metrics at different speeds using the optimal CP for each subject: MAPE, Pearson's correlation coefficient, and residuals analysis.

Optimal Contact Pressure (mmHg)			
Speed [km/h]	**MAPE [%] (σ)**	**ρ [−]**	**Residuals μ ± σ [bpm]**
3	2.38 ± 2.99	0.94	0 ± 5
6	2.07 ± 2.58	0.93	1 ± 4
8	4.90 ± 3.09	0.83	−1 ± 9

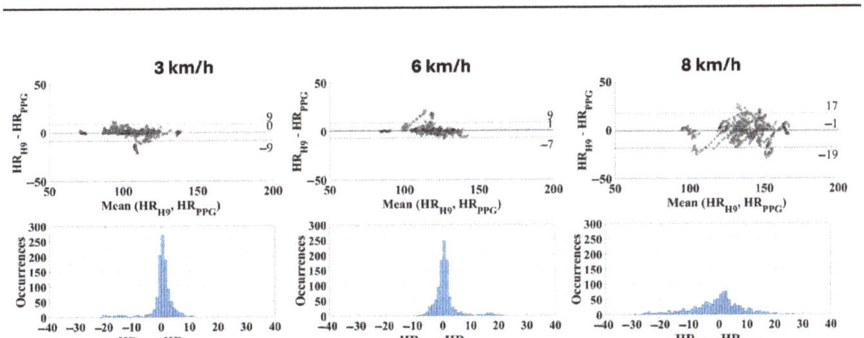

Figure 8. Bland–Altman plots and distribution of deviations related to the optimal contact pressure for each subject (bin width: 1 bpm).

4. Discussion and Conclusions

In this paper the authors evaluated the effect of different contact pressure values on the quality of the data gathered through a wrist-band wearable sensor. In particular, the analysis was performed on the heart rate (HR) values derived from a test (prototypal bracelet) device and a reference (Polar H9 chest strap) device, collecting photoplethysmographic (PPG) and electrocardiographic (ECG) signals, respectively. It is worth mentioning that the ECG-based chest strap, which is the most widely-adopted reference system, does not provide information regarding its accuracy during physical activities; on the other hand, it is beyond doubt that a medical-grade 12-lead ECG device would be the best choice as the gold standard. However, since medical certification is beyond the aim of this work at present, the Polar H9 device can be considered suitable (additionally, it has been extensively employed in the literature as a gold standard instrument [42,43]), and any uncertainty has been assumed to be included in the expanded uncertainty estimated in the analyses.

An extensive metrological evaluation was performed to evaluate the effective performance of the proposed PPG-based prototype in the measurement of the HR during different intensities of physical activity. First of all, it is crucial to underline the adequateness of the tightening system, which allowed the substantial maintenance of the contact pressure during the tests without loosening issues. Plus, limited variations in blood pressure values were observed before and after the dynamic exercises, but no further considerations were explored on the impact of blood pressure and physical activity intensities [44].

Observing the synchronized signal portions (Figure 5), it is immediately evident that the contact pressure directly affects the agreement between the HR values derived from the test and reference devices, leading to increasing measurement differences when the sensor adherence is not adequate and, thus, to a worsening signal quality. Consequently, this leads to a higher measurement uncertainty, which can undermine the reliability of the usage of PPG-based devices during certain exercises. This becomes particularly relevant in sport applications and in the context of wellbeing assessments, which are among the main application fields for wearable sensors [45,46].

The results in terms of the performance metrics (i.e., the mean average percentage error and correlation coefficient—Tables 2 and 3) show that the intermediate contact pressure (CP), i.e., 60 mmHg, is the optimal choice considering all the activity levels (i.e., walking/running at 3, 6, and 8 km/h). In fact, this CP value consistently allowed the achievement of the lowest MAPE, with average values ranging from 3.36% to 6.83% depending on the walking speed/activity intensity level, versus 4.99–9.12% for CP = 20 mmHg and 4.37–11.85% for CP = 75 mmHg. This was similar for the correlation coefficient; the CP = 60 mmHg always resulted in a value higher than 0.70 (achieving 0.88 at 3 km/h). This result is in line with the findings of the previous work of some of the authors [21], where a CP of 54 mmHg was identified as optimal, with a MAPE ranging from 2.4% to 3.8% and a Pearson's correlation coefficient of 0.81–0.95 at different physical activity intensities (corresponding to HR values of 90 bpm—low intensity, 120 bpm—medium intensity, and 140 bpm—high intensity). Such results confirm that the proposed methodology allows us to obtain accurate results, with a MAPE < 5% and a strong linear correlation with the reference device. This cannot be stated for the activity speed of 8 km/h, which on the contrary leads to excessively impactful motion artifacts, impairing the quality of the results.

However, in the previous study [21] no contact pressure values higher than 54 mmHg were analyzed and the different levels were chosen by asking the participants to wear the bracelet as a smartwatch with different tightening levels. Thus, the objective of the present study was to assess the metrological performance of the PPG sensor with pressures higher than 54 mmHg, chosen empirically based on the previous study. Although theoretically the optimal contact pressure should be equal to the transmural pressure [9], a recent study [47] reported that the PPG signal responses may be dependent on the wavelength and on the skin and vasculature properties as the CP changes. Thus, the results, in addition to confirming those previously obtained, demonstrate that 75 mmHg is an excessive pressure that causes a significant decrease in precision at different speeds (Figure 7), especially at 8 km/h.

These considerations are directly reflected in the analysis of the measurement differences (residuals), as observed in Table 4. The distribution is tighter with a CP equal to 60 mmHg and a walking/running speed up to 6 km/h (Figure 7). Precision and trueness reached their best values at the same CP of 60 mmHg for any intensity of physical activity, as seen in Table 4 and Figure 6. Specifically, the confidence intervals (coverage factor $k = 1.96$) varied from $[-12, 10]$ bpm at 3 km/h to $[-23, 23]$ bpm at 8 km/h, with a mean value of -1 bpm and 0 bpm, respectively. At 8 km/h the residuals seem not to have a defined distribution, appearing almost casual.

Evaluating the effect of the contact pressure on the quality of the PPG signal during activity execution is extremely interesting, since PPG-based wearables are notably prone to motion artifacts. Studies in the literature [20] consider the HR measurement as reliable when the mean absolute percentage error (MAPE) is lower than 5% and the linear correlation with the reference value is strong (i.e., $\rho > 0.7$). Upon this basis, all the values measured at a walking speed of 3 km/h can be considered reliable (MAPE within the range 3.36–4.99, $\rho \geq 0.77$), regardless of the CP (Table 2). On the other hand, when running is considered, the results cannot be considered acceptable at a CP of 20 mmHg (MAPE = 6.11%, $\rho = 0.62$), but they are acceptable at higher CP values (best with a CP of 60 mmHg, with MAPE = 3.62% and $\rho = 0.80$, whereas with CP = 75 mmHg MAPE is 4.73% and ρ is 0.75, and this still considered reliable). At 8 km/h none of the CP values allow us to obtain accurate HR values; this can be attributed to the high variability in the measurement results, as can be observed in the Bland–Altman plot (Figure 6, right). If a specific CP value is considered, the MAPE increases with the activity level, as expected, due to possible motion artifacts. On the other hand, the value of the correlation coefficient decreases due to a worsening correlation with the reference values. It is feasible to assume that at the highest contact pressure (i.e., 75 mmHg) and at the highest exercise intensity (i.e., 8 km/h) the effect of the tightening, combined with the lower damping, has a compressive effect on the capillary bed, causing a significant decrease in the signal quality.

However, it is worth mentioning that any given contact pressure is not necessarily the optimal one for a different subject; this is attributable to the intrinsic inter-subject variability found in the wrist morphological characteristics and in the cardiovascular properties. In fact, if individualized optimal values are considered, the measurement uncertainty significantly decreases (Table 5) from 15 bpm (with a CP equal to 60 mmHg) to 8 bpm when considering a running speed of 6 km/h (coverage factor $k = 2$). At the same time, the accuracy (or rather, trueness) is optimal, with a mean residuals value equal to 1 bpm (instead of 0 bpm when considering the same running speed). This is reflected also in the distribution of the residuals (Figure 8), which appears very high and tight up to 6 km/h (standard deviation equal to 5 bpm and 4 bpm at 3 km/h and 6 km/h, respectively); indeed, at 8 km/h the measurement uncertainty is almost doubled (standard deviation of 9 bpm).

In conclusion, it is evident that the contact between the sensor and the skin is a crucial aspect in PPG-based sensors and a system capable of optimizing the contact pressure depending on individual characteristics would be beneficial and could also widen the applications of such devices. The possibility of personalizing the sensors can improve the measurement accuracy and this could be widely exploited in fields like indoor comfort assessment (also during movements) and sport applications (where motion is unavoidable and needs to be thoroughly considered without impairing the accurate measurement of physiological signals), just as examples. In future, it would be interesting also to extend the study to subjects with different skin tones, hence covering most of the Fitzpatrick scale; it is worth underlining that the skin tone assessment can be performed on different body areas provided that they are protected from sunlight and hence representative of the effective skin phototyping of the subject (e.g., other works in the literature use the inner part of the upper arm) [48].

These results and the considerations that can be inferred are relevant to the fine-tuning of a wearable system capable of accurately measuring the physiological signals that can be exploited to provide an insight into the overall status of the subject, while also reflecting their personal perception of the indoor comfort in a certain living environment. This can contribute to the improvement of the quality of living and in future could be extended also to clinical settings (this would require a clinical research study for validation purposes and marketing of the device, according to national and international guidelines and recommendations, e.g., EU Regulation 536/2014—Clinical Trial Regulation (CTR) [49] and EU Directive 2001/83 [50], among others). Additionally, it is worth noting that the results from such experimental tests can be exploited for the optimization of the hardware design of wearable sensors capable of auto-tightening the bracelet around the wrist, hence improving the contact pressure between the PPG sensor and the skin and hence, the quality of the measured data. In turn, these data can inform PCMs that will be better performing and consequently more effective in the enhancement of indoor well-being, as well as in the building energy consumption through the integration of the monitoring data analysis within the building management and control systems.

Author Contributions: Conceptualization, M.A., L.D., L.S., G.C., and F.S.; methodology, M.A., L.D., L.S., G.D., C.G., G.C., and F.S.; validation, L.S. and G.C.; formal analysis, L.A., C.G., and F.S.; investigation, L.S., F.B., L.A., and C.G.; resources, M.A. and G.D.; data curation, L.D., L.S., L.A., C.G., and G.C.; writing—original draft preparation, F.B., G.C., and F.S.; supervision, M.A., L.S., L.D., and F.S.; funding acquisition, M.A., G.C., F.B., F.S., L.S., and L.D. All authors have read and agreed to the published version of the manuscript.

Funding: M.A., G.C., F.B. and F.S. were co-funded by the Italian Ministry of Research through the WEPOP (Prot. 2022RKLB3J) "WEarable PlatfOrm for Personalized comfort" project, within the PRIN 2022 program. L.D. was supported by the Italian Ministry of Education, University and Research "Sensoristica intelligente, infrastrutture e modelli gestionali per la sicurezza di soggetti fragili" (4FRAILTY) project (PON R&I ARSO1_00345).

Institutional Review Board Statement: The study was conducted in accordance with the Declaration of Helsinki and approved by the Ethics Committee of the University of Palermo (51594/23) and by the Ethics Committee of Università Politecnica delle Marche.

Informed Consent Statement: Written informed consent was obtained from all subjects involved in the study.

Data Availability Statement: Data will be made available upon request to the corresponding Author.

Conflicts of Interest: The authors declare no conflicts of interest.

References

1. Life Expectancy and Healthy Life Expectancy. Available online: https://www.who.int/data/gho/data/themes/mortality-and-global-health-estimates/ghe-life-expectancy-and-healthy-life-expectancy (accessed on 1 November 2024).

2. Maresova, P.; Javanmardi, E.; Barakovic, S.; Barakovic Husic, J.; Tomsone, S.; Krejcar, O.; Kuca, K. Consequences of chronic diseases and other limitations associated with old age—A scoping review. *BMC Public Health* **2019**, *19*, 1431. [CrossRef] [PubMed]
3. Baig, M.M.; GholamHosseini, H.; Moqeem, A.A.; Mirza, F.; Lindén, M. A Systematic Review of Wearable Patient Monitoring Systems—Current Challenges and Opportunities for Clinical Adoption. *J. Med. Syst.* **2017**, *41*, 115. [CrossRef] [PubMed]
4. Lopreite, M.; Misuraca, M.; Puliga, M. An analysis of the thematic evolution of ageing and healthcare expenditure using word embedding: A scoping review of policy implications. *Socioecon. Plann. Sci.* **2023**, *87*, 101600. [CrossRef]
5. Cosoli, G.; Mansi, S.A.; Pigliautile, I.; Pisello, A.L.; Revel, G.M.; Arnesano, M. Enhancing personal comfort: A machine learning approach using physiological and environmental signals measurements. *Measurement* **2023**, *217*, 113047. [CrossRef]
6. Abd-Alrazaq, A.; AlSaad, R.; Shuweihdi, F.; Ahmed, A.; Aziz, S.; Sheikh, J. Systematic review and meta-analysis of performance of wearable artificial intelligence in detecting and predicting depression. *NPJ Digit. Med.* **2023**, *6*, 84. [CrossRef] [PubMed]
7. Liu, P.; Song, Y.; Yang, X.; Li, D.; Khosravi, M. Medical intelligence using PPG signals and hybrid learning at the edge to detect fatigue in physical activities. *Sci. Rep.* **2024**, *14*, 16149. [CrossRef] [PubMed]
8. Loh, H.W.; Xu, S.; Faust, O.; Ooi, C.P.; Barua, P.D.; Chakraborty, S.; Tan, R.S.; Molinari, F.; Acharya, U.R. Application of photoplethysmography signals for healthcare systems: An in-depth review. *Comput. Methods Programs Biomed.* **2022**, *216*, 106677. [CrossRef] [PubMed]
9. Scardulla, F.; Cosoli, G.; Spinsante, S.; Poli, A.; Iadarola, G.; Pernice, R.; Busacca, A.; Pasta, S.; Scalise, L.; D'Acquisto, L. Photoplethysmograhic sensors, potential and limitations: Is it time for regulation? A comprehensive review. *Measurement* **2023**, *218*, 113150. [CrossRef]
10. Almarshad, M.A.; Saiful Islam, M.; Al-Ahmadi, S.; BaHammam, A.S. Diagnostic Features and Potential Applications of PPG Signal in Healthcare: A Systematic Review. *Healthcare* **2022**, *10*, 547. [CrossRef] [PubMed]
11. Are Your Home and Your Belongings Internet-Connected? Products Eurostat News-Eurostat. Available online: https://ec.europa.eu/eurostat/web/products-eurostat-news/-/ddn-20210225-1 (accessed on 29 April 2025).
12. Tamura, T.; Maeda, Y.; Sekine, M.; Yoshida, M. Wearable Photoplethysmographic Sensors—Past and Present. *Electronics* **2014**, *3*, 282–302. [CrossRef]
13. Haque, C.A.; Hossain, S.; Kwon, T.H.; Kim, K.D. Comparison of Different Methods to Estimate Blood Oxygen Saturation using PPG. In Proceedings of the 2021 International Conference on Information and Communication Technology Convergence (ICTC), Jeju Island, Republic of Korea, 20–22 October 2021; pp. 792–794. [CrossRef]
14. León-Valladares, D.; Barrio-Mateu, L.A.; Cortés-Carmona, N.; Fuentes-Lizana, G.; Cabanas, A.M.; Latorre-Progulakis, K.; Fuentes-Guajardo, M.; Martín-Escudero, P. Determining factors of pulse oximetry accuracy: A literature review. *Rev. Clínica Española (Engl. Ed.)* **2024**, *224*, 314–330. [CrossRef]
15. Man, P.K.; Cheung, K.L.; Sangsiri, N.; Shek, W.J.; Wong, K.L.; Chin, J.W.; Chan, T.T.; So, R.H.Y. Blood Pressure Measurement: From Cuff-Based to Contactless Monitoring. *Healthcare* **2022**, *10*, 2113. [CrossRef] [PubMed]
16. Kuwalek, P.; Burlaga, B.; Jesko, W.; Konieczka, P. Research on methods for detecting respiratory rate from photoplethysmographic signal. *Biomed. Signal Process. Control* **2021**, *66*, 102483. [CrossRef]
17. Ferizoli, R.; Karimpour, P.; May, J.M.; Kyriacou, P.A. Arterial stiffness assessment using PPG feature extraction and significance testing in an in vitro cardiovascular system. *Sci. Rep.* **2024**, *14*, 2024. [CrossRef] [PubMed]
18. Afzali Arani, M.S.; Costa, D.E.; Shihab, E. Human Activity Recognition: A Comparative Study to Assess the Contribution Level of Accelerometer, ECG, and PPG Signals. *Sensors* **2021**, *21*, 6997. [CrossRef] [PubMed]
19. Cosoli, G.; Antognoli, L.; Veroli, V.; Scalise, L. Accuracy and Precision of Wearable Devices for Real-Time Monitoring of Swimming Athletes. *Sensors* **2022**, *22*, 4726. [CrossRef] [PubMed]
20. Hwang, S.; Seo, J.O.; Jebelli, H.; Lee, S.H. Feasibility analysis of heart rate monitoring of construction workers using a photoplethysmography (PPG) sensor embedded in a wristband-type activity tracker. *Autom. Constr.* **2016**, *71*, 372–381. [CrossRef]
21. Scardulla, F.; D'Acquisto, L.; Colombarini, R.; Hu, S.; Pasta, S.; Bellavia, D. A Study on the Effect of Contact Pressure during Physical Activity on Photoplethysmographic Heart Rate Measurements. *Sensors* **2020**, *20*, 5052. [CrossRef] [PubMed]
22. Sirkiä, J.-P.; Panula, T.; Kaisti, M. Investigating the impact of contact pressure on photoplethysmograms. *Biomed. Eng. Adv.* **2024**, *7*, 100123. [CrossRef]
23. Lambert Cause, J.; Solé Morillo, Á.; García-Naranjo, J.C.; Stiens, J.; da Silva, B. The Impact of Contact Force on Signal Quality Indices in Photoplethysmography Measurements. *Appl. Sci.* **2024**, *14*, 5704. [CrossRef]
24. Ray, D.; Collins, T.; Woolley, S.; Ponnapalli, P. A Review of Wearable Multi-wavelength Photoplethysmography. *IEEE Rev. Biomed. Eng.* **2021**, *16*, 136–151. [CrossRef] [PubMed]
25. Yan, L.; Hu, S.; Alzahrani, A.; Alharbi, S.; Blanos, P. A Multi-Wavelength Opto-Electronic Patch Sensor to Effectively Detect Physiological Changes against Human Skin Types. *Biosensors* **2017**, *7*, 22. [CrossRef]

26. Han, S.; Roh, D.; Park, J.; Shin, H. Design of Multi-Wavelength Optical Sensor Module for Depth-Dependent Photoplethysmography. *Sensors* **2019**, *19*, 5441. [CrossRef] [PubMed]
27. Krishnan, S.; Argüello-Prada, E.J.; Ferney, J.; García, C. Machine Learning Applied to Reference Signal-Less Detection of Motion Artifacts in Photoplethysmographic Signals: A Review. *Sensors* **2024**, *24*, 7193. [CrossRef] [PubMed]
28. Pandey, R.K.; Chao, P.C.P. A Low-Power 54 μW Adaptive Analog Front-End with Adaptive Intensity Control for an Organic PPG Sensor in Wearable Devices. In Proceedings of the 2021 IEEE Sensors, Sydney, Australia, 31 October–3 November 2021; pp. 1–4. [CrossRef]
29. Scardulla, F.; Riggi, C.; Diana, G.; D'Acquisto, L. Preliminary Analisys on the Effect of Skin Temperature on Photoplethysmographic Signal. In Proceedings of the 2024 IEEE International Workshop on Metrology for Living Environment, Chania, Greece, 12–14 June 2024; pp. 6–10. [CrossRef]
30. Bent, B.; Goldstein, B.A.; Kibbe, W.A.; Dunn, J.P. Investigating sources of inaccuracy in wearable optical heart rate sensors. *Npj Digit. Med.* **2020**, *3*, 18. [CrossRef] [PubMed]
31. Cosoli, G.; Antognoli, L.; Scalise, L. Methods for the metrological characterization of wearable devices for the measurement of physiological signals: State of the art and future challenges. *MethodsX* **2023**, *10*, 102038. [CrossRef] [PubMed]
32. Kaur, A.; Arora, A. A Review on Reliability, Security and Future Aspects of Photoplethysmography. In Proceedings of the 2019 2nd International Conference on Intelligent Computing, Instrumentation and Control Technologies (ICICICT), Kannur, India, 5–6 July 2019; pp. 1480–1485. [CrossRef]
33. Consumer Technology Association (CTA). Physical Activity Monitoring for Heart Rate (ANSI/CTA-2065). 2018. Available online: https://shop.cta.tech/products/physical-activity-monitoring-for-heart-rate (accessed on 29 September 2022).
34. Takahashi, H.; Yamashita, S.; Yakura, N. Validating the Accuracy of Omron HEM-1026 (HCR-1901T2/HCR-1902T2): Blood Pressure Monitoring Device at Home, According to the Universal Protocol. *Vasc. Health Risk Manag.* **2024**, *20*, 333. [CrossRef] [PubMed]
35. Romagnoli, S.; Ripanti, F.; Morettini, M.; Burattini, L.; Sbrollini, A. Wearable and Portable Devices for Acquisition of Cardiac Signals while Practicing Sport: A Scoping Review. *Sensors* **2023**, *23*, 3350. [CrossRef] [PubMed]
36. Martín-Escudero, P.; Cabanas, A.M.; Dotor-Castilla, M.L.; Galindo-Canales, M.; Miguel-Tobal, F.; Fernández-Pérez, C.; Fuentes-Ferrer, M.; Giannetti, R. Are Activity Wrist-Worn Devices Accurate for Determining Heart Rate during Intense Exercise? *Bioengineering* **2023**, *10*, 254. [CrossRef] [PubMed]
37. Morresi, N.; Casaccia, S.; Revel, G.M. Metrological characterization and signal processing of a wearable sensor for the measurement of heart rate variability. In Proceedings of the 2021 IEEE International Symposium on Medical Measurements and Applications (MeMeA), Lausanne, Switzerland, 23–25 June 2021; pp. 1–6. [CrossRef]
38. Pollreisz, D.; TaheriNejad, N. Detection and Removal of Motion Artifacts in PPG Signals. *Mob. Netw. Appl.* **2022**, *27*, 728–738. [CrossRef]
39. Liang, Y.; Elgendi, M.; Chen, Z.; Ward, R. Analysis: An optimal filter for short photoplethysmogram signals. *Sci. Data* **2018**, *5*, 180076. [CrossRef]
40. Tan, L.; Jiang, J. *Digital Signal Processing: Fundamentals and Applications*; Academic Press: Cambridge, MA, USA, 2018; ISBN 0128150726.
41. Lapitan, D.G.; Rogatkin, D.A.; Molchanova, E.A.; Tarasov, A.P. Estimation of phase distortions of the photoplethysmographic signal in digital IIR filtering. *Sci. Rep.* **2024**, *14*, 6546. [CrossRef] [PubMed]
42. Li, Y.X.; Huang, J.L.; Yao, X.Y.; Mu, S.Q.; Zong, S.X.; Shen, Y.F. A ballistocardiogram dataset with reference sensor signals in long-term natural sleep environments. *Sci. Data* **2024**, *11*, 1091. [CrossRef] [PubMed]
43. Baek, S.; Ha, Y.; Park, H. won Accuracy of Wearable Devices for Measuring Heart Rate During Conventional and Nordic Walking. *PMR* **2021**, *13*, 379–386. [CrossRef] [PubMed]
44. Pescatello, L.S.; Kulikowich, J.M. The aftereffects of dynamic exercise on ambulatory blood pressure. *Med. Sci. Sports Exerc.* **2001**, *33*, 1855–1861. [CrossRef] [PubMed]
45. De Fazio, R.; Mastronardi, V.M.; De Vittorio, M.; Visconti, P. Wearable Sensors and Smart Devices to Monitor Rehabilitation Parameters and Sports Performance: An Overview. *Sensors* **2023**, *23*, 1856. [CrossRef] [PubMed]
46. Yang, Y. Application of wearable devices based on artificial intelligence sensors in sports human health monitoring. *Meas. Sens.* **2024**, *33*, 101086. [CrossRef]
47. Lambert Cause, J.; Solé Morillo, Á.; da Silva, B.; García-Naranjo, J.C.; Stiens, J. Evaluating Vascular Depth-Dependent Changes in Multi-Wavelength PPG Signals Due to Contact Force. *Sensors* **2024**, *24*, 2692. [CrossRef] [PubMed]
48. Gupta, V.; Sharma, V.K. Skin typing: Fitzpatrick grading and others. *Clin. Dermatol.* **2019**, *37*, 430–436. [CrossRef] [PubMed]

49. Regulation-536/2014-EN-EUR-Lex. Available online: https://eur-lex.europa.eu/eli/reg/2014/536/oj/eng (accessed on 5 July 2025).
50. Directive-2001/83-EN-EUR-Lex. Available online: https://eur-lex.europa.eu/eli/dir/2001/83/oj/eng (accessed on 5 July 2025).

Disclaimer/Publisher's Note: The statements, opinions and data contained in all publications are solely those of the individual author(s) and contributor(s) and not of MDPI and/or the editor(s). MDPI and/or the editor(s) disclaim responsibility for any injury to people or property resulting from any ideas, methods, instructions or products referred to in the content.

sensors

MDPI

Article

Enhancing Wearable Fall Detection System via Synthetic Data

Minakshi Debnath, Sana Alamgeer, Md Shahriar Kabir and Anne H. Ngu *

Department of Computer Science, Texas State University, San Marcos, TX 78666-4684, USA; stg60@txstate.edu (M.D.); sana.alamgeer@txstate.edu (S.A.); cpi12@txstate.edu (M.S.K.)
* Correspondence: angu@txstate.edu

Abstract: Deep learning models rely heavily on extensive training data, but obtaining sufficient real-world data remains a major challenge in clinical fields. To address this, we explore methods for generating realistic synthetic multivariate fall data to supplement limited real-world samples collected from three fall-related datasets: SmartFallMM, UniMib, and K-Fall. We apply three conventional time-series augmentation techniques, a Diffusion-based generative AI method, and a novel approach that extracts fall segments from public video footage of older adults. A key innovation of our work is the exploration of two distinct approaches: video-based pose estimation to extract fall segments from public footage, and Diffusion models to generate synthetic fall signals. Both methods independently enable the creation of highly realistic and diverse synthetic data tailored to specific sensor placements. To our knowledge, these approaches and especially their application in fall detection represent rarely explored directions in this research area. To assess the quality of the synthetic data, we use quantitative metrics, including the Fréchet Inception Distance (FID), Discriminative Score, Predictive Score, Jensen–Shannon Divergence (JSD), and Kolmogorov–Smirnov (KS) test, and visually inspect temporal patterns for structural realism. We observe that Diffusion-based synthesis produces the most realistic and distributionally aligned fall data. To further evaluate the impact of synthetic data, we train a long short-term memory (LSTM) model offline and test it in real time using the SmartFall App. Incorporating Diffusion-based synthetic data improves the offline F1-score by 7–10% and boosts real-time fall detection performance by 24%, confirming its value in enhancing model robustness and applicability in real-world settings.

Keywords: fall detection; time-series data; synthetic data generation; Diffusion; video extraction

1. Introduction

Falls are a leading cause of injury and one of the primary contributors to unintentional deaths among adults aged 65 and older worldwide [1,2]. This high incidence highlights the urgent need for effective prevention and intervention strategies to protect the health and well-being of the elderly population. In response, research on fall detection technologies has grown substantially, particularly in the use of wearable devices such as smartwatches and inertial measurement unit (IMU) sensors. These devices are increasingly favored due to their portability, affordability, and non-intrusive nature [3].

There is also an increasing interest in leveraging deep learning models for learning the dynamical behavior of falls and the accurate detection of fall events using a wearable device [4]. However, in many physiological processes like the onset of falls, the dynamics are exceedingly complex, and deep learning techniques are not effective, especially when given a limited amount of data for training. Moreover, a fall is a rare event, and it is impossible to collect a large amount of fall data. In the past, researchers have embarked on collecting simulated fall data from young adults in a controlled environment. The data collection process is costly and laborious. In the context of our SmartFall multi-modal (SmartFallMM) dataset (the work is under review), it took a whole month of effort to collect data from 40 people. This excludes the effort needed to implement suitable data collection systems for various types of IMU sensors, as well as the effort in recruiting the participants. Therefore, integrating deep learning techniques into these systems faces a significant challenge: the severe shortage of real human fall data for training. To address this, synthetic data is increasingly being used [5,6] to overcome this limitation. Data augmentation techniques are a widely recognized method for mitigating the problem of small datasets. By artificially expanding the available data through various transformations and modifications, these techniques help improve the robustness and generalizability of deep learning models.

Recently, generative AI has gained prominence in research, particularly in the field of synthetic data generation across various domains, including images and time-series data. Techniques such as Generative Adversarial Networks (GANs) [7–9], Variational Autoencoders (VAEs) [10], and Diffusion models [11,12] are increasingly utilized for creating synthetic time-series data. These methods are capable of generating data that closely resembles real-world data by learning complex patterns and structures from existing data. In particular, the Diffusion-based generative AI model has emerged recently as a popular tool for data generation, demonstrating exceptional performance across a variety of applications.

Platforms like YouTube with video datasets of people's daily life activities, including free fall (i.e., the downward motion phase during a fall when the body experiences minimal support and acceleration approaches gravity [13]) can be another source for obtaining authentic fall data. Unlike the majority of currently available fall datasets, which rely on simulated falls, leveraging video content from public video sources could provide a more accurate representation of real-world falls, leading to better training data for fall detection systems. Pose estimation has become a cornerstone in analyzing human motion across various industries, including sports and healthcare [14]. By harnessing machine learning and computer vision, these techniques can extract skeletal data from videos, offering detailed insights into body movement. Using pose estimation techniques on video content from publicly accessible video sources, researchers might be able to create realistic fall data that closely mirrors actual human movements.

However, there is no comparative study yet on the impact of synthetic data on the performance of fall detection models trained with a mix of real and synthetic fall data, particularly within real-world applications. In general, researchers tend to evaluate the quality of generated data using similarity metrics such as principal component analysis (PCA) and t-distributed stochastic neighbor embedding (t-SNE) [15], which capture distributional resemblance but not necessarily functional performance. Therefore, to validate both the practical utility and quality of the generated data, it is crucial to demonstrate that models trained with synthetic data can achieve high F1-scores in real-world scenarios.

To address the persistent challenge of limited real-world fall data, this paper introduces a novel hybrid framework that combines three complementary approaches: traditional augmentation techniques, Diffusion-based generative modeling, and video-based pose estimation. This paper presents an experimental study of the effectiveness of these techniques for generating multivariate synthetic fall data. Additionally, it explores different metrics to evaluate the quality of generated synthetic data. We examine and compare several methods for generating and augmenting fall data, including three established time-series data augmentation techniques, such as jittering, magnitude warping, and rotation, a Diffusion-based generative AI technique, as well as a novel approach involving pose estimation from video recordings of falls in older adults.

The main contributions of this paper are as follows:

- The implementation of two distinct synthetic data generation techniques for time-series data, including video-based extraction (pose estimation) and generative AI-based (Diffusion) generation, where we used traditional data augmentation methods as a baseline for performance comparison.
- A comparative evaluation of fall detection model performance, incorporating both real and synthetic data across three different public fall datasets using three different synthetic data generation techniques.
- A demonstration that synthetic data produced using the Diffusion approach and pose estimation from video recordings can enhance the fall detection model's performance in real-world scenarios using the SmartFall App.
- An exploration of various quantitative evaluation metrics provides insights into the quality and utility of synthetic data for effective model training.

The remainder of this paper is structured into five main sections. Section 2 reviews recent advancements in synthetic data generation, including the use of generative AI techniques such as GANs, Diffusion models, and pose estimation, as well as general data augmentation methods. Section 3 details the overall methodology, covering the synthetic data generation processes, the datasets employed, the deep learning model used for fall detection, and an explanation of the experimental setup. Section 4 outlines three synthetic fall data generation methods and compares their suitability for fall detection. Section 5 presents the results of both offline and real-time evaluations. Finally, Section 6 offers conclusions, and Section 7 discusses potential directions for future research. This paper is an extended version of the short paper presented at AIME 2024 [16].

2. Related Work

The generation of synthetic time-series data for accelerometer-based fall detection systems has been a significant area of research. Various generative techniques have been proposed to address this challenge, each with its strengths and limitations. In this section, we discuss the techniques that fall within the scope of this study.

2.1. Generative Adversarial Networks (GANs)

Generative Adversarial Networks (GANs) have been widely used for synthetic time-series data generation. For example, TimeGAN [7,9] integrates the strengths of both GANs and Recurrent Neural Networks (RNNs) to capture temporal dependencies in data. Zhang et al. [17] also applied GANs to learn the conditional probability distribution of essential features in the real dataset and generate samples based on the learned distribution. Other frameworks, like TSGM [10] with GANs and Variational Autoencoders (VAEs) and DoppelGANger [18] with GANs only, offered flexibility for generating synthetic data with complex dependencies.

In the medical domain also, GANs have been employed to generate synthetic medical health records, which are used to train machine learning models for disease diagnosis and treatment prediction. For example, the paper [8] showcases the use of GANs to create synthetic medical records that maintain the temporal and sequential patterns found in real datasets. This technique addresses significant challenges, such as patient privacy concerns and the scarcity of large, diverse datasets for training machine learning models. Another method, MedGAN [19], used an autoencoder and GANs to generate synthetic patient records close to the real data of patients.

However, GANs are known to suffer from instability during training and mode collapse, where the generator produces limited diversity in outputs. Also, GAN-based models often require large amounts of labeled data and extensive computational resources, which are not always available for fall detection datasets. These issues can be particularly problematic when aiming to generate a variety of scenarios necessary for effective fall detection systems. The complexity and extensive tuning required for GANs can be a hindrance, especially when dealing with the high-dimensional and noisy nature of accelerometer data.

2.2. Diffusion Models

Diffusion models have recently emerged as a powerful tool for generating synthetic time-series data. These models effectively handle complex data distributions in domains like healthcare and finance because they can capture intricate patterns and dependencies within the data. For example, the TS-Diffusion model [11] effectively generates high-fidelity and diverse time-series data by leveraging Neural Ordinary Differential Equations (ODEs). Similarly, Diffusion-TS [20] introduces interpretability by incorporating Fourier-based loss terms, allowing it to uncover time-series components like seasonality and trends. Tian et al. [12] proposed a method that used Denoising Diffusion Probabilistic Models (DDPM) to generate diverse and realistic synthetic time-series data of Electronic Health Records (EHRs).

Despite their strengths, Diffusion-based methods face challenges in capturing subtle and rapid motion patterns critical for accelerometer-based fall detection. For example, TS-Diffusion struggles with representing short, transient events, potentially overlooking key biomechanical details. Although Diffusion-TS enhances interpretability through frequency-domain analysis, it often fails to reconstruct abrupt, non-stationary movements typical of falls. In contrast, DDPMs show promise by effectively modeling small datasets and dynamic temporal structures, making them better suited for realistic fall data generation.

2.3. Pose Estimation Techniques

Pose estimation methods have enabled the analysis of human motion in various fields, such as sports, healthcare, and animation. These methods are extensively utilized to extract kinematics or skeleton data from videos by leveraging advancements in computer vision and machine learning to track human body joints and generate detailed motion data. For instance, recent work [14] has focused on improving 2D human pose estimation across rare camera views using synthetic data, particularly in sports applications, by incorporating diverse datasets that include video footage from platforms like YouTube.

Despite the widespread application of pose estimation for capturing complex motion data, there remains a significant gap in datasets specifically targeting fall detection scenarios. While pose estimation is often used to extract skeleton data, no study has demonstrated converting this data into accelerometer data for fall detection. Addressing this gap requires collecting and generating fall-specific data, using pose estimation to ensure the data reflects real-world fall events involving the wrist joints, used in an accelerometer format to enhance the training of fall detection systems.

2.4. Augmentation Techniques

Various augmentation techniques have been employed to generate synthetic time-series data to enhance the training and performance of machine learning models [21]. For example, Dynamic Time Warping (DTW) [22] allows for the creation of new time series by averaging existing series with varying weights. Other techniques like window warping and suboptimal warping [23] manipulate the time-series data to simulate different temporal patterns. Jittering, scaling, permutation, and frequency-based augmentations [24] are also commonly used to introduce variability and simulate real-world noise and conditions.

Despite the advantages of augmentation techniques, generating synthetic data, which accurately captures the variety and complexity of fall scenarios, remains challenging. Falls can vary greatly in direction, speed, and impact, and it can be difficult for standard augmentation techniques to simulate these nuances effectively. Additionally, there is a lack of studies demonstrating the reliability of these techniques specifically for fall data, which highlights the importance of evaluating their impact on machine learning models. Conducting further research into these techniques is crucial for validating their effectiveness and enhancing the robustness of fall detection systems.

3. Methodology

Figure 1 illustrates the complete workflow of the proposed methodology. A detailed explanation of each component in the workflow is provided below.

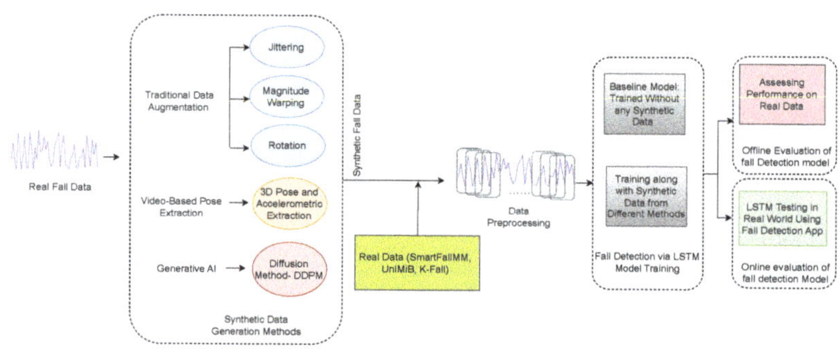

Figure 1. Schematic representation of the overall approach.

3.1. Datasets

We utilized three publicly available fall-related datasets for various synthetic data generation techniques, along with a video dataset and supplementary fall recordings sourced from YouTube. The public datasets are SmartFallMM (https://github.com/txst-cs-smartfall/SmartFallMM-Dataset) (accessed on 29 May 2025) data collected in our laboratory, the UniMiB [25], and the K-Fall [26]. These datasets consist of various simulated falls and daily activities carried out by healthy young adults. The video dataset [27] includes recordings of older adults falling in a long-term care facility but with limited availability. To supplement this, we scraped YouTube for "falls of elderly people" videos. It is worth pointing out that the YouTube-sourced videos were not used as a benchmark dataset but rather as a supplementary source to enhance training diversity. Only videos with clearly observable fall dynamics were used to ensure data validity.

The UniMib dataset was collected to enhance human activity recognition and fall detection using data from wearable devices, such as smartphones. This dataset was collected using a Samsung Galaxy Nexus I9250 smartphone, equipped with a Bosch BMA220 triaxial low-g acceleration sensor. Additionally, the smartphone's built-in microphone recorded audio signals at 8000 Hz, aiding in data annotation. During data collection, participants placed the smartphone in their front trouser pockets, alternating between the left and right sides. Acceleration and audio data were recorded using a custom mobile application and stored in separate files on the smartphone. For data annotation, claps were performed by an external subject before and after each activity or fall to guide the participant and to minimize movement artifacts that could interfere with the recorded sensor data. A 3 s signal window centered around each peak was used for training. The sampling rate of this dataset is 50Hz. The data was collected from 30 participants, aged between 18 and 60. It covers a wide range of activities, including 9 types of daily life activities (ADLs) and 8 different types of falls.

The K-Fall [26] dataset was collected for detecting pre-fall, fall, and post-fall events. It includes data from 32 healthy young adults who performed 21 daily activities and 15 simulated falls, with each activity repeated 5 times. Data collection utilized a custom-designed system that synchronizes sensor data with high-frequency video. A nine-axis inertial sensor was attached to the lower back of participants, recording data from a three-axis accelerometer, gyroscope, and magnetometer at a frequency of 100 Hz. This sensor also provided orientation measurements in Euler angles with data integrated and filtered using an extended Kalman filter. The data was transmitted via Bluetooth to a Raspberry Pi 4, which managed the recording process using a custom graphical user interface (GUI) and handled synchronization between the sensor and a Raspberry Pi HQ camera. The camera captured video at 90 frames per second to provide synchronized visual data. The video recordings have been primarily used for accurate labeling of fall events. For this study, only the accelerometer data was used.

SmartFallMM is a multi-modal dataset collected in our laboratory, involving both older and young adults. The dataset includes data from 24 older adults between the age of 60 and 93 and 16 young adults aged between 20 and 30. For this paper, we utilized data from 12 young adult participants (7 male and 5 female). At the time of this experimental study, there were only 12 younger people data that were labeled. Participants were instructed to simulate five types of falls (front, back, left, right, and rotational) on an air mattress and perform nine types of ADLs, with five repetitions each. Three Azure Kinect cameras recorded all falls and ADLs, capturing the skeletal posture associated with each activity for cross-modal learning research and accurately labeling the beginning and ending of all trials. The Meta Sensor (from MBIENT lab) was positioned on the right wrist, the Huawei watch on the left wrist, the Nexus smartphone on the right hip, and another Meta Sensor on the left hip. This setup allowed us to collect data from four crucial joints of the human body during various movements. This data collection process was approved by the Institutional Review Board (IRB 9461) at Texas State University. Only the accelerometer data from the Huawei Smartwatch was used to generate synthetic fall data.

The final dataset we used includes a collection of videos [27] featuring older adults experiencing falls in long-term care facilities, as well as short clips of falls of older adults from YouTube. The YouTube videos capture candid fall scenarios, providing an alternative, realistic representation of falls in everyday environments. We used 24 videos from YouTube in this study and 34 videos from the two long-term care facilities. These falls occurred exclusively in shared spaces like lounges, dining rooms, and hallways within the facilities.

For the SmartFallMM and UniMiBand K-Fall public fall datasets, all the existing fall samples in the datasets were utilized for synthetic data generation. In the video extraction method, we utilized all available fall videos. The original real fall datasets were then expanded by progressively adding synthetic fall data until a balance was achieved between fall and no-fall instances. This strategy ensures even distribution within the datasets, enhancing the reliability of the subsequent analysis.

3.2. Fall Detection Learning Model

We employed a simple long short-term memory (LSTM) deep learning model, which has been shown to be effective for time-series data due to its capability to learn temporal dynamics. Our LSTM model's input layer has 3 nodes corresponding to the accelerometer's x, y, and z vectors, with an input shape of (W, 3, 64), where W represents the window size set to 128, capturing around four seconds of data, and 64 is the batch size. The training process utilizes the Binary Cross Entropy (BCE) loss function with the ADAM optimizer. This model has been successfully deployed and tested in our SmartFall App [28], where it demonstrated better performance compared to the 1D CNN, Gradient Boosting, and Random Forest methods [29]. Figure 2 represents the basic LSTM architecture.

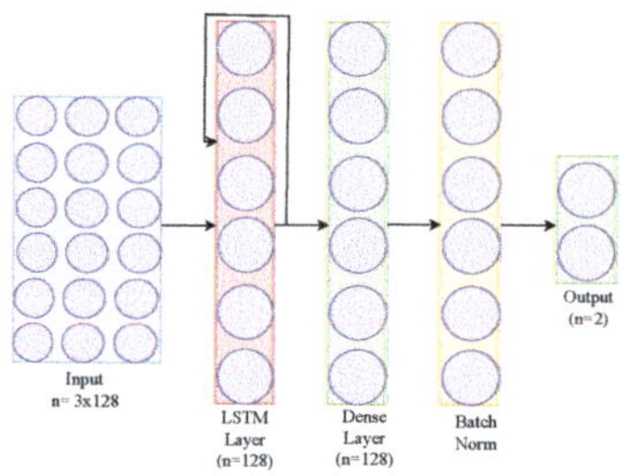

Figure 2. Basic LSTM architecture.

This LSTM model consists of four layers, including an LSTM layer, a Batch Normalization layer, and one dense layer with the number of neurons matching the window-size parameter.

3.3. Data Preprocessing, Training, and Evaluation

In the data preprocessing phase, we utilized the sliding window technique to divide the dataset into overlapping windows, applying a step size of 10, which introduces ten new data points into each new window. For consistency across all experiments, the window size was set to 128. This size was chosen based on prior research [28], which demonstrated that it effectively captures the necessary fall patterns for LSTM models. Our experiments focused exclusively on the basic LSTM model, which was tested under various training scenarios. Baseline models were trained using only the original datasets, i.e., without the addition of any synthetic data. The dataset was split into training, validation, and test sets in a 70/20/10 ratio. A 5-fold validation approach was implemented, with each fold consisting of data from 20% of individuals for validation, 10% for testing, and 70% for training. These baseline models provided a reference for assessing the impact of synthetic data.

To evaluate the effectiveness of synthetic data, we trained new LSTM models using a combination of the original training dataset and synthetic data generated by the various methods described in Section 4. Across all three datasets, SmartFallMM, UniMiB, and K-Fall, we generated and included synthetic data only for the fall class, without synthesizing non-fall or activities of daily living (ADL) data. This decision was driven by the fact that real-world data already contains a substantially larger amount of non-fall samples compared to fall events. To prevent class imbalance, we carefully determined the number of synthetic fall samples added to the training data for each dataset, ensuring that the final training set maintained a balanced or nearly balanced distribution between fall and non-fall events (approximately a 50/50 ratio). Validation and testing were performed exclusively on real data to fairly assess the model's generalization performance.

To address potential concerns about preserving extreme variations in fall dynamics, we ensured that no statistical outlier removal or normalization was applied to the original fall data used for training any of the synthetic data generation methods. This design choice was intentional to retain exceptionally high kinetic and kinematic patterns, such as abrupt accelerations or sharp impacts, that often characterize real-world fall events.

The experimental results are presented in the Results Section. The effectiveness of the synthetic data was evaluated using standard metrics such as precision, recall, F1-score, and accuracy to determine which method for generating synthetic data was most effective in addressing the lack of real fall data.

We also tested the most effective model trained with the addition of synthetic data in a real-world environment using the SmartFall App. This real-time assessment involved three student participants, with approval from IRB 9461 at Texas State University. Each participant wore a smartwatch running the SmartFall App, simulated five different types of falls on a queen-size air mattress, and carried out a specified list of seven activities of daily living (ADLs). Both true positive and false positive predictions were documented during the evaluation. In our real-world testing of the fall detection app, the participants performed five repetitions of each of five different fall types, totaling fifteen falls, alongside seven daily life activities to assess false alarm rates. For the LSTM model trained with SmartFallMM data combined with Diffusion-based synthetic data (SF w DF), the system achieved high recall (0.97), missing none or at most one fall, and produced approximately four false positive alerts during daily activities. Similarly, for the model incorporating video-extracted synthetic data (SF w VE), recall was slightly lower (0.95), with around one missed fall and about five false positives. These results demonstrate strong sensitivity in detecting falls, though further work is needed to reduce false alarms and improve practical deployment.

4. Synthetic Data Generation

This section presents the three distinct strategies we employed to generate synthetic fall data: traditional data augmentation techniques, video-based extraction through pose estimation, and Diffusion-based generative modeling. Table 1 summarizes the key characteristics, strengths, and constraints of each of these synthetic data generation techniques, providing insight into their suitability for fall detection applications.

Table 1. Comparison of traditional, video-based, and AI-based synthetic data generation methods.

Method Type	Techniques Used	Key Features	Advantages	Limitations
Traditional Data Augmentation	(1) Jittering (2) Magnitude Warping (3) Rotation	(1) Applies noise, scaling, and rotation (2) Simulates signal variation (3) Enhances model robustness	(1) Simple and computationally efficient (2) Fast to integrate into training (3) Improves generalization to small perturbations	(1) Limited diversity in output (2) Cannot synthesize new or unseen fall dynamics
Video-Based Pose Extraction	(1) 2D-to-3D keypoint estimation (2) Derivation of velocity and acceleration signals	(1) Converts video motion into sensor-like signals (2) Captures fall dynamics from real-world footage	(1) Bridges vision and sensor modalities (2) Generates realistic joint-level signals (3) Cost-effective alternative to sensor data collection	(1) Manual labeling effort required (2) Privacy concerns in video sourcing (3) Background clutter may affect accuracy
DDPM-Based Synthetic Generation	(1) Diffusion models using U-Net (2) Score-based generative training	(1) Learns temporal patterns in real data (2) Generates high-fidelity synthetic sequences	(1) Produces diverse and naturalistic data (2) Captures subtle temporal variations (3) Supports privacy-preserving data synthesis	(1) High computational resource demands (2) Requires careful hyperparameter tuning

4.1. Basic Data Augmentations

Data augmentation has become a widely adopted technique in computer vision, but its application in time-series analysis is still relatively underexplored. In our study, we implemented three specific data augmentation methods: jittering [30], magnitude warping [30], and rotation [21].

Jittering involves adding random Gaussian noise to time-series data. By introducing small, random perturbations, jittering helps simulate real-world variations and imperfections that may not be present in the original dataset. This augmentation makes the model more robust to noisy data, improving its ability to generalize to new, unseen data while preserving the core patterns in the time series.

Magnitude warping adjusts the amplitude of the time-series data by applying a scaling factor derived from a cubic spline curve. The cubic spline curve modifies the magnitude of the entire time series in a smooth and controlled manner. This technique allows for the exploration of different signal strengths and ranges, which can be crucial for models that need to perform well across various magnitudes of data, effectively helping the model learn to handle amplitude variations.

Rotation simulates different sensor placements by applying rotations to the time-series data. For example, data recorded from a sensor on the left wrist might be rotated to mimic the sensor being placed on the right wrist. This augmentation introduces variability in sensor orientation without changing the actual labels associated with the data. It helps the model learn to recognize patterns despite differences in how data is collected, making the model more adaptable to different sensor configurations and improving its generalization to diverse scenarios.

4.2. *Diffusion Method—DDPM*

Denoising Diffusion Probabilistic Models (DDPMs) [15,31] represent a class of generative AI models that have demonstrated remarkable success in synthesizing high-quality data across domains, such as images and audio.

We have integrated a Diffusion model with the U-Net architecture adapted from previous work [15]. Initially developed for image analysis, the U-Net architecture, as illustrated in Figure 3, has been adapted for time-series data by employing one-dimensional (1D) convolutional layers with a kernel size of 7 and padding of 3, enabling it to capture crucial temporal dependencies in the data.

The time-series input is first processed with RMSNorm to ensure stable training. The network architecture, comprising ResNet blocks and Linear Attention units, executes downsampling and upsampling operations to refine features and preserve temporal information. Each block incorporates time and sinusoidal positional embeddings, enabling the model to effectively respond to Diffusion timesteps and sequence positions.

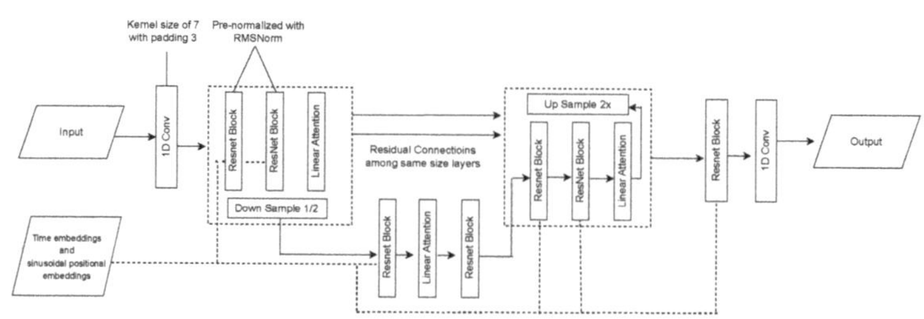

Figure 3. Schematic illustration of the U-Net architecture adapted for time-series data in the Diffusion model.

Li's original model [15] faced difficulties in effectively handling accelerometer data with varying lengths across different fall scenarios, which made it challenging to standardize the input for the model. Additionally, the model lacked a reliable normalization method, which could lead to inconsistencies during training. To address these issues, we introduced a padding strategy during the preprocessing stage. Specifically, we padded each sample with its last value, which allowed us to standardize the input size across all data points without discarding any important information. This modification not only stabilized the model's performance but also ensured that no critical details were lost during data preparation. Equations (1) and (2) detail the strategies we used.

Adjusting Trial Lengths:

$$L_{\text{new}} = \left\lceil \frac{L_{\max}}{8} \right\rceil \times 8 \tag{1}$$

Padding Operation:

$$x' = \text{pad}(x, L_{\text{new}} - L_{\text{current}}, v) \tag{2}$$

where

- x is the original trial data;
- $L_{\max}$, L_{current}, and L_{new} are the max, current, and new lengths, respectively;
- pad denotes the padding operation;
- v is the value used for padding, the last element of x.

To address normalization challenges, particularly in training stability with deep networks, we introduced RMSNorm [31] within the ResNet blocks of the U-Net structure. This normalization technique contributes to stable training dynamics. The use of scaling activation magnitudes improves convergence speed and model performance with variable-length time-series data. Implementing padding strategies and RMSNorm ensures consistent handling of variable-length data, optimizing the generation of synthetic time-series data and enhancing fall detection models.

4.3. Extraction of Fall Data from Video via Pose Estimation

In 3D pose estimation from video, deep learning methods, particularly neural networks, are widely used. Researchers utilize large datasets, including real-world data, to improve how well models perform across different situations. A common approach is using 2D-to-3D estimation techniques, which rely on 2D keypoints to make computations more efficient and allow for real-time results. There is also a trend toward end-to-end systems that directly predict 3D poses, making the process more straightforward. We have implemented the methodology outlined in [32,33]. To accurately extract fall data, we processed 34 publicly available videos from [27] and 24 additional YouTube videos. Our method involved cropping the video frames to isolate the falling person, focusing on reducing extraction time and highlighting relevant data. We retained 1 to 2 s of footage before and after the fall. Each video was adjusted for resolution and brightness as needed.

Using 3D pose estimation, we extracted the positions of 17 joints from the human skeleton in each video. For synthetic data generation, we focused on specific joint positions depending on the target dataset. For instance, to match the SmartFallMM dataset, we extracted accelerometer data from the left wrist joint. For the UniMiB dataset, we extracted data from both the left and right hips. Acceleration data was derived from 3D keypoints by calculating the velocity from position changes and then the acceleration from velocity changes. To simulate accelerometer signals more precisely, we computed joint-specific acceleration data from the extracted 3D coordinates by calculating discrete second-order differences across consecutive frames. For example, considering the left wrist joint, let $\mathbf{p}(f) = [x(f), y(f), z(f)]$ denote the 3D positional coordinates at frame f. The acceleration vector $\mathbf{a}(f)$ is estimated using the discrete second derivative:

$$\mathbf{a}(f) = \frac{\mathbf{p}(f+1) - \mathbf{p}(f)}{\Delta t^2} \tag{3}$$

where Δt is the time interval between video frames ($\Delta t = 1/46\,\text{s}$). Each axis component, $a_x(f)$, $a_y(f)$, and $a_z(f)$, was computed independently and scaled to produce acceleration values in m/s^2.

Figure 4 illustrates the methodology for obtaining the accelerometer readings from the joint positions in the video. This approach yielded approximately 54 fall samples from the videos. To ensure data quality, we addressed occlusions by discarding samples where key joints (e.g., the left wrist or hips) were not reliably tracked due to motion blur, low resolution, or brief occlusion during the fall. Additionally, we applied temporal smoothing to correct minor discontinuities in joint trajectories. Only sequences with stable and continuous joint tracking across the fall window were retained for synthetic accelerometer extraction.

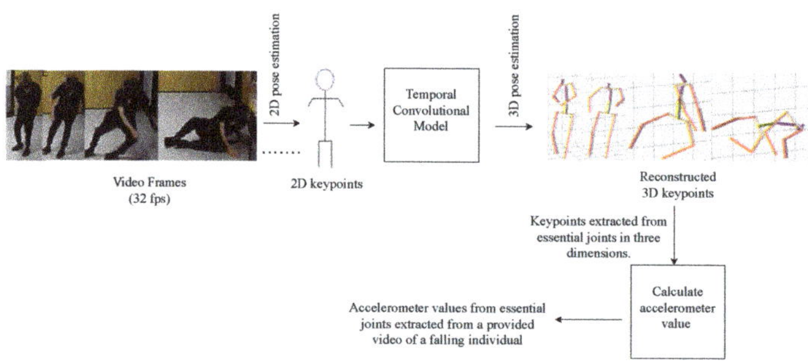

Figure 4. Pipeline for extracting synthetic accelerometer data from video frames. Colored segments in the 3D skeletons represent different body parts (e.g., arms, legs, torso), with distinct colors used to aid visual interpretation. The dots indicate the presence of additional frames in the video sequence.

4.4. Evaluation Metrics for Synthetic Data Quality

To evaluate the quality of the synthetic time-series data, we used the following metrics:

- **FID Score**: Measures the difference in contextual representations of synthetic and real samples, based on the Fréchet distance between feature distributions. The score ranges from 0 to ∞, where lower values indicate higher similarity and thus better quality [34].
- **Discriminative Score**: Evaluates how well a binary classifier can distinguish real from synthetic data. Typically expressed as an accuracy percentage ranging from 50% to 100%. A lower score, close to 50%, indicates higher similarity between real and synthetic data, as it implies the classifier cannot reliably differentiate them [9].
- **Predictive Score**: Measures the utility of synthetic data for forecasting future time steps. It is often reported as a prediction error metric (e.g., RMSE) or as accuracy for classification tasks. The score ranges from 0 to ∞ (for error metrics), with lower values indicating better predictive utility [9].
- **Jensen–Shannon Divergence (JSD)**: Quantifies the similarity between the probability distributions of real and synthetic data. JSD values range from 0 to 1, where 0 indicates identical distributions, and higher values indicate greater divergence [35].
- **Kolmogorov–Smirnov (KS) Test:** A non-parametric statistical test used to compare two samples and determine whether they come from the same distribution. Unlike normality tests that only assess fit to a specific distribution, the K–S test directly compares empirical distributions, making it especially useful for evaluating model outputs or validating synthetic data against real samples [36]. The test outputs a D-statistic ranging from 0 to 1, where lower values suggest greater similarity between distributions. A corresponding p-value indicates statistical significance; higher p-values imply no significant difference between the distributions.

5. Results

For consistency and clarity throughout the Results Section, we use the following abbreviations for datasets and augmentation methods: SF refers to SmartFallMM, UM denotes UniMiB, and KF represents K-Fall. Regarding the synthetic data generation and augmentation methods, Jit stands for jittering, VE denotes video extraction, MW represents magnitude warping, and Ro corresponds to rotation. These abbreviations will be used consistently in all the result tables and discussions. Additionally, to ensure completeness and benchmark against state-of-the-art methods, we include a comparison with Diffusion-TS, which is a recent generative AI-based approach.

5.1. Quality Evaluation Results for Synthetic Data

Table 2 presents a comprehensive comparison of the evaluation metrics for the three datasets across multiple methods, the Context-FID Score, Discriminative Score, Predictive Score, Jensen–Shannon Divergence (JSD), and Kolmogorov–Smirnov statistic (KS), where lower values indicate better performance. The bolded values in the table indicate the best performance for each score. The proposed Diffusion (ours) model achieves the lowest Context-FID scores (e.g., 0.409 on SmartFallMM and 1.523 on UniMiB), demonstrating better contextual fidelity. The model achieves notably low discriminative scores (e.g., 0.704 on SmartFallMM and 0.500 on K-Fall), substantially outperforming Jit (0.833 and 0.921) and MW (0.801 and 0.879). This suggests that the generated data more accurately captures the statistical and temporal dynamics of real accelerometer signals. Moreover, the model achieves the lowest JSD values across all the datasets, 0.056 on SmartFallMM, 0.009 on UniMiB, and 1.610 on K-Fall, indicating closer distributional alignment with real data than other approaches, such as Diffusion-TS (0.241, 0.874, and 0.458, respectively) and Jit (0.295, 0.981, and 0.598). This alignment suggests a better preservation of the underlying statistical properties essential for realistic synthetic data.

For KS, across all three datasets, our proposed Diffusion-based generator consistently achieves the smallest average $\bar{D}$: 0.142 on SmartFallMM and 0.095 on UniMiB, indicating that its synthetic accelerometer data most closely match the real distributions. On the K-Fall dataset, the video-extracted method performs slightly better than proposed Diffusion model ($\bar{D} = 0.321$ vs. $\bar{D} = 0.345$). Jit (0.314–0.434), MW (0.211–0.377), and Ro (0.188–0.459) yield substantially higher $\bar{D}$ values, underscoring their inability to capture true sensor statistics. Notably, UniMiB proves easiest to model (all methods $\bar{D} < 0.13$), whereas K-Fall remains the hardest, with even the best method exceeding $\bar{D} = 0.32$. These results demonstrate that learned generative approaches, particularly our Diffusion variant, offer better distributional fidelity compared to traditional augmentations, especially on complex real-world time-series signals. In summary, the proposed Diffusion (ours) model consistently outperforms all the competing methods across all the datasets and metrics, highlighting the robustness and generalizability in generating better-quality, real-like, and task-relevant synthetic data for fall detection applications.

Figures 5–7 provide a comparative analysis of real and synthetic fall data, focusing on Diffusion-generated and video-extracted synthetic data, the two best-performing generation methods, across the three datasets. Each dataset comprises two comparison scenarios: "Real vs. Synthetic (Diffusion)" and "Real vs. Synthetic (Video Extracted)."

We analyzed the statistical distribution of the real and synthetic data to assess their similarity. In all the plots shown in this figure, the x-axis represents accelerometer values collected from wearable devices such as smartwatches, depending on the specific dataset. These values measure acceleration along different axes (X, Y, and Z) and are typically expressed in meters per second squared (m/s^2). Higher absolute values indicate stronger movements or impacts detected by the wearable sensor. The y-axis indicates the frequency of these values. These plots illustrate how well the synthetic data reflects the statistical distribution of real data, highlighting the similarity in their normal distributions. In the SmartFallMM, UniMiB, and K-Fall datasets, the Diffusion-generated synthetic fall data closely match the distribution of the real fall data, indicating a strong alignment. Notably, the synthetic data derived from YouTube videos via video extraction also demonstrated a similar distributional pattern to those in the benchmark datasets, supporting the validity of these videos as a viable supplementary data source. While the video-extracted synthetic data exhibits a narrower distribution range, particularly in the accelerometer values along the x, y, and z axes, compared to the real fall data, this indicates that the synthetic data captures a more focused spectrum of fall dynamics. This focused representation can be useful for emphasizing key fall characteristics while still providing valuable insights into the dynamics of falls. Overall, these results suggest that the synthetic data effectively replicates the statistical variability observed in real falls across all three datasets.

Table 2. Comparison of evaluation metrics for different methods across multiple datasets. Bold values indicate the best (lowest) score achieved for each metric and dataset combination.

Evaluation Metrics	Methods	SmartFallMM	UniMiB	K-Fall
Context-FID Score (↓)	Proposed Diffusion	**0.409**	**1.523**	**2.753**
	Diffusion-TS	0.722	2.192	3.179
	Video Extracted	0.910	2.654	3.854
	Jit	0.845	2.500	3.620
	MW	0.872	2.712	3.730
	Ro	0.801	2.410	3.500
Discriminative Score (↓)	Proposed Diffusion	0.704	1.620	**0.500**
	Diffusion-TS	**0.500**	**0.321**	1.155
	Video Extracted	0.910	1.875	0.882
	Jit	0.833	1.743	0.921
	MW	0.801	1.692	0.879
	Ro	0.850	1.730	0.901
Predictive Score (↓)	Proposed Diffusion	**0.735**	**0.745**	**0.759**
	Diffusion-TS	0.773	0.976	2.636
	Video Extracted	0.862	1.112	1.983
	Jit	0.820	1.045	1.901
	MW	0.793	1.089	2.001
	Ro	0.834	1.099	2.090
JSD (↓)	Proposed Diffusion	**0.056**	**0.009**	1.610
	Diffusion-TS	0.241	0.874	**0.458**
	Video Extracted	0.342	1.020	0.602
	Jit	0.295	0.981	0.598
	MW	0.310	1.001	0.610
	Ro	0.280	0.998	0.590
KS Score (↓)	Proposed Diffusion	**0.142**	**0.095**	0.345
	Diffusion-TS	0.231	0.112	0.347
	Video Extracted	0.150	0.103	**0.321**
	Jit	0.314	0.112	0.434
	MW	0.211	0.105	0.377
	Ro	0.188	0.121	0.459

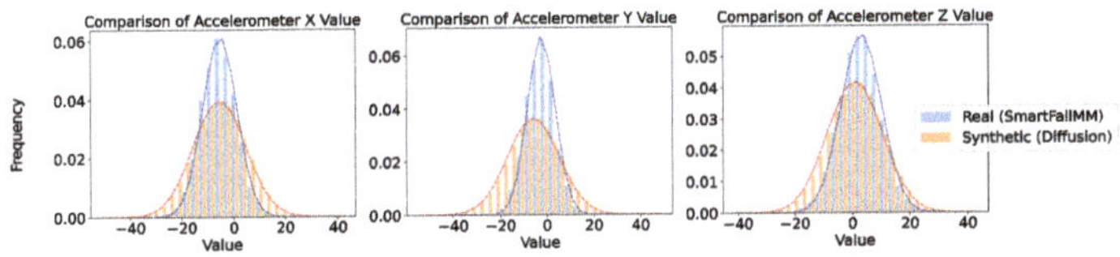

(**a**) SF vs. SF with Diffusion

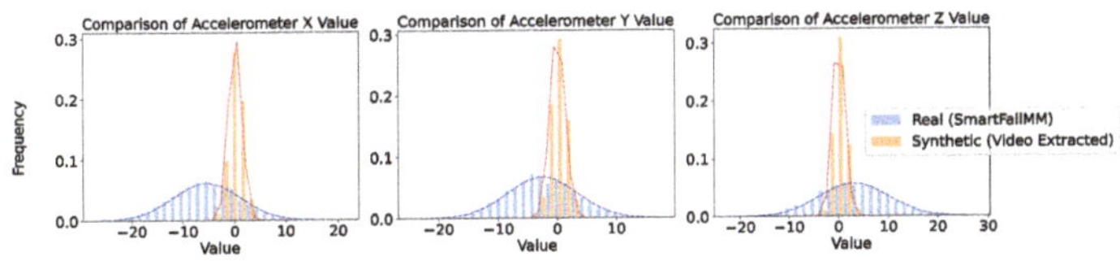

(**b**) SF vs. SF with Video Extracted

Figure 5. Comparison of real and generated data for SmartFallMM. Blue represents the real data distribution, while orange denotes the synthetic data distribution.

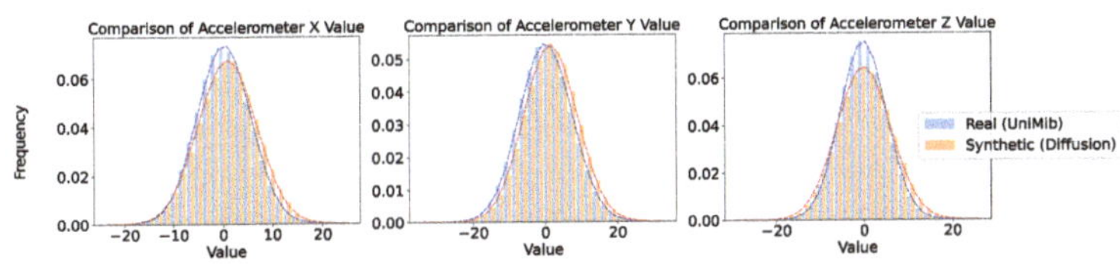

(**a**) UM vs. UM with Diffusion

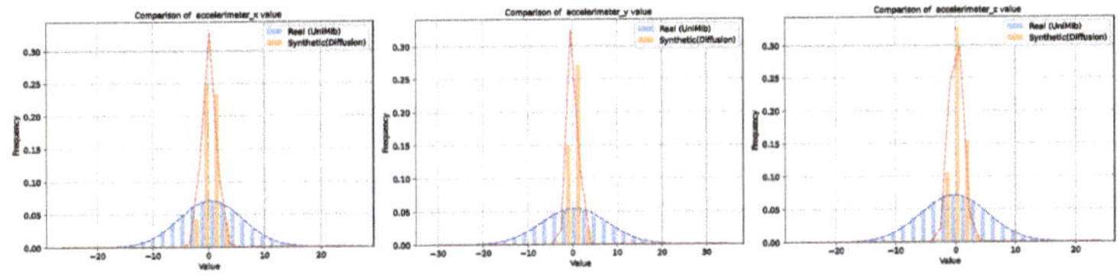

(**b**) UM vs. UM with Video Extracted

Figure 6. Comparison of real and generated data for UniMiB. Blue represents the real data distribution, while orange denotes the synthetic data distribution.

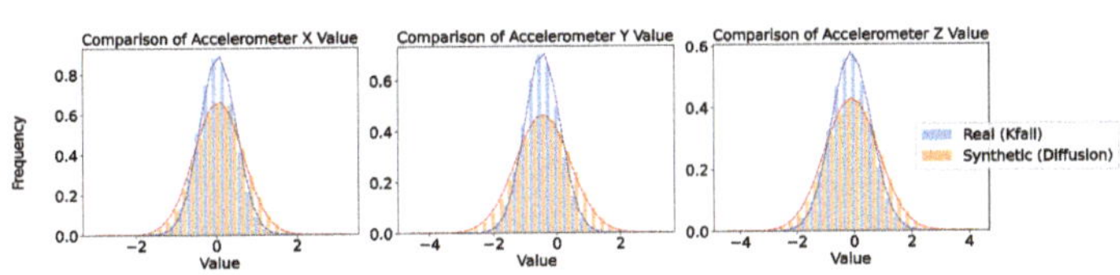

(**a**) KF vs. KF with Diffusion

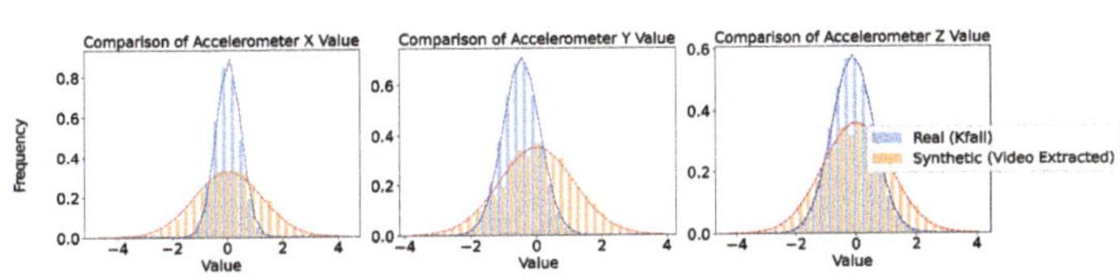

(**b**) KF vs. KF with Video Extracted

Figure 7. Comparison of real and generated data for K-Fall. Blue represents the real data distribution, while orange denotes the synthetic data distribution.

5.2. Offline Evaluation of Fall Detection Model

Table 3 presents an overview of the comparative evaluation performed on the three datasets, incorporating the synthetic data generated using three different methods. The values highlighted in bold and italics correspond to the highest and second-highest improvements, respectively, observed in the fall detection task. In this analysis, we present the average results from 5-fold validation, which involves conducting five experiments for each dataset, both with and without synthetic data. The average precision, recall, F1-score, and accuracy from these experiments are reported for each dataset. To enhance clarity, we refer to SmartFallMM as SF, UniMiB as UM, K-Fall as KF, video extraction from older home people along with YouTube videos as VE, jittering as Jit, magnitude warping as MW, and rotation as Ro. The results obtained by solely using the SmartFallMM dataset (SF) without any data augmentation yield an F1-score of 0.72 and an accuracy of 0.77. After applying synthetic data generated from various techniques, the most promising result obtained for SF is by incorporating the combination of the VE method, achieving an F1-score of 0.84, which indicates an improvement of almost 12%.

Using Diffusion-generated data also showed improved performance with an F1-score of 0.80 than the baseline result. These results demonstrate that Diffusion and video extraction-based methods are both effective in synthesizing accelerometer data for fall detection. They enhance the diversity of our dataset by covering a broader range of fall scenarios. We also evaluated and compared our findings using two other public datasets: UniMiB and K-Fall. Both K-Fall and UniMiB exhibited performance improvements when utilizing Diffusion-generated data and the video extraction method. Specifically, UniMiB showed an approximate 8% improvement, while K-Fall demonstrated a 12% increase compared to the baseline results. Diffusion-generated data demonstrated improvements across all three datasets. Also, when additional video data was incorporated (YouTube-scraped video) using the video extraction method, the results improved compared to using only the video dataset from the long-term facility presented in our previous work [16]. This suggests a promising approach by including more videos of elderly individuals from sources like YouTube to extract skeleton data and augment real-world fall data.

As observed, the model trained with synthetic data from the proposed Diffusion method consistently outperforms the baseline across all the datasets. Compared to the baseline, the proposed method improves the F1-score by 11.11% on SF (from 0.72 to 0.80), 13.51% on UM (from 0.74 to 0.84), and 16.22% on KF (from 0.74 to 0.86). Similarly, the accuracy increases by 6.49% on SF (from 0.77 to 0.82), 7.89% on UM (from 0.76 to 0.82), and 10% on KF (from 0.80 to 0.88). These consistent and substantial gains reinforce the effectiveness of the proposed Diffusion method in enhancing fall detection performance. While the video-extracted (VE) method also performs competitively (e.g., highest F1-score on SF and UM), the proposed Diffusion method demonstrates a more balanced and consistently high performance across all the metrics and datasets. Based on these results, we recognize both approaches as effective for improving fall detection performance in our framework.

These offline-evaluation results not only validate the effectiveness of our methods but also emphasize the role of offline evaluation in guiding real-world deployment strategies. In fall detection, where real-world fall data is both rare and ethically difficult to obtain, consistent improvements across multiple datasets suggest that the model is learning robust, transferable patterns. These findings establish a foundation for real-time deployment and guide further validation efforts in operational settings.

Table 3. Fall detection model performance with synthetic data from various generation techniques across three datasets. Bold and italic values represent the highest and second-highest improvements observed in the fall detection task.

Datasets	Precision	Recall	F1-Score	Accuracy
SF	0.79	0.68	0.72	0.77
SF w DF	0.82	0.81	*0.80*	*0.82*
SF w VE	0.95	0.78	**0.86**	**0.88**
SF w Jit	0.78	0.66	0.73	0.71
SF w MW	0.82	0.64	0.72	0.71
SF w Ro	0.75	0.68	0.70	0.70
UM	0.71	0.75	0.74	0.76
UM w DF	0.79	0.83	*0.84*	*0.82*
UM w VE	0.82	0.88	**0.86**	**0.84**
UM w Jit	0.70	0.72	0.74	0.76
UM w MW	0.73	0.76	0.75	0.76
UM w Ro	0.69	0.72	0.71	0.73
KF	0.90	0.60	0.74	0.80
KF w DF	0.92	0.82	**0.86**	**0.88**
KF w VE	0.90	0.78	0.83	0.85
KF w Jit	0.80	0.78	0.79	0.80
KF w MW	0.85	0.72	0.80	0.81
KF w Ro	0.82	0.81	0.81	0.82

5.3. Real-Time Evaluation of Fall Detection Model

Table 4 showcases the real-world evaluation results for the top-performing offline model. Real-world testing was conducted with the SmartFall App, which is designed to operate with data sensed from a wearable watch. The presented values represent average performance across three participants, with the bold values showing the highest gain. Initially, we evaluated the SmartFall App using a basic LSTM model trained solely on real SmartFallMM data (SF), which achieved an F1-score of 0.62. We then evaluated the top models trained with combinations of synthetic and real data, following the protocol described in Section 3.3. These include models trained using Diffusion-generated data (SF w DF) and video-extracted data (SF w VE).

The best real-world performance was observed from the model trained with Diffusion-generated data, achieving an F1-score of 0.86, which is an improvement of approximately 24%. The model trained with VE data also performed well, yielding an F1-score of 0.82 (a 20% improvement). These results closely mirror the trends observed in our offline evaluation. Specifically, the model trained with Diffusion-generated data showed an F1-score improvement of 11.11% offline (from 0.72 to 0.80) and 24% in real-world testing (from 0.62 to 0.86). Similarly, the video extraction method yielded an offline F1-score gain of 19.44% and a corresponding 20% boost in real-time performance. This strong correlation between offline and online results supports the practical utility of offline evaluation as a reliable indicator of real-world effectiveness, especially when robust synthetic data generation techniques are applied.

Table 4. Real-world results using SmartFFall App. Bold values represent the highest gain in the performance.

Dataset	Precision	Recall	F1-Score
SF	0.56	0.75	0.62
SF w DF	0.78	0.97	**0.86**
SF w VE	0.72	0.95	**0.82**

5.4. Limitations

Despite the promising results, this study has a few limitations. First, the current model does not account for individual-specific movement patterns, which may affect generalizability across diverse populations. This highlights the need for personalization techniques that tailor the model to each user's activity profile. Second, our data pipeline relies solely on accelerometer input, omitting additional sensor modalities such as gyroscope or barometric data that could enhance contextual understanding and improve fall detection accuracy. Third, the video extraction process, though effective, depends on manual frame selection, which is time-consuming and may introduce human bias. This limits scalability and consistency, calling for automated frame selection strategies. Finally, the real-world testing reported in Table 4 was conducted on a limited sample of only three younger participants and involved simulated falls. While this provides a proof of concept for evaluating the utility of synthetic data, it does not fully reflect the movement characteristics of the intended elderly population. Testing on actual elderly falls is ethically and practically infeasible. Future work should focus on expanding the participant base and exploring methods to bridge this population gap through personalization and data-driven adaptation techniques.

6. Conclusions

Data scarcity poses a critical barrier to the development of robust fall detection systems, primarily due to the ethical and logistical constraints of collecting real-world fall events. To address this challenge, we propose two advanced synthetic data generation strategies: a Denoising Diffusion Probabilistic Model (DDPM) and a video-based pose estimation approach. These methods are systematically compared with conventional augmentation techniques, such as jittering, magnitude warping, and rotation. To elaborate, our Diffusion-based approach adopts the DDPM framework to generate synthetic fall data that closely mimics the statistical and temporal characteristics of real falls. On the other hand, the video extraction technique leverages the pose estimation technique from computer vision to capture fall events from videos and extract key motion patterns and sensor data. The distinct contribution of this work lies in the integration of the proposed methods, which are underexplored in fall detection, and their validation through both offline benchmarking and real-world (online) deployment. This dual validation ensures that the generated data not only aligns statistically with real-world signals but also translates effectively into practical, deployable fall detection systems.

In offline evaluations, across all datasets, including SmartFallMM, UniMiB, and K-Fall, the results showed that models trained with a mix of real and synthetic data outperformed those trained with real data alone, highlighting the value of synthetic data in enhancing model robustness and generalization. Specifically, the inclusion of DDPM-based and video-extracted synthetic data led to more accurate and reliable fall detection models.

Notably, our top-performing synthetic data generation methods for the SmartFallMM dataset, which are the Diffusion-based and video-extracted techniques, also demonstrated promising real-world performance, making it effective for practical deployment in fall detection systems. Specifically, the Diffusion-based method improved the offline F1-score on SmartFallMM by 11.11% (from 0.72 to 0.80), which corresponded to a 24% gain in the real-world F1-score (from 0.62 to 0.86) when tested using the SmartFall App. This strong alignment underscores the practical merit of offline evaluation as a reliable predictor of real-world performance.

7. Future Work

While our proposed methods demonstrated substantial improvements in both offline performance and real-world deployment, further enhancements are necessary for adoption in high-stakes environments, such as eldercare or clinical monitoring. To this end, our future work will focus on addressing the limitations by first exploring the personalization techniques that adapt the model to individual user behavior, the integration of additional sensor modalities (e.g., gyroscope, barometer, or ambient context), and the refinement of the synthetic data pipeline itself. In particular, based on our findings that the DDPM Diffusion-based and video extraction methods yielded the best results, we aim to refine the video extraction process by focusing on relevant frames only, as suggested by Votel and Li [37], who successfully used three frames per video for the training of a fall detection model. Extracting frames manually is time-consuming and labor-intensive, so we intend to explore automatic frame selection methods, such as the one proposed by Liu et al. [38], which automatically identifies images with fall activity. This approach could streamline the data preprocessing phase and potentially enhance the performance of the fall detection system by using only the most pertinent information from each video.

Secondly, we aim to further explore other video extraction techniques for generating synthetic fall data. Recording the real-world fall data from older adults poses significant challenges because of privacy issues, making it essential to explore alternative data sources. One such source is generative AI tools like Sora [39], an AI platform with text-to-video capabilities, which can produce videos up to one minute long. By leveraging videos of elderly individuals falling from AI tools, we can further enhance our synthetic datasets and improve fall detection models.

In parallel, we intend to explore the cross-modal data generation techniques such as text-to-motion methods IMUGPT [40] and T2M-GPT [41], which generate skeleton data-based motions from given text prompts. Those cross-model generation techniques are built upon a large human motion dataset.

Furthermore, future research could explore integrating fall-type classification into the synthetic data pipeline and extending these techniques, originally developed for fall detection, to other domains, such as sports injury prevention and rehabilitation monitoring. This could broaden the research's impact by enhancing predictive models and monitoring systems across diverse movement-related applications.

Finally, while our study focuses on synthetic data generation using Diffusion models and video-based pose estimation, we acknowledge the critical role of gold-standard datasets based on 3D motion capture and artificially induced falls using programmable systems, like Motek. Such systems, equipped with built-in safety mechanisms, have recently enabled the collection of high-fidelity fall data through controlled perturbations. We believe that incorporating such gold-standard benchmarks will be essential in future studies to validate and further refine synthetic data pipelines, offering a complementary path alongside generative AI approaches.

Author Contributions: Conceptualization, A.H.N. and M.D.; methodology, M.D.; software, M.D. and M.S.K.; validation, A.H.N., M.D., and S.A.; formal analysis, A.H.N., M.D., and S.A.; investigation,

M.D. and S.A.; resources, A.H.N.; data curation, M.D. and M.S.K.; writing—original draft preparation, M.D. and S.A.; writing—review and editing, M.D., S.A., and A.H.N.; visualization, M.D.; supervision, A.H.N.; project administration, A.H.N.; funding acquisition, A.H.N. All authors have read and agreed to the published version of the manuscript.

Funding: This research was funded by the National Science Foundation (NSF) under the Smart and Connected Health (SCH) Program grant number 21223749.

Institutional Review Board Statement: The study protocols were approved by the Texas State University's Institutional Review Board (IRB), with protocol numbers #9613 for the movement study.

Informed Consent Statement: Informed consent was obtained from all subjects involved in this study.

Data Availability Statement: We are happy to share the synthetic fall data generated in this study for research purposes upon request.

Conflicts of Interest: The authors declare no conflicts of interest.

References

1. Matos-Carvalho, J.P.; Correia, S.D.; Tomic, S. Sensitivity Analysis of LSTM Networks for Fall Detection Wearable Sensors. In Proceedings of the 2023 6th Conference on Cloud and Internet of Things (CIoT), Lisbon, Portugal, 20–22 March 2023; pp. 112–118. [CrossRef]
2. Sherin Zebiah, S.; Shalomy, A.V.; Kachhap, J.; Tete, N.; Nancy, R.; Ananthi, A.; Prasanna, J.; Subathra, M.; George, S. Human Fall Detection Using Machine Learning and Deep Learning Techniques: A Survey. In Proceedings of the 2023 4th International Conference on Signal Processing and Communication (ICSPC), Coimbatore, India, 23–24 March 2023; pp. 253–257. [CrossRef]
3. Maray, N.; Ngu, A.H.; Ni, J.; Debnath, M.; Wang, L. Transfer learning on small datasets for improved fall detection. *Sensors* **2023**, *23*, 1105. [CrossRef]
4. Islam, M.M.; Tayan, O.; Islam, M.R.; Islam, M.S.; Nooruddin, S.; Nomani Kabir, M.; Islam, M.R. Deep Learning Based Systems Developed for Fall Detection: A Review. *IEEE Access* **2020**, *8*, 166117–166137. [CrossRef]
5. Le, T.A.; Baydin, A.G.; Zinkov, R.; Wood, F. Using synthetic data to train neural networks is model-based reasoning. In Proceedings of the 2017 International Joint Conference on Neural Networks (IJCNN), Anchorage, AK, USA; 14–19 May 2017; pp. 3514–3521. [CrossRef]
6. Zherdev, D.; Zherdeva, L.; Agapov, S.; Sapozhnikov, A.; Nikonorov, A.; Chaplygin, S. Producing Synthetic Dataset for Human Fall Detection in AR/VR Environments. *Appl. Sci.* **2021**, *11*, 1938. [CrossRef]
7. Juneja, T.; Bajaj, S.B.; Sethi, N. Synthetic Time Series Data Generation Using Time GAN with Synthetic and Real-Time Data Analysis. In Proceedings of the International Conference on Recent Innovations in Computing, Jammu, India, 26–27 October 2023; Singh, Y., Verma, C., Zoltán, I., Chhabra, J.K., Singh, P.K., Eds.; Springer : Singapore, 2023; pp. 657–667.
8. Ashrafi, N.; Schmitt, V.; Spang, R.P.; Möller, S.; Voigt-Antons, J.N. Protect and Extend—Using GANs for Synthetic Data Generation of Time-Series Medical Records. *arXiv* **2024**, arXiv:2402.14042. [CrossRef]
9. Yoon, J.; Jarrett, D.; van der Schaar, M. Time-series Generative Adversarial Networks. In Proceedings of the Advances in Neural Information Processing Systems, Vancouver, BC, Canada, 8–14 December 2019; Volume 32.
10. Nikitin, A.; Iannucci, L.; Kaski, S. TSGM: A Flexible Framework for Generative Modeling of Synthetic Time Series. *arXiv* **2024**, arXiv:2305.1156.
11. Li, Y. TS-Diffusion: Generating Highly Complex Time Series with Diffusion Models. *arXiv* **2023**, arXiv:2311.03303. [CrossRef]
12. Tian, M.; Chen, B.; Guo, A.; Jiang, S.; Zhang, A.R. Reliable Generation of EHR Time Series via Diffusion Models. *arXiv* **2023**, arXiv:2310.15290. [CrossRef]
13. Noury, N.; Fleury, A.; Rumeau, P.; Bourke, A.; Laighin, G.O.; Rialle, V.; Lundy, J. Fall detection—Principles and Methods. In Proceedings of the 2007 29th Annual International Conference of the IEEE Engineering in Medicine and Biology Society, Lyon, France, 22–26 August 2007; pp. 1663–1666. [CrossRef]
14. Purkrabek, M.; Matas, J. Improving 2D Human Pose Estimation in Rare Camera Views with Synthetic Data. In Proceedings of the 2024 IEEE 18th International Conference on Automatic Face and Gesture Recognition (FG), Istanbul, Turkey, 27–31 May 2024; IEEE: Piscataway, NJ, USA , 2024. [CrossRef]
15. Li, X. Mitigating Data Shortage in Biomedical Signal Analysis: An Investigation into Transfer Learning and Generative Models. Ph.D. Dissertation, Texas State University, San Marcos, TX, USA, 2023.

16. Debnath, M.; Kabir, M.S.; Ni, J.; Ngu, A.H.H. The Impact of Synthetic Data on Fall Detection Application. In Proceedings of the Artificial Intelligence in Medicine, Salt Lake City, UT, USA, 9–12 July 2024; Finkelstein, J., Moskovitch, R., Parimbelli, E., Eds.; Springer: Cham, Switzerland, 2024; pp. 204–209.
17. Zhang, C.; Kuppannagari, S.R.; Kannan, R.; Prasanna, V.K. Generative Adversarial Network for Synthetic Time Series Data Generation in Smart Grids. In Proceedings of the 2018 IEEE International Conference on Communications, Control, and Computing Technologies for Smart Grids (SmartGridComm), Aalborg, Denmark, 29–31 October 2018; pp. 1–6. [CrossRef]
18. Lin, Z.; Jain, A.; Wang, C.; Fanti, G.; Sekar, V. Using GANs for Sharing Networked Time Series Data: Challenges, Initial Promise, and Open Questions. In Proceedings of the ACM Internet Measurement Conference, ACM, 2020, IMC'20, Virtual Event, 27–29 October 2020. [CrossRef]
19. Choi, E.; Biswal, S.; Malin, B.; Duke, J.; Stewart, W.F.; Sun, J. Generating Multi-label Discrete Patient Records using Generative Adversarial Networks. *arXiv* **2018**, arXiv:1703.06490. [CrossRef]
20. Yuan, X.; Qiao, Y. Diffusion-TS: Interpretable Diffusion for General Time Series Generation. *arXiv* **2024**, arXiv:2403.01742. [CrossRef]
21. Um, T.T.; Pfister, F.M.J.; Pichler, D.; Endo, S.; Lang, M.; Hirche, S.; Fietzek, U.; Kulić, D. Data augmentation of wearable sensor data for Parkinson's disease monitoring using convolutional neural networks. In Proceedings of the 19th ACM International Conference on Multimodal Interaction, ACM, 2017, ICMI'17, Glasgow, UK, 13–17 November 2017. [CrossRef]
22. Fawaz, H.I.; Forestier, G.; Weber, J.; Idoumghar, L.; Muller, P.A. Data augmentation using synthetic data for time series classification with deep residual networks. *arXiv* **2018**, arXiv:1808.02455. [CrossRef]
23. Flores, A.; Tito-Chura, H.; Apaza-Alanoca, H. Data Augmentation for Short-Term Time Series Prediction with Deep Learning. In Proceedings of the Intelligent Computing, Virtually, 15–16 July 2021; Arai, K., Ed.; Springer: Cham, Switzerland, 2021; pp. 492–506.
24. Iglesias, G.; Talavera, E.; González-Prieto, A.; Mozo, A.; Gómez-Canaval, S. Data Augmentation techniques in time series domain: A survey and taxonomy. *Neural Comput. Appl.* **2023**, *35*, 10123–10145. [CrossRef]
25. Micucci, D.; Mobilio, M.; Napoletano, P. UniMiB SHAR: A Dataset for Human Activity Recognition Using Acceleration Data from Smartphones. *Appl. Sci.* **2017**, *7*, 1101. [CrossRef]
26. Yu, X.; Jang, J.; Xiong, S. A large-scale open motion dataset (KFall) and benchmark algorithms for detecting pre-impact fall of the elderly using wearable inertial sensors. *Front. Aging Neurosci.* **2021**, *13*, 692865. [CrossRef]
27. Robinovitch, S. *Falls Experienced by Older Adult Residents in Long-Term Care Homes*; Databrary New York University: New York, NY, USA, 2018. [CrossRef]
28. Ngu, A.H.; Metsis, V.; Coyne, S.; Srinivas, P.; Salad, T.; Mahmud, U.; Chee, K.H. Personalized Watch-Based Fall Detection Using a Collaborative Edge-Cloud Framework. *Int. J. Neural Syst.* **2022**, *32*, 2250048. . [CrossRef] [PubMed]
29. Mauldin, T.R.; Ngu, A.H.; Metsis, V.; Canby, M.E. Ensemble Deep Learning on Wearables Using Small Datasets. *ACM Trans. Comput. Healthc.* **2021**, *2*, 5. [CrossRef]
30. Nikitin, A.; Iannucci, L.; Kaski, S. TSGM—A Flexible Framework for Synthetic Time Series Generative Modeling. *Adv. Neural Inf. Process. Syst.* **2024** , *37*, 129042–129061.
31. Ho, J.; Jain, A.; Abbeel, P. Denoising Diffusion Probabilistic Models. In Proceedings of the Advances in Neural Information Processing Systems 33 (NeurIPS 2020), Virtual, 6–12 December 2020; Larochelle, H., Ranzato, M., Hadsell, R., Balcan, M., Lin, H., Eds.; Curran Associates, Inc.: Sydney, Australia, 2020; Volume 33, pp. 6840–6851.
32. Liu, J.; Rojas, J.; Li, Y.; Liang, Z.; Guan, Y.; Xi, N.; Zhu, H. A graph attention spatio-temporal convolutional network for 3D human pose estimation in video. In Proceedings of the 2021 IEEE International Conference on Robotics and Automation (ICRA), Xi'an, China, 30 May–5 June 2021; IEEE: Piscataway, NJ, USA, 2021; pp. 3374–3380.
33. Alamgeer, S.; Souissi, Y.; Ngu, A.H.H. AI-Generated Fall Data: Assessing LLMs and Diffusion Model for Wearable Fall Detection. *arXiv* **2025**, arXiv:2505.04660. [CrossRef]
34. Paul, J.; Bohlke-Schneider, M.; Mercado, P.; Kapoor, S.; Singh, N.R.; Flunkert, V.; Gasthaus, J.; Januschowski, T. PSA-GAN: Progressive Self Attention GANs for Synthetic Time Series. *arXiv* **2022**, arXiv:2108.00981. [CrossRef]
35. Menéndez, M.; Pardo, J.; Pardo, L.; Pardo, M. The Jensen-Shannon Divergence. *J. Frankl. Inst.* **1997**, *334*, 307–318. [CrossRef]
36. Wikipedia Contributors. Kolmogorov–Smirnov Test—Wikipedia, The Free Encyclopedia. 2024. Available online: https://en.wikipedia.org/wiki/Kolmogorov (accessed on 19 May 2025).
37. Votel, R.; Na Li, G.R. Next-Generation Pose Detection with MoveNet and TensorFlow.js. *Tensorflow Blog*, 17 May 2021. Available online: https://blog.tensorflow.org/2021/05/next-generation-pose-detection-with-movenet-and-tensorflowjs.html (accessed on 29 May 2025).
38. Khan, H.; Ullah, I.; Shabaz, M.; Omer, M.F.; Usman, M.T.; Guellil, M.S.; Koo, J. Visionary vigilance: Optimized YOLOV8 for fallen person detection with large-scale benchmark dataset. *Image Vis. Comput.* **2024**, *149*, 105195. [CrossRef]

39. OpenAI. Sora System Card. 2024. Available online: https://openai.com/index/sora-system-card/ (accessed on 27 May 2025).
40. Leng, Z.; Bhattacharjee, A.; Rajasekhar, H.; Zhang, L.; Bruda, E.; Kwon, H.; Plötz, T. IMUGPT 2.0: Language-Based Cross Modality Transfer for Sensor-Based Human Activity Recognition. *arXiv* **2024**, arXiv:2402.01049. [CrossRef]
41. Zhang, J.; Zhang, Y.; Cun, X.; Huang, S.; Zhang, Y.; Zhao, H.; Lu, H.; Shen, X. T2M-GPT: Generating Human Motion from Textual Descriptions with Discrete Representations. *arXiv* **2023**, arXiv:2301.06052. [CrossRef]

Disclaimer/Publisher's Note: The statements, opinions and data contained in all publications are solely those of the individual author(s) and contributor(s) and not of MDPI and/or the editor(s). MDPI and/or the editor(s) disclaim responsibility for any injury to people or property resulting from any ideas, methods, instructions or products referred to in the content.

sensors

MDPI

Article

Can a Commercially Available Smartwatch Device Accurately Measure Nighttime Sleep Outcomes in Individuals with Knee Osteoarthritis and Comorbid Insomnia? A Comparison with Home-Based Polysomnography

Céline Labie [1,2], Nils Runge [1,2,3], Zosia Goossens [2,3], Olivier Mairesse [3], Jo Nijs [2,4,5], Anneleen Malfliet [2,6,7], Dieter Van Assche [1,8], Kurt de Vlam [8,9], Luca Menghini [10,11], Sabine Verschueren [1,†] and Liesbet De Baets [1,2,12,13,*,†]

1 Musculoskeletal Rehabilitation Research Group, Department of Rehabilitation Sciences, Faculty of Movement and Rehabilitation Sciences, KU Leuven, 3001 Leuven, Belgium; celine.labie@kuleuven.be (C.L.); nils.arno.andreas.runge@vub.be (N.R.); dieter.vanassche@kuleuven.be (D.V.A.); sabine.verschueren@kuleuven.be (S.V.)
2 Pain in Motion Research Group (PAIN), Department of Physiotherapy, Human Physiology and Anatomy, Faculty of Physical Education and Physiotherapy, Vrije Universiteit Brussel, 1050 Brussels, Belgium; zosia.goossens@vub.be (Z.G.); jo.nijs@vub.be (J.N.); anneleen.malfliet@vub.be (A.M.)
3 Brain, Body and Cognition, Faculty of Psychology and Educational Sciences, Vrije Universiteit Brussel, 1090 Brussels, Belgium; olivier.mairesse@vub.be
4 PijnPraxis.be Private Practice for Pain Physiotherapy, 3970 Leopoldsburg, Belgium
5 Unit of Physiotherapy, Department of Health and Rehabilitation, Institute of Neuroscience and Physiology, Sahlgrenska Academy, University of Gothenburg, SE-405 30 Gothenburg, Sweden
6 Chronic Pain Rehabilitation, Department of Physical Medicine and Physiotherapy, University Hospital Brussels, 1050 Brussels, Belgium
7 Research Foundation Flanders, 1000 Brussels, Belgium
8 Division of Rheumatology, University Hospitals Leuven, 3000 Leuven, Belgium; kurt.devlam@uzleuven.be
9 Skeletal Biology & Engineering Research Center, Department of Development and Regeneration, KU Leuven, 3000 Leuven, Belgium
10 Department of General Psychology, University of Padova, 35131 Padova, Italy; luca.menghini-1@unitn.it
11 Human Inspired Technology Research Centre, University of Padova, 35131 Padova, Italy
12 Department of Physical and Rehabilitation Medicine, University Hospitals Leuven, 3212 Leuven, Belgium
13 Leuven Algologic Center, University Hospitals Leuven, 3212 Leuven, Belgium
* Correspondence: liesbet.debaets@kuleuven.be
† These authors contributed equally to this work.

Highlights

What are the main findings?

- Fitbit Sense demonstrates high accuracy and sensitivity for detecting sleep in individuals with knee osteoarthritis and insomnia.
- The device shows limited ability to differentiate quiet wakefulness from sleep and less precision in classifying specific sleep stages.

What is the implication of the main finding?

- Fitbit Sense can be a useful complementary tool for monitoring general sleep duration, timing, and regularity in this population.
- Sleep stage and fragmentation data should be interpreted with caution, as agreement with polysomnography decreases under more disrupted sleep conditions.

Abstract: Sleep is a vital physiological process for recovery and health. In people with knee osteoarthritis (OA), disrupted sleep is common and linked to worse clinical outcomes. Commercial sleep trackers provide an accessible option to monitor sleep in this population, but their accuracy for detecting sleep, wake, and sleep stages remains uncertain. This study compared nighttime sleep data from polysomnography (PSG) and Fitbit Sense in individuals with knee OA and insomnia. Data were collected from 53 participants (60.4% women, mean age 51 ± 8.2 years) over 62 nights using simultaneous PSG and Fitbit recording. Fitbit Sense showed high accuracy (85.76%) and sensitivity (95.95%) for detecting sleep but lower specificity (50.96%), indicating difficulty separating quiet wakefulness from sleep. Agreement with PSG was higher on nights with longer total sleep time, higher sleep efficiency, shorter sleep onset, and fewer awakenings, suggesting better performance when sleep is less fragmented. The device showed limited precision in classifying sleep stages, often misclassifying deep and REM sleep as light sleep. Despite these issues, Fitbit Sense may serve as a useful complementary tool for monitoring sleep duration, timing, and regularity in this population. However, sleep stage and fragmentation data should be interpreted cautiously in both clinical and research settings.

Keywords: knee osteoarthritis; sleep; insomnia; wearable; polysomnography; accuracy

1. Introduction

Sleep is a universal behavior essential for recovery and health regulation. In chronic pain, it not only reflects symptom burden but also actively contributes to pain persistence and impaired functioning [1]. Knee osteoarthritis (OA) is a leading cause of chronic pain and disability [2], with increasing evidence linking sleep disturbances to poorer disease outcomes [3–8]. Monitoring sleep is necessary to better understand these associations, and although research on self-reported sleep in OA populations is well established, studies using objective sleep measures remain limited [9–11]. Preliminary evidence suggests that prolonged nocturnal awakenings are predictive of next-day pain intensity [12], and alterations in sleep staging have also been documented [13–15]. Moreover, individuals with knee OA and comorbid insomnia display irregular sleep–wake patterns, further disrupting endogenous activity–rest rhythms [16].

This limited research on objective sleep outcomes in knee OA highlights the need for accessible, reliable sleep tracking tools for research and clinical practice [17–20]. While PSG is the gold standard, it requires a costly, burdensome setup in-laboratory or a home-based setup and trained sleep scorers. Research-grade actigraphy devices are recommended for estimating sleep in individuals with insomnia [21]. They detect sleep–wake states but have limited ability to assess sleep stages [22]. Consumer sleep tracking (CST) devices, including smartwatches, finger rings, and smart straps, offer a scalable and non-invasive alternative for monitoring sleep in real-world settings. These devices are popular because of their lower cost and greater acceptability. Using sensors, such as accelerometry and photoplethysmography-based sensors, CST devices estimate sleep outcomes, potentially providing an affordable, user-friendly solution for large-scale sleep data collection in free-living conditions [23]. For individuals with chronic pain, CST may support self-management by increasing awareness of behavioral sleep patterns and their interaction with symptom fluctuations. Clinicians may benefit from these data to tailor interventions and monitor treatment responses over time. Moreover, CST's ability to concurrently capture other behavioral health-related indicators, such as physical activity and stress-related metrics, supports a more integrated understanding of lifestyle factors contributing to pain and recovery [24,25]. These devices not only enable continuous monitoring but also hold promise for advancing personalized healthcare applications, particularly in the prevention and management of chronic conditions such as knee OA and insomnia [26–28].

Despite these advantages and opportunities, the accuracy and performance of CST devices remain largely unknown, especially in clinical populations, emphasizing the necessity for thorough evaluation in research and practice [20]. Standardized protocols for evaluating CST device performance have recently been developed [29], and studies have assessed their accuracy across diverse populations, including healthy individuals [30,31] and those with sleep disorders [32,33], Huntington's disease [34], and psychiatric conditions [35,36]. To date, no research has investigated the quality performance of CST devices in populations with chronic pain conditions. In individuals with knee OA, pain-related sleep disruptions, irregular sleep–wake cycles, and altered movement patterns may further compromise CST accuracy, underscoring the need for population-specific performance testing. This study addresses this gap by assessing the performance of the Fitbit Sense for nighttime sleep outcomes compared with home-based polysomnography in individuals with knee OA and comorbid insomnia.

2. Materials and Methods

Data were collected from a large randomized controlled trial (RCT) investigating the effects of cognitive–behavioral therapy for insomnia integrated with evidence-based physiotherapy in people with knee OA and insomnia (ClinicalTrials.gov NCT05387473—registration date 24 May 2022) [37]. For this study, data from the screening procedure and post-intervention assessment were used. This study was approved by the ethics committees of the University Hospitals Leuven and the University Hospital Brussels (S66306) and conducted in compliance with the Helsinki Declaration [38]. All participants provided written informed consent before enrolment in this study. The analyses were conducted following Menghini et al.'s standardized framework and guidelines for CST performance evaluation [29,39]. This study's findings were reported and written up following the guidelines provided by the 22-item checklist from the STROBE Statement for cross-sectional studies [40].

2.1. Participants

Participants were recruited from the community areas around Brussels and Leuven (Belgium) between May 2022 and August 2024. Participants were initially screened by phone to assess their eligibility. Included participants were adults aged ≥ 45 years reporting average knee pain $\geq 3/10$ on most days of the week for at least 3 months, with KOA as classified by the American College of Rheumatology [41], and diagnosed with insomnia using the Diagnostic and Statistical Manual of Mental Disorders, fifth Edition criteria [42]. Exclusion criteria included: (1) body mass index > 30 kg/m^2; (2) knee replacement; (3) known pre-diagnosed sleep disorder such as sleep apnea or periodic limb movement disorders; (4) rheumatological or neurological conditions; (5) cancer diagnosis or treatment in the last year; (6) receiving cholinesterase inhibitors; (7) receiving supervised knee exercise therapy or knee joint infiltration in previous 6 months; (8) presence of an external factor limiting opportunity to sleep (e.g., newborn). Participants underwent at-home polysomnography to screen for intrinsic sleep disorders such as sleep apnea or periodic limb movement disorders, which would result in exclusion from the trial. However, for this study, all screened individuals at the moment of the screening procedure were considered for inclusion, even if a sleep disorder was detected during the PSG screening night. Furthermore, data collected during the post-intervention follow-up assessment were also considered for this study.

2.2. Measurements

2.2.1. Procedure and Data Collection

The sleep assessment was performed during one night between 30 August 2022 and 21 November 2024, and was part of the at-home screening procedure or post-intervention assessment of the original RCT. An experienced researcher—specifically trained in installing the CST and PSG devices—visited the participants' homes in the evening hours to install both devices and inform the participants of their appropriate use. Participants were sleeping with a CST device, the Fitbit Sense (Fitbit®, San Francisco, CA, USA—Firmware and software version: FITBIT OS 5.3.1—Version 44.128.6.17) and simultaneously underwent at-home PSG measurement using the Alice PDX portable sleep diagnostic system (Philips Respironics, Murrysville, PA, USA). Both devices were used and worn as per the manufacturer's instructions and specifications. The device was placed on the non-dominant wrist and a finger's width above the styloid process of the ulna, and was worn until the next morning.

For this one-night assessment, the timing and duration of their sleep were self-selected and unrestricted, and participants were instructed to adhere to their habitual sleep schedule. Participants were asked to refrain from caffeine after lunch, smoking in the evening, and alcohol consumption as well as intensive sport for the whole day before the evaluation. Additionally, participants were asked to take their regular medications and not to initiate new pharmacological treatments on the day of the recording.

Around the time of the sleep assessment, participants received an invitation via email to digitally complete self-reported questionnaires for sociodemographic, pain-related, and sleep-related factors using the UZ Leuven REDCap online system (REDCap 12.4.13; Vanderbilt University, Nashville, TN, USA) [37].

2.2.2. Data Processing

The time in bed (TIB) period was defined as the interval between lights-off (participant getting in bed with the intention to sleep—start initiated by the participant pressing a button on the PSG) and lights-on times (participant getting out of bed with the intention to start the day—end initiated by turning the PSG device off). These events were primarily determined by the participant pressing a button on the PSG; however, when the button was not pressed, an alternative approach was used. Specifically, lights-off was estimated based on the PSG-detected transition from standing to lying down closest to the sleep diary-reported time (used on 34 nights, compared with 30 nights where the button was pressed). Similarly, lights-on was determined by the transition from lying down to standing when the button press was missing (used on 8 nights, compared with 56 nights where the button was pressed). The exported epochs and the TIB period were manually synchronized to the PSG's lights-off/on times.

Fitbit Sense Sleep Outcomes and Processing

The Fitbit Sense was selected for this clinical validation study based on its superior performance compared with other Fitbit models [30,43], the unavailability of the Fitbit Sense 2 at the time, and cost considerations that allowed for an adequate sample size within the original RCT design. Fitbit Sense tracks motion, heart rate, heart rate variability, and respiratory rate via a 3-axis accelerometer (sampling rate of 100 Hz) and optical plethysmography, respectively (sampling rate disclosed by Fitbit). Collected sensor data are processed via a device- and company-specific algorithm [44]. The Fitbit Sense's proprietary sleep stage classification algorithm integrates motion, heart rate variability, and respiratory rate data to provide information on sleep versus wake states and sleep staging (i.e., wake; light sleep as equivalent of PSG N1 + N2; deep sleep as equivalent of PSG N3; and REM sleep). These data can be extracted from the Fitbit web interface (Small Steps Labs LLC, San Diego, CA, USA) once the device is synchronized with the interface. For this study, the algorithm sensitivity option for the 'normal setting' was selected. No additional custom thresholds or smoothing beyond the default Fitbit Sense algorithm settings were applied.

For each recording, Fitabase provides sleep stage data, including timestamps, in 30 s epochs to allow epoch-by-epoch (EBE) analysis.

PSG Sleep Outcomes, Processing, and Scoring

PSG was used as a gold standard reference measure to evaluate sleep outcomes obtained by the CST device. The overnight home-based PSG assessment was conducted in a free-living monitoring environment, where factors such as sound and temperature were not controlled. The following parameters were recorded: electroencephalography (EEG) (electrodes placement on frontal midline (Fz) and central midline (Cz), each referenced to the left mastoid (M1)), electrooculography (EOG) (left and right), electromyography (EMG) chin and tibialis (left and right), oronasal flowmetry (pressure-based airflow with snore detection through a nasal cannula and thermistor), oxygen saturation (SpO2, finger probe, Oximetry board Nonin), pulse rate (from the oximeter probe) and body position (up, supine, prone, left, and right). The PSG electrode sites were measured and applied according to the International 10–20 System of EEG Electrode Placement. PSG data were collected and recorded at a sampling rate of 250 Hz, with bandpass filters applied as follows: ECG (0.318 Hz to 81 Hz), EEG (0.318 Hz to 35 Hz), and EMG (9.7 Hz to 86 Hz). After visual inspection of recordings, manual scoring software (Sleepware 3G, Philips Respironics, Murrysville, PA, USA) was used for scoring sleep versus wake states, sleep stages, and events following the American Academy of Sleep Medicine Manual [45] (version 2.4 released April 2017). Each 30 s epoch of PSG data was categorized as wake, N1, N2, N3, or REM sleep by two independent raters, who were blinded to the CST device's output [39]. The inter-rater reliability, measured by Cohen's Kappa, was 68%. Discrepancies were addressed through discussion, with a third rater consulted for any unresolved disagreements. Raw data of scored recordings were exported as xls-files to allow EBE analysis. For the purpose of this study, outcomes of interest were the same sleep outcomes as derived from the Fitbit Sense (TST, SE, SOL, and WASO), and duration (in minutes) spent for each sleep stage. PSG stages N1 and N2 were merged into one category representing light sleep in the device, and N3 was referred to as deep sleep.

2.3. Data Selection

Data from participants who met the inclusion criteria from this study and with no missing sleep data were used. PSG recordings were defined as failed if scoring was not possible because of poor signal quality, loosening of material, other technical issues, or when no signals were recorded at all. For the device, nights with poor signal quality (no sleep stages output) or nights with no recorded sleep episode data were excluded. Available eligible nights from screening and post-intervention assessment were selected, allowing multiple nights per participant.

2.4. Statistical Analysis

Outcomes were calculated based on the recommended standards for evaluating CST devices, utilizing the open-source R code from Menghini et al. (2021) [29]. The performance evaluation followed this standardized framework, which included discrepancy analysis, Bland–Altman plots, and EBE analysis. EBE analysis was conducted using 30 s epochs, following the temporal synchronization of Fitbit Sense with PSG data, with all analyses limited to the lights-off-to-lights-on period. The sample size was based on the number of nights for which valid, concurrent data were available from both Fitbit Sense and PSG. A significance threshold of 0.05 was applied for all tests. All analyses were carried out using R software version 4.3.2 (R Foundation for Statistical Computing). No log transformations or normalization procedures were applied to data before analysis.

2.4.1. Descriptives

Descriptive statistics were generated for all demographic and baseline characteristics. The normality of their distributions was assessed using histograms, Q–Q plots, and Kolmogorov–Smirnov tests. Percentages were calculated for categorical variables. For the self-reported questionnaires and sleep outcomes, the mean (standard deviation) was reported for normally distributed data, while the median (interquartile range) was reported for non-normally distributed data. A threshold of 1.5 times the interquartile range was applied to the discrepancies between Fitbit Sense and PSG across all sleep outcomes to identify outliers.

2.4.2. Discrepancy Analysis and Bland–Altman Plot

Discrepancy analysis evaluated bias and limits of agreement (LOAs) between sleep outcomes derived from Fitbit Sense and PSG. Bias was determined by subtracting PSG values from Fitbit Sense values, where positive differences indicate overestimation by Fitbit Sense and negative differences indicate underestimation. These discrepancies, assessed at the individual level, estimate systematic bias and random measurement error (95% LOAs) using Bland–Altman plots. The analysis involved testing key assumptions: (1) whether bias remains constant or varies with measurement size (proportional bias), (2) whether the variability of differences is uniform (homoscedasticity) or changes in measurement size (heteroscedasticity), and (3) whether differences follow a normal or non-normal distribution. To assess proportional bias and heteroscedasticity for each sleep outcome, linear regression was used. When proportional bias was significant ($p < 0.05$), indicating changes in bias with measurement size, discrepancies were adjusted accordingly by modeling the bias as a linear function of PSG-derived values, allowing for the estimation of conditional bias at specific PSG levels rather than assuming a constant mean difference across the measurement range. Similarly, if heteroscedasticity was detected, LOAs were modeled based on measurement size to account for variability in the differences. The significance of Fitbit Sense over- and underestimation was assessed based on the 95% confidence intervals of the bias.

2.4.3. EBE Analysis

To assess performance metrics for each sleep stage, the outcomes were dichotomized, comparing the sleep stage of interest against all others. Error matrices comparing the Fitbit Sense outcomes to PSG were generated for each participant and the whole group. Specifically, the number of epochs classified as light sleep, deep sleep, REM sleep, or wake by both methods was computed. The average proportion of correct (sensitivity) and incorrect (specificity) Fitbit Sense classifications relative to PSG classifications was reported using a proportional error matrix, along with the corresponding standard deviation and 95% confidence intervals. The traditional definitions of sensitivity (i.e., the ability to correctly classify sleep epochs) and specificity (i.e., the ability to correctly classify wake epochs) have been adapted to account for multi-stage sleep classification [20,39]. Instead of a binary sleep–wake categorization (0 = wake, 1 = sleep), classifications were expanded to include four distinct stages: 0 = wake, 1 = "light" sleep, 2 = "deep" sleep, and 3 = REM sleep. Sensitivity was calculated for each sleep stage, representing Fitbit Sense's ability to correctly identify a given PSG stage (wake, "light," "deep," or REM), while specificity measured its ability to correctly classify all other PSG stages. For further details on these definitions, see Menghini et al. [29]. Cohen's kappa (κ), which measures classification agreement beyond chance on a scale from 0 to 1, and the prevalence-adjusted bias-adjusted kappa (PABAK) coefficient were computed for each sleep stage. Additionally, the positive predictive value (the proportion of epochs identified as a target stage by Fitbit Sense that are also classified as that stage by PSG) and negative predictive value (the proportion of epochs not classified as a target stage by Fitbit Sense that are also not classified as that stage by PSG) were calculated for each sleep stage.

2.4.4. Sensitivity Analysis

Several sensitivity analyses were conducted to assess the robustness of the findings. The first evaluated the performance and accuracy of this study device using all available data points, including outliers. The second focused on data from participants with only a single night of recordings. The third compared baseline demographic characteristics between included and excluded participants, using either a t-test or its nonparametric equivalent, depending on data distribution.

3. Results

3.1. Characteristics of Participants

A total of 155 eligible individuals with knee OA and insomnia were identified for this study. Of these, 13 participants withdrew before assessment due to lack of interest, medical interventions, or inability to follow the study protocol (e.g., time constraints). After checking for valid concurrent nights, 105 home-based sleep assessments remained. During data processing, 41 nights were excluded because of recording failures from the PSG or study device. From the remaining 64 nights, a total of 14 outliers were identified using the predefined criterion of 1.5 times the interquartile range for the discrepancies between the device and PSG across all sleep outcomes. Outliers were further examined through visual inspection of data, allowing for a more comprehensive assessment. Two participants accounted for half of these outliers and were removed from the analysis because of multiple unlikely values. All other data points were cross-checked with original records, considered valid, and included in the final analysis. The final number of nights included for the Fitbit–PSG comparison was 62 nights (44 from the baseline screening and 18 from the post-intervention assessment) from 53 participants (Figure 1).

Participants were predominantly older adults (61 years $\pm$ 8.2; 98.1% White-Caucasian; 60.4% female; BMI of 24.7 $\pm$ 2.6 kg/m^2) with long-standing knee pain (mean duration: 9.7 $\pm$ 9.7 years) of moderate severity (NRS: 5.2 $\pm$ 1.9). A total of 86.5% experienced night-time pain, and 84.9% of participants reported one or more comorbidities other than knee pain and insomnia. Average insomnia duration was 11.1 $\pm$ 11 years, with 96.1% reporting poor sleep quality (PSQI > 5) and 66.7% meeting criteria for clinical insomnia (ISI $\geq$ 15). The detailed demographics and baseline characteristics of the 53 participants studied are presented in Table 1. PSG-derived sleep outcomes are reported in Table 2. The number of participants for each outcome measure is provided in Supplementary Material Table S1.

Table 1. Demographics and baseline characteristics of the participants in this study. Categorical variables are reported as frequencies.

Demographics	Mean (SD)/N	Median (IQR)	Range/Percentage
Age, years	61.0 (8.2)	63 (13.5)	45–78
Female	32		60.4
BMI, kg/m^2	24.7 (2.6)		20.2–30.5
White-Caucasian	52		98.1
Postsecondary education	46		86.8
Currently working	23		43.4
Lives alone	17		30.1
Smoking	5		9.3

Table 1. *Cont.*

Demographics	Mean (SD)/N	Median (IQR)	Range/Percentage
Baseline characteristics			
Pain duration, years	9.7 (9.7)	5 (12.7)	0.8–45
Pain NRS	5.2 (1.9)	5 (3)	2–9
Pain at night	7.0		86.5
KOOS pain	54.2 (14.2)		13.9–91.7
KOOS Function	62.6 (13.9)	63.2 (13.2)	23.5–94.1
KOOS Quality of life	39.3 (15.7)		0–81.3
≥1 Comorbidities	45		84.9
BPI Severity	5 (2.1)	5 (2.3)	1.3–14
BPI Interference	3.8 (1.8)	3.6 (3.3)	1–7.4
CSI	38 (12.7)		12–68
HADS Anxiety	6.3 (3.3)	6 (4)	2–15
Moderate to severe symptoms	6		12
HADS Depression	5.2 (3.7)	4.5 (6)	0–16
Moderate to severe symptoms	4		8
Sleep characteristics			
Sleep problem duration, years	11.1 (11.0)	8 (14.5)	0.6–54
Sleep medication use	12		22.6
ISI	15.8 (4.1)	17 (4)	5–24
No Insomnia	2		3.9
Subclinical insomnia	15		29.4
Clinical insomnia, moderate	31		60.8
Clinical insomnia, severe	3		5.9
PSQI	10.2 (2.9)	10 (2)	4–17
Poor sleep quality (PSQI > 5)	49		96.1
BFS Mental Fatigue	29.6 (23.6)	25 (41.7)	0–83.3
BFS Physical Fatigue	34 (21.9)	33.3 (33.3)	0–83.3
ESS	8.4 (4.1)	8 (4)	1–20
Excessive daytime sleepiness (ESS > 10)	17		32
AHI (h^{-1})	9 (9.7)	5.1 (9)	0.5–45.7
Moderate (AHI ≥ 15)	10		16.1
Severe (AHI ≥ 30)	3		4.8
PLMSI (h^{-1})	8.5 (18)	1.3 (7)	0–100.2
Moderate (PLMSI ≥ 25)	5		8.1
Severe (PLMSI ≥ 50)	2		1.2

NRS = Numeric Rating Scale; KOOS = Knee Disability and Osteoarthritis Outcome Score; BPI = Brief Pain Inventory; CSI = Central Sensitization Inventory; HADS = Hospital Anxiety and Depression Scale; ISI = Insomnia Severity Index; PSQI = Pittsburgh Sleep Quality Index; BFS = Brugmann Fatigue Scale; ESS = Epworth Sleepiness Scale; AHI = Apnea–Hypopnea Index; PLMSI = Periodic Limb Movement Sleep Index.

Table 2. PSG-derived sleep outcomes for the participants in this study.

	Mean (SD)	Median (IQR)	Range
TIB (min)	477 (60)	482 (81)	318–593
TST (min)	373 (63)	387 (76)	213–489
SE (%)	78 (8)	79 (11)	51–91
SOL (min)	20 (18)	17 (20)	1–114
WASO (min)	84 (37)	80 (52)	23–169
Light			
Duration (min)	241 (50)	241 (61)	126–339
Percentage	65 (10)	66 (13)	36–85
Deep			
Duration (min)	47 (30)	45 (38)	2–160
Percentage	13 (8)	11 (10)	1–41
REM			
Duration (min)	85 (30)	86 (38)	0–146
Percentage	22 (6)	23 (7)	0–34

TIB = Total time in bed; TST = Total sleep time; SE = Sleep efficiency; SOL = Sleep onset latency; WASO = Wake after sleep onset; SD = Standard deviation; IQR = Interquartile range.

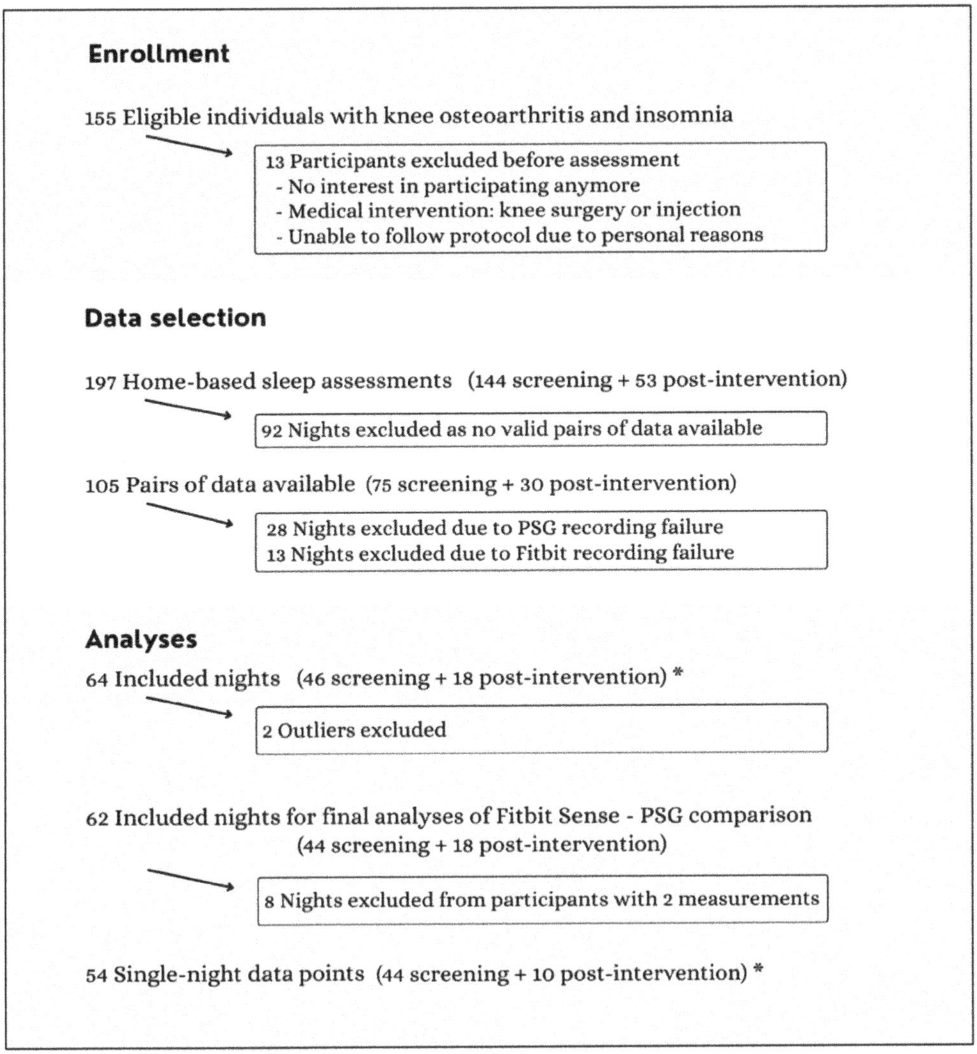

Figure 1. Flow chart illustrating the study enrollment, data selection, and analyses. * Additional analyses are available in Supplemental Material.

3.2. *Fitbit Sense–PSG Comparison*

3.2.1. Discrepancy Analysis

Individual-level discrepancies were calculated by subtracting PSG-derived outcomes from Fitbit-derived outcomes for each subject. Figure 2 presents the summary statistics and distribution for each sleep outcome.

Group-level discrepancies were assessed using mean bias (Fitbit Sense–PSG), mean absolute difference, and standard error. A summary of group-level differences between the Fitbit Sense and PSG is provided in Table 3.

Table 3. Group-level differences between Fitbit Sense and PSG. The positive mean bias indicates overestimation by Fitbit Sense, while the negative mean bias indicates underestimation. The mean absolute difference captures the average deviation magnitude regardless of direction. The standard error represents the variability of the mean bias estimate, with smaller values indicating more consistent measurements across observations.

	Mean Bias ± SD (95% CI)	Mean Absolute Difference	Mean Standard Error (95% CI)
TST (min)	37.05 ± 32.04 (28.97, 45.13)	40.81	0.52 (39.86–41.85)
SE (%)	7.89 ± 6.95 (6.13, 9.65)	8.65	0.11 (8.42–8.88)
SOL (min)	−5.8 ± 22.5 (−11.52, −0.08)	13.1	0.4 (12.3–13.8)
WASO (min)	−31.23 ± 31.45 (−39.22, −23.24)	35.79	0.37 (35.05–36.53)
Light (min)	22.29 ± 53.17 (8.74, 35.84)	45.63	0.67 (44.28–46.98)
Deep (min)	13.74 ± 33.23 (5.34, 22.14)	40.81	0.44 (39.93–41.69)
REM (min)	1.02 ± 36.13 (−10.15, 12.19)	28.08	0.43 (27.21–28.95)

PSG = Polysomnography; TIB = Total time in bed; TST = Total sleep time; SE = Sleep efficiency; WASO = Wake after sleep onset; REM = Rapid-eye-movement; SD = Standard deviation; CI = Confidence interval.

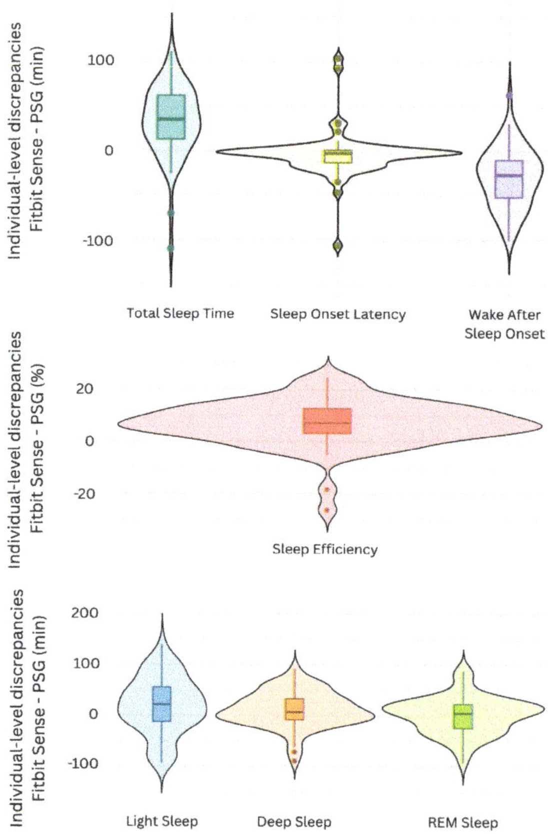

Figure 2. Distribution of individual-level discrepancies between Fitbit Sense and PSG for each sleep outcome. The boxplot elements include a colored horizontal line indicating the median, a box representing the interquartile range (first to third quartile), and whiskers extending to the upper and lower adjacent values. These elements are consistently applied across all violin plots, which further illustrate the data distribution by showing the density of values across different ranges. Data points beyond 1.5 times the interquartile range are represented by colored dots, highlighting potential outliers within the sample. PSG = Polysomnography; REM = Rapid-eye-movement.

The results of Bland–Altman analysis showed a significant proportional bias across all sleep outcomes, with the magnitude and direction of this bias varying according to the range of PSG-derived values (Table 4). These biases were modeled using linear regression, and the corresponding Bland–Altman plots illustrate how the bias changes across the range of measurements (Figure 3).

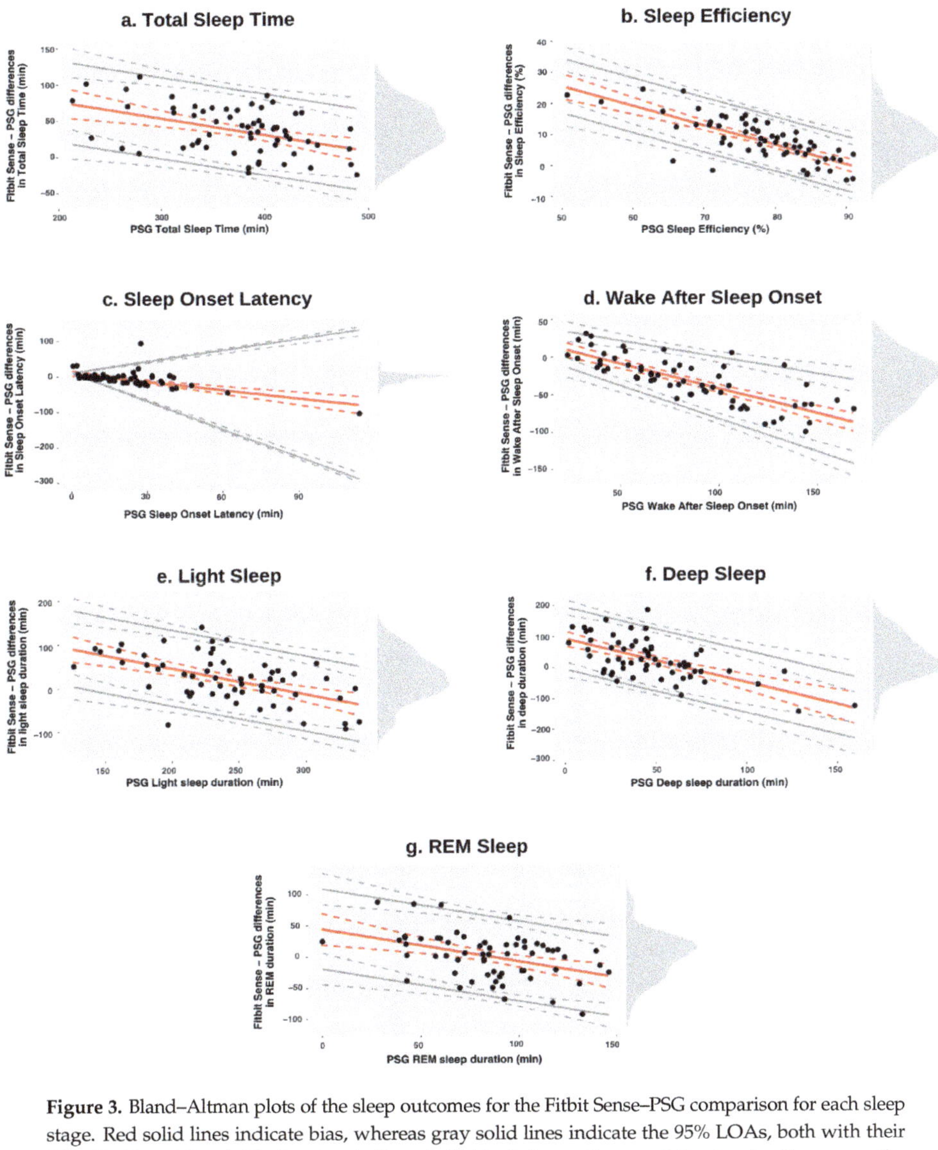

Figure 3. Bland–Altman plots of the sleep outcomes for the Fitbit Sense–PSG comparison for each sleep stage. Red solid lines indicate bias, whereas gray solid lines indicate the 95% LOAs, both with their 95% CIs (dotted lines). Black points indicate individual observations, and the density diagram on the right side of each plot represents the distribution of the Fitbit Sense–PSG differences. The zero point on the *y*-axis denotes perfect agreement between the two methods, while values above and below zero indicate overestimation and underestimation, respectively, in comparison to PSG. A diagonal trend in the mean bias line indicates a considerable proportional bias, whereas diverging limits of agreement suggest notable heteroscedasticity. PSG = Polysomnography; REM = Rapid-eye-movement.

Table 4. Group-level discrepancies, bias, and LOAs between Fitbit Sense and PSG. If proportional bias was observed, a linear model was used to predict discrepancies based on the corresponding PSG outcomes, with 95% confidence intervals reported for the model's intercept (b0) and slope (b1). In cases of heteroscedasticity, another linear model was applied to predict the absolute residuals of the initial model using PSG-derived measures, with 95% confidence intervals provided for the intercept (c0) and slope (c1).

	Fitbit Sense Mean, ±SD	PSG Mean, ±SD	Proportional Bias, 95% CI	LOA Lower, 95% CI	LOA Upper, 95% CI
TST (min)	409.90 ± 56.40	372.85 ± 63.34	123.37 − 0.23 × ref b0 = [79.46, 167.28], b1 = [−0.35, −0.12]	bias − 55.83 bias − [49.46, 64.79]	bias + 55.83 bias + [49.46, 64.79]
SE (%)	85.87 ± 5.41	77.99 ± 8.35	57.54 − 0.64 × ref b0 = [46.69, 68.4], b1 = [−0.78, −0.5]	bias − 8.78 bias − [6.96, 10.98]	bias + 8.78 bias + [6.96, 10.98]
SOL (min)	14.52 ± 17.53	20.34 ± 17.77	10.75 − 0.82 × ref b0 = [4.02, 17.49], b1 = [−1.07, −0.56]	bias − ref × 1.85 bias − ref × [1.7, 1.93]	bias + ref × 1.85 bias + ref × [1.7, 1.93]
WASO (min)	53.00 ± 22.67	84.23 ± 37.50	25.12 − 0.67 × ref b0 = [13.09, 37.16], b1 = [−0.8, −0.54]	bias − 2.46 (7.64 + 0.09 × ref) c0 = [0.91, 14.36], c1 = [0.02, 0.16]	bias + 2.46 (7.64 + 0.09 × ref) c0 = [0.91, 14.36], c1 = [0.02, 0.16]
Light (min)	263.27 ± 48.35	240.98 ± 49.70	166.66 − 0.6 × ref b0 = [110.36, 222.96], b1 = [−0.83, −0.37]	bias − 86.34 bias − [71.84, 103.48]	bias + 86.34 bias + [71.84, 103.48]
Deep (min)	60.90 ± 26.93	47.16 ± 30.10	47.19 − 0.71 × ref b0 = [34.99, 59.38], b1 = [−0.93, −0.49]	bias − 49.92 bias − [41.52, 59.66]	bias + 49.92 bias + [41.52, 59.66]
REM (min)	85.72 ± 35.64	84.70 ± 30.01	45.03 − 0.52 × ref b0 = [19.85, 70.21], b1 = [−0.8, −0.24]	bias − 63.88 bias − [55.04, 75.85]	bias + 63.88 bias + [55.04, 75.85]

TIB = Total time in bed; TST = Total sleep time; SE = Sleep efficiency; SOL = Sleep onset latency, WASO = Wake after sleep onset; REM = Rapid-eye-movement; SD = Standard deviation; PSG = Polysomnography; LOA = Limit of agreement; CI = Confidence interval.

TST and SE exhibited significant negative proportional biases, indicating that the device tended to overestimate these outcomes, particularly at higher PSG-derived values. For TST, the overestimation decreased as PSG values surpassed 450 min, beyond which the bias became statistically non-significant (Figure 3a). At an average PSG-derived TST of 373 min, the device overestimated TST by 37.58 min. Similarly, SE was overestimated for nights with high PSG-derived SE values, with the bias becoming non-significant when PSG-based SE exceeded 85% (Figure 3b). A negative proportional bias was also found for *SOL and WASO* (Figure 3c,d). However, whereas TST and SE discrepancies showed larger deviations from zero for smaller PSG-derived values, the opposite pattern was found for SOL and WASO, showing larger deviations for longer PSG-derived SOL and WASO values. The bias became statistically non-significant for lower PSG-derived values (approximately 13 min and 50 min or less, respectively). Notably, the variability in differences decreased as PSG-derived SOL and WASO values declined (i.e., heteroscedasticity), suggesting improved agreement at lower SOL and WASO levels. *Light sleep* duration was also overestimated by the device, particularly when PSG-derived values were below 250 min. For values exceeding this threshold, the bias became non-significant (Figure 3e). However, the wide LOAs associated with this measure indicate considerable variability and poor agreement with PSG, highlighting the need for caution when interpreting light sleep estimates. For both *deep sleep and REM sleep* durations, the average bias appeared to be close to zero near the mean PSG-derived values. However, the bias became more evident at the lower and upper extremes of the PSG measurement range (Figure 3f,g).

3.2.2. Epoch-by-Epoch Analysis

Light sleep (73 ± 10%) and REM sleep (60 ± 27%) showed the highest correct classification rates. A substantial proportion (37 ± 15%) of wake epochs were misclassified as light sleep, indicating difficulties for Fitbit Sense in distinguishing wakefulness from light sleep. Furthermore, 46 ± 25% of deep sleep epochs and 33 ± 25% of REM sleep epochs were misclassified as light sleep, indicating difficulties in differentiating between those stages too.

The detailed group-level proportional error matrix from the EBE analysis is presented in Table 5 and provides insights into systematic misclassifications and the overall performance of the classification. Additionally, the absolute error matrix, which considers the total epoch count in a given classification category without differentiating between subjects, is provided in the Supplementary Material Table S2.

Table 5. Group-level proportional error matrix reporting the group average proportion of epochs in each sleep stage as mean (standard deviation) [95% confidence intervals]. Each cell of the matrix displays the average proportion of epochs assigned to each classification category, along with the corresponding standard deviation and 95% confidence intervals. This matrix quantifies the proportion of epochs assigned to each sleep stage by Fitbit Sense relative to PSG.

PSG Stage	Fitbit Sense Wake	Fitbit Sense Light	Fitbit Sense Deep	Fitbit Sense REM
Wake	0.51 (0.15) [0.47, 0.55]	0.37 (0.15) [0.33, 0.4]	0.02 (0.04) [0.01, 0.03]	0.11 (0.1) [0.08, 0.13]
Light	0.04 (0.03) [0.04, 0.05]	0.73 (0.10) [0.7, 0.75]	0.13 (0.07) [0.11, 0.15]	0.10 (0.07) [0.08, 0.12]
Deep	0.02 (0.06) [0.01, 0.03]	0.46 (0.25) [0.39, 0.52]	0.49 (0.27) [0.42, 0.55]	0.03 (0.08) [0.01, 0.05]
REM	0.04 (0.05) [0.02, 0.05]	0.33 (0.25) [0.26, 0.39]	0.03 (0.06) [0.02, 0.05]	0.60 (0.27) [0.54, 0.67]

PSG = Polysomnography; REM = Rapid-eye-movement.

The Fitbit Sense demonstrated distinct performance patterns across sleep stages when evaluated against PSG using stage-specific group-level EBE metrics.

Wake Detection

Fitbit Sense demonstrated high accuracy (85.76 ± 5.49%) and very high specificity (95.95 ± 2.9%) for wake detection, indicating reliable identification of non-wake epochs.

However, sensitivity was moderate (50.96 ± 15.46%), reflecting a significant under-detection of wake, often misclassified as light sleep. The agreement with PSG was moderate to substantial (κ = 0.51 ± 0.13; PABAK = 0.72 ± 0.11), confirming reliable wake detection despite this underestimation.

Light Sleep

Light sleep showed the lowest accuracy (67.22 ± 7.85%) and specificity (62.62 ± 13.32%), indicating frequent misclassification of other stages as light sleep. While sensitivity was highest for light sleep (72.50 ± 9.98%), the positive predictive value was moderate (66.51 ± 11.98%), suggesting overestimation of light sleep at the expense of deeper stages. The agreement with PSG was low (κ = 0.34 ± 0.15).

Deep Sleep

Deep sleep detection had high accuracy (87.65 ± 4.57%) and specificity (91.76 ± 4.41%), but low sensitivity (48.75 ± 27.24%), showing frequent misclassification of deep sleep as light sleep. The positive predictive value was low (39.65 ± 25.94%), highlighting substantial false positives. Despite low κ (0.34 ± 0.23), prevalence-adjusted agreement was substantial (PABAK = 0.75 ± 0.10).

REM Sleep

REM sleep accuracy was high (85.23 ± 6.02%) with good specificity (91.76 ± 4.41%). Sensitivity was moderate (60.41 ± 26.51%), indicating a notable proportion of REM epochs were misclassified, mostly as light sleep. The positive predictive value was modest (56.62 ± 18.09%) and agreement metrics showed moderate reliability (κ = 0.47 ± 0.23; PABAK = 0.70 ± 0.12).

Details of the group-level EBE metrics are provided in Table 6. These EBE metrics are reported as averages ± SD across subjects. Detailed visualizations of individual-level EBE metrics are available in the Supplementary Material Figure S1. Additionally, group-level EBE metrics based on the absolute error matrix, which considers the total epoch count per condition without differentiating between subjects, are also available in the Supplementary Material Table S3.

Table 6. Group-level EBE metrics assessing Fitbit Sense agreement for each sleep stage detection. EBE agreement metrics are reported for each sleep stage, compared against all other possible classifications, and averaged out for all participants. The Fitbit Sense was evaluated against the corresponding reference epochs from the PSG. Sensitivity represents the percentage of epochs identified as a specific sleep stage by PSG that are correctly detected as that stage by Fitbit Sense. Data is reported as mean (standard deviation) [95% confidence intervals]. Values approaching 1.0 represent higher accuracy for the given metric.

Stage	Accuracy	Sensitivity	Specificity	PPV	NPV	Kappa	PABAK
Wake	85.76 (5.49) [84.44, 87.17]	50.96 (15.46) [47.14, 54.72]	95.95 (2.90) [95.27, 96.71]	76.91 (14.22) [73.49, 80.51]	87.05 (7.08) [85.3, 88.82]	0.51 (0.13) [0.48, 0.55]	0.72 (0.11) [0.69, 0.74]
Light	67.22 (7.85) [65.33, 69.19]	72.50 (9.98) [70.05, 75.07]	62.62 (13.32) [59.36, 65.93]	66.51 (11.98) [63.62, 69.50]	68.87 (10.70) [66.18, 71.48]	0.34 (0.15) [0.31, 0.38]	0.34 (0.16) [0.31, 0.38]
Deep	87.65 (4.57) [86.55, 88.82]	48.75 (27.24) [42.04, 55.61]	91.76 (4.41) [90.70, 92.86]	39.65 (25.94) [33.23, 45.94]	94.44 (4.91) [93.32, 95.73]	0.34 (0.23) [0.28, 0.39]	0.75 (0.09) [0.73, 0.78]
REM	85.23 (6.02) [83.75, 86.74]	60.41 (26.51) [53.82, 67.12]	90.79 (5.27) [89.54, 92.15]	56.62 (18.09) [52.29, 61.15]	91.39 (6) [89.94, 92.91]	0.47 (0.23) [0.41, 0.53]	0.70 (0.12) [0.67, 0.74]

EBE = Epoch-by-epoch; PSG = Polysomnography; REM = Rapid-eye-movement; PPV = Positive predictive value; NPV = Negative predictive value; PABAK = Prevalence and bias-adjusted kappa.

3.3. Sensitivity Analyses

The results of the first sensitivity analysis, which examined performance and accuracy over 64 nights including all data points with outliers, are presented in Supplementary Material Figure S2 and Table S4–S6. The second sensitivity analysis, based on 54 nights from participants with only a single night of data, is described in Supplementary Material Figure S3 and Table S7–S9. The third sensitivity analysis compared baseline characteristics between included and excluded participants. No significant demographic differences were found, except for age, with excluded participants being slightly younger. These results are available in Supplementary Material Table S10.

4. Discussion

This is the first study to assess the performance of the Fitbit Sense for detecting sleep outcomes compared with PSG under free-living conditions in individuals with knee OA and insomnia. The device demonstrates strong sleep vs. wake detection, with high accuracy (85.76%) and sensitivity (95.95%) for identifying sleep epochs. These results align with prior performance evaluation studies [43,46,47] and support its utility for general sleep monitoring in this clinical population. The high sleep sensitivity suggests the device reliably detects sleep periods, making it suitable for tracking overall sleep duration and parameters such as timing, regularity, and chronotype in individuals with knee OA and insomnia.

Our results also show Fitbit Sense–PSG agreement varied depending on the size of the measurement. The agreement was highest on nights with longer sleep (TST > 450 min), more time asleep (SE > 85%), faster sleep onset (SOL < 13 min), and fewer awakenings (WASO < 50 min). Under these conditions, discrepancies were minimal or non-significant, suggesting robust performance when sleep is less fragmented. These findings align with previous Fitbit–PSG comparisons in healthy [30,31] and clinical populations, including insomnia [32], obstructive sleep apnea [33], Huntington's disease [34], major depressive disorders [36], and psychiatric disorders [35]. As in prior studies, sleep duration, light sleep, and REM were overestimated, while the duration of nocturnal awakening was underestimated. However, unlike earlier studies reporting underestimation of deep sleep [30–36], the device overestimated this stage. Similar overestimations of deep sleep by other CSTs [30,31] suggest such discrepancies may reflect device-specific algorithms or population-specific factors.

Despite strengths, the device had limited wake detection accuracy with a moderate wake sensitivity of 50.96%. While performing better than earlier Fitbit models (13.10–69.76%) [30,32,33,47–50], this highlights a persistent challenge for CSTs: distinguishing quiet wakefulness from sleep. Reliance on actigraphy and heart rate sensors likely contributes to the misclassification of motionless wake as sleep. When compared with the newer Fitbit Sense 2 in young healthy adults, two studies reported modest improvements in sleep stage sensitivity. Sensitivities for light, REM, and deep sleep using Sense 2 were 77.34% and 78.00%, 67.10% and 61.70%, and 68.12% and 67.30%, respectively, versus 72.50%, 48.75%, and 60.41% in our study [46,51]. These differences may reflect updated algorithms and enhanced sensors, including improved heart rate variability and skin temperature tracking, possibly supporting more accurate sleep staging. These differences may also reflect population-related variability, as our sample differs from the healthy young adults assessed in prior studies. Sleep staging and wearable sensor performance may vary across populations with differing health conditions and sleep disturbances, influencing the accuracy of CSTs, which generally perform better in healthy individuals. For example, Kang et al. (2017) reported lower accuracy of the Fitbit Flex in individuals with insomnia compared with good sleepers, suggesting sleep disturbances may impair device performance [52]. Similarly, a study using the same CST as ours found slightly lower agreement with PSG in individuals with psychiatric disorders, with light, deep, and REM sleep sensitivities of 68%, 44%, and 49%, respectively, less than those observed in our sample [35].

In individuals with knee OA, pain-related awakenings involving subtle repositioning due to discomfort may lack sufficient wrist movement and remain undetected by CSTs [8]. Pain- and insomnia-related arousals (brief EEG-defined awakenings) may fall below CST sensitivity thresholds, leading to underreported nocturnal awakenings and fragmented sleep, key features of knee OA-related sleep disturbance [13–15,53]. Similarly, individuals with insomnia often lie awake without movement, making these wakeful periods prone to misclassification as sleep. Beyond the sleep–wake distinction, CSTs face difficulties in accurately identifying sleep staging. The device exhibited high specificity across stages, but lower sensitivity for wake, deep, and REM sleep, with many epochs misclassified as light sleep. This imprecision complicates clinical interpretation, potentially masking true sleep disturbances in individuals with knee OA and insomnia. Pain in knee OA may fragment sleep cycles via micro-arousals, disrupting stage continuity. Consequently, CST algorithms may label ambiguous epochs as light sleep. Since these devices already have difficulties detecting REM and deep sleep in healthy individuals, underrepresentation of these stages in individuals with knee OA and insomnia could exacerbate the overestimation of light sleep and underestimation of restorative sleep.

4.1. Clinical Implications

This study underscores the clinical relevance and limitations of the Fitbit Sense for sleep monitoring in individuals with knee OA and comorbid insomnia. The device offers an accessible way to monitor sleep-derived parameters, including sleep duration, timing, regularity, and chronotype, which are increasingly recognized as modifiable factors influencing pain-related outcomes. For clinicians and researchers, such data may help contextualize fluctuations in symptom severity and functional limitations, offering a more continuous and objective view compared with self-reported questionnaires, which may be subject to recall bias and limited by periodic assessments. For example, patterns of irregular sleep timing or short sleep duration may inform the delivery of components within cognitive–behavioral therapy for insomnia (CBT-I), such as sleep scheduling or stimulus control, and may also help guide behavioral strategies in individuals with chronic pain, where improving sleep regularity and alignment could contribute to better symptom management. For patients, increased visibility into their sleep patterns may foster awareness of unhelpful behaviors (e.g., inconsistent sleep schedules or extended time in bed) and encourage sleep-promoting practices, such as adherence to regular routines or time in bed restrictions. This feedback loop between data and behavior may be relevant in populations where pain-related hyperarousal (i.e., difficulty falling asleep due to pain-induced arousal) or unhelpful beliefs about sleep (e.g., unrealistic expectations such as needing eight hours of sleep to function the next day). However, the device's reduced accuracy in detecting wakefulness, estimating sleep efficiency, and classifying sleep stages suggests it is not suitable for diagnostic decision-making or detailed assessment of sleep architecture. Instead, CSTs should supplement clinical assessments, subjective evaluation, and validated tools such as sleep diaries or, when needed, polysomnography (PSG) [54].

These limitations emphasize the importance of user education and careful interpretation. The observed variability in device accuracy across the magnitude of nighttime sleep outcomes (e.g., duration of nocturnal awakenings) suggests that the device may be less precise for individuals with more severe sleep disruptions. Clinicians and users should consider this potential limitation, as accuracy may decrease with more extreme sleep patterns, affecting the interpretation of results. Without guidance, CST data may result in false reassurance or increased anxiety, potentially exacerbating both sleep disturbances and pain. Incorporating patient-reported outcomes alongside objective data enables clinicians to identify and interpret discrepancies such as reports of non-restorative sleep despite apparently adequate total sleep time, within the patient's broader clinical picture, as CSTs are unable to capture the complex and context-dependent factors that influence sleep quality and quantity. Understanding the limitations of CSTs is essential for effective and appropriate use. Importantly, the continuous feedback from sleep trackers can trigger orthosomnia (increased sleep-related anxiety that may lead to insomnia), especially in patients already prone to hyperarousal [55]. Heightened focus on nightly data may reinforce excessive monitoring and cognitive preoccupation with sleep. To prevent this, clinicians should encourage adaptive interpretation of CST data, emphasizing longer-term trends rather than nightly fluctuations. This should occur within a biopsychosocial framework that considers emotional regulation, physical activity, and cognitive patterns contributing to sleep and pain outcomes.

4.2. Future Research

Future research should include normative comparisons for disorder-specific effects and assess intra-individual variability over multiple nights to detect natural fluctuations and treatment-related changes. Research involving larger, more diverse cohorts with rigorous control of potential confounders—including comorbidities (e.g., obesity and chronic diseases), medications affecting sleep architecture and behavior, and environmental factors such as geographical living environments and household context—is essential to better understand their impact on wearable device performance and sleep outcomes.

Longitudinal studies are needed to evaluate CST stability during interventions, especially in older adults with knee OA. CST–PSG agreement should be assessed based on clinically meaningful thresholds rather than perfect agreement, focusing on trend detection. Physiological states, including inflammation, autonomic dysregulation, or metabolic shifts, may influence CST accuracy by altering sleep architecture or sensor responsiveness. Given the high rate of multimorbidity in knee OA populations [56,57], comorbidities such as cardiovascular disease, diabetes, and depression should be accounted for in device validation.

Despite limitations, CSTs support patient engagement and self-monitoring, potentially improving adherence to behavioral sleep interventions. Future studies should examine how patients interpret CST data, especially in cases of mismatch with subjective experience, to minimize misperceptions, reduce orthosomnia risk, and optimize clinical integration [58].

4.3. Strengths and Limitations

This study presents several strengths. It focuses on a clinically relevant and understudied population and evaluates sleep tracking under free-living conditions using gold-standard home PSG. The methodology adhered to established scoring guidelines, including double scoring of PSG data, and examined both global and stage-specific sleep parameters.

However, several limitations should be acknowledged. The findings are specific to the Fitbit Sense and its proprietary algorithms, which lack transparency and limit both generalizability to other consumer sleep technology (CST) devices and reproducibility of results. This lack of algorithmic openness remains a key limitation in advancing evidence-based use of wearables in clinical and research settings. Future work should prioritize collaboration with manufacturers, adoption of open-source algorithms, and population-specific calibration to improve the accuracy, reproducibility, and trustworthiness of wearable-derived sleep measures. Additionally, only the default "normal" sensitivity mode was tested, and it remains unclear how alternative settings (e.g., "sensitive") might impact classification accuracy, highlighting the need for future research to compare performance across modes.

Second, this study focused solely on nighttime sleep, excluding daytime sleep episodes or napping, which may differ physiologically. In addition, the use of a single night of in-home PSG may not fully capture night-to-night variability in sleep, particularly in individuals with insomnia. While this approach aligns with most validation studies of CST devices, it limits the reliability of assessing typical sleep patterns and device performance across multiple nights. Third, inter-rater variability in PSG scoring may have introduced random error but reflects typical practice and likely led to conservative estimates of agreement. Fourth, although the device's limitations in detecting wakefulness may vary by sleep pattern, stratified analyses by insomnia severity or the presence of comorbid sleep disorders were not performed because of limited sample sizes. Fifth, the predominantly White sample, with only one African American participant, may restrict generalizability given known sensor performance differences related to skin tone [59].

5. Conclusions

This study provides important evidence on the performance of the Fitbit Sense for sleep monitoring in individuals with KOA and comorbid insomnia. While the device performs well in detecting overall sleep under free-living conditions, limitations remain in identifying wakefulness and accurately classifying sleep stages. These findings indicate that, although not appropriate for diagnostic use, the Fitbit Sense may be a useful complementary tool for tracking sleep-derived parameters, including sleep duration, timing, regularity, and chronotype, as well as analyzing individual variability and longitudinal trends at both personal and population levels. The observed variability in device accuracy across the magnitude of nighttime sleep outcomes (e.g., duration of nocturnal awakenings) underscores the importance of personalized application and cautious interpretation.

Supplementary Materials: The following supporting information can be downloaded at: https://www.mdpi.com/article/10.3390/s25154813/s1, Supplementary Table S1: Number of participants for each outcome measure; Supplementary Table S2: Group-level absolute error matrix; Supplementary Figure S1: Individual EBE metrics for each considered stage represented in boxplots; Supplementary Table S3: Group-level absolute error matrix assessing Fitbit Sense agreement for each sleep stage detection; Supplementary Table S4: Group-level discrepancies, bias and LOAs between Fitbit Sense and PSG for 64 nights; Supplementary Figure S2: Bland–Altman plots of the sleep outcomes for Fitbit Sense–PSG comparison for each sleep stage; Supplementary Table S5: Group-level proportional error matrix reporting the group average proportion of epochs in each sleep stage; Supplementary Table S6: Group-level EBE metrics assessing Fitbit Sense agreement for each sleep stage detection; Supplementary Table S7: Group-level discrepancies, bias and LOAs between Fitbit Sense and PSG for 64 nights; Supplementary Figure S3: Bland–Altman plots of the sleep outcomes for Fitbit Sense–PSG comparison for each sleep stage; Supplementary Table S8: Group-level proportional error matrix reporting the group average proportion of epochs in each sleep stage; Supplementary Table S9: Group-level EBE metrics assessing Fitbit Sense agreement for each sleep stage detection; Supplementary Table S10: baseline characteristics comparison between included and excluded participants.

Author Contributions: Conceptualization, C.L. and L.D.B.; methodology, C.L., L.D.B. and L.M.; validation, C.L., N.R., Z.G. and L.M.; formal analysis, C.L.; investigation, C.L., N.R. and L.D.B.; data curation, C.L.; writing—original draft preparation, C.L. and L.D.B.; writing—review and editing, C.L., N.R., Z.G., O.M., J.N., A.M., D.V.A., K.d.V., L.M., S.V. and L.D.B.; visualization, C.L.; supervision, L.D.B., S.V. and J.N.; project administration, L.D.B.; funding acquisition, O.M., J.N., A.M., D.V.A., K.d.V., S.V. and L.D.B. All authors have read and agreed to the published version of the manuscript.

Funding: This study was funded by a research grant, and the Fitbit Sense devices were acquired independently, with no participation from Fitbit. This study received financial support from the Applied Biomedical Research Program of the Research Foundation Flanders (FWO-TBM project no. TBM2021–T000521 N–54252), as part of the PREMEO trial ("Towards Precision Medicine for Osteoarthritis: Added Value of Cognitive Behavioural Therapy for Insomnia"). Additional funding was provided by the Strategic Research Program SRP90 ("Pain Never Sleeps: Unravelling the Sleep–Pain Interaction in Patients with Chronic Pain"), supported by the Research Council of the Vrije Universiteit Brussel, Brussels, Belgium. The funding bodies had no role in the review of this work.

Institutional Review Board Statement: This study was conducted in accordance with the Declaration of Helsinki and was approved by the ethical committees of the University Hospitals Leuven and the University Hospital Brussels (S66306) and was conducted in compliance with the Helsinki Declaration.

Informed Consent Statement: Informed consent was obtained from all subjects involved in this study.

Data Availability Statement: All data are available upon reasonable request to the corresponding author.

Acknowledgments: We would like to thank all participants, co-investigators, and the study team for their meaningful contributions to this project. A large language model was used to assist with language checking after the manuscript was written. All content was reviewed and approved by the authors.

Conflicts of Interest: Jo Nijs and the Vrije Universiteit Brussel received lecturing/teaching fees from various professional associations and educational organizations. The other authors have no potential conflicts of interest to report.

Abbreviations

The following abbreviations are used in this manuscript:

OA	Osteoarthritis
PSG	Polysomnography
CST	Consumer sleep tracking
RCT	Randomized controlled trial
TIB	Time in bed
EBE	Epoch-by-epoch
EEG	Electroencephalography
EOG	Electrooculography
EMG	Electromyography
TST	Total sleep time
SE	Sleep efficiency
SOL	Sleep onset latency
WASO	Wake after sleep onset
SD	Standard deviation
IQR	Interquartile range
LOAs	Limits of agreement
PABAK	Prevalence-adjusted bias-adjusted kappa
NRS	Numeric Rating Scale
KOOS	Knee Disability and Osteoarthritis Outcome Score
BPI	Brief Pain Inventory
CSI	Central Sensitization Inventory
HADS	Hospital Anxiety and Depression Scale
PSQI	Pittsburgh Sleep Quality Index
ISI	Insomnia Severity Index
BFS	Brugmann Fatigue Scale
ESS	Epworth Sleepiness Scale
AHI	Apnea–Hypopnea Index

PLMSI	Periodic Limb Movement Sleep Index
REM	Rapid-eye-movement
CI	Confidence interval

References

1. Husak, A.J.; Bair, M.J. Chronic Pain and Sleep Disturbances: A Pragmatic Review of Their Relationships, Comorbidities, and Treatments. *Pain Med.* **2020**, *21*, 1142–1152. [CrossRef]
2. Steinmetz, J.D.; Culbreth, G.T.; Haile, L.M.; Rafferty, Q.; Lo, J.; Fukutaki, K.G.; Cruz, J.A.; Smith, A.E.; Vollset, S.E.; Brooks, P.M. Global, regional, and national burden of osteoarthritis, 1990–2020 and projections to 2050: A systematic analysis for the Global Burden of Disease Study 2021. *Lancet Rheumatol.* **2023**, *5*, e508–e522. [CrossRef]
3. Thorlund, J.B.; Skou, S.T.; Roos, E.M.; Grønne, D.T.; Vægter, H.B. 264—How common is insomnia among patients with knee and hip osteoarthritis? A cross-sectional study using data from the Good Life with osteoArthritis in Denmark (GLA:D®) register. *Osteoarthr. Cartil.* **2024**, *32*, S194. [CrossRef]
4. Jacob, L.; Smith, L.; Konrad, M.; Kostev, K. Association between sleep disorders and osteoarthritis: A case–control study of 351,932 adults in the UK. *J. Sleep Res.* **2021**, *30*, e13367. [CrossRef] [PubMed]
5. Liu, M.; McCurry, S.M.; Belza, B.; Dobra, A.; Buchanan, D.T.; Vitiello, M.V.; Von Korff, M. Effects of osteoarthritis pain and concurrent insomnia and depression on health care use in a primary care population of older adults. *Arthritis Care Res.* **2019**, *71*, 748–757. [CrossRef] [PubMed]
6. Taylor, S.S.; Hughes, J.M.; Coffman, C.J.; Jeffreys, A.S.; Ulmer, C.S.; Oddone, E.Z.; Bosworth, H.B.; Yancy, W.S.; Allen, K.D. Prevalence of and characteristics associated with insomnia and obstructive sleep apnea among veterans with knee and hip osteoarthritis. *BMC Musculoskelet. Disord.* **2018**, *19*, 79. [CrossRef]
7. Silva, A.; Mello, M.T.; Serrão, P.R.; Luz, R.P.; Ruiz, F.; Bittencourt, L.R.; Tufik, S.; Mattiello, S.M. Influence of obstructive sleep apnea in the functional aspects of patients with osteoarthritis. *J. Clin. Sleep Med.* **2018**, *14*, 265–270. [CrossRef]
8. Woolhead, G.; Gooberman-Hill, R.; Dieppe, P.; Hawker, G. Night pain in hip and knee osteoarthritis: A focus group study. *Arthritis Care Res.* **2010**, *62*, 944–949. [CrossRef]
9. Jung, J.H.; Seok, H.; Choi, S.J.; Bae, J.; Lee, S.H.; Lee, M.H.; Kim, J.-H.; Song, G.G. The association between osteoarthritis and sleep duration in Koreans: A nationwide cross-sectional observational study. *Clin. Rheumatol.* **2018**, *37*, 1653–1659. [CrossRef]
10. Park, H.-M.; Kwon, Y.-J.; Kim, H.-S.; Lee, Y.-J. Relationship between sleep duration and osteoarthritis in middle-aged and older women: A nationwide population-based study. *J. Clin. Med.* **2019**, *8*, 356. [CrossRef]
11. Ni, J.; Zhou, W.; Cen, H.; Chen, G.; Huang, J.; Yin, K.; Sui, C. Evidence for causal effects of sleep disturbances on risk for osteoarthritis: A univariable and multivariable Mendelian randomization study. *Osteoarthr. Cartil.* **2022**, *30*, 443–450. [CrossRef]
12. Whibley, D.; Braley, T.J.; Kratz, A.L.; Murphy, S.L. Transient Effects of Sleep on Next-Day Pain and Fatigue in Older Adults With Symptomatic Osteoarthritis. *J. Pain* **2019**, *20*, 1373–1382. [CrossRef] [PubMed]
13. Yeung, W.K.; Morgan, K.; Mckenna, F. Comparison of sleep structure and psychometric profiles in patients with fibromyalgia, osteoarthritis and healthy controls. *J. Sleep Res.* **2018**, *27*, 292–300. [CrossRef] [PubMed]
14. Doherty, M.; Smith, J. Elusive'alpha-delta'sleep in fibromyalgia and osteoarthritis. *Ann. Rheum. Dis.* **1993**, *52*, 245. [CrossRef]
15. Leigh, T.; Hindmarch, I.; Bird, H.; Wright, V. Comparison of sleep in osteoarthritic patients and age and sex matched healthy controls. *Ann. Rheum. Dis.* **1988**, *47*, 40–42. [CrossRef] [PubMed]
16. Spira, A.P.; Runko, V.T.; Finan, P.H.; Kaufmann, C.N.; Bounds, S.C.; Liu, L.; Buenaver, L.F.; McCauley, L.M.; Ancoli-Israel, S.; Smith, M.T. Circadian rest/activity rhythms in knee osteoarthritis with insomnia: A study of osteoarthritis patients and pain-free controls with insomnia or normal sleep. *Chronobiol. Int.* **2015**, *32*, 242–247. [CrossRef]
17. *Chronic Pain (Primary and Secondary) in Over 16s: Assessment of All Chronic Pain and Management of Chronic Primary Pain;* NICE Guideline NG193; National Institute for Health and Care Excellence (NICE): London, UK, 2021.
18. Edwards, R.R.; Schreiber, K.L.; Dworkin, R.H.; Turk, D.C.; Baron, R.; Freeman, R.; Jensen, T.S.; Latremoliere, A.; Markman, J.D.; Rice, A.S. Optimizing and accelerating the development of precision pain treatments for chronic pain: IMMPACT review and recommendations. *J. Pain* **2023**, *24*, 204–225. [CrossRef]
19. Lim, D.C.; Najafi, A.; Afifi, L.; Bassetti, C.L.; Buysse, D.J.; Han, F.; Högl, B.; Melaku, Y.A.; Morin, C.M.; Pack, A.I. The need to promote sleep health in public health agendas across the globe. *Lancet Public Health* **2023**, *8*, e820–e826. [CrossRef]
20. de Zambotti, M.; Goldstein, C.; Cook, J.; Menghini, L.; Altini, M.; Cheng, P.; Robillard, R. State of the science and recommendations for using wearable technology in sleep and circadian research. *Sleep* **2023**, *47*, zsad325. [CrossRef]
21. Smith, M.T.; McCrae, C.S.; Cheung, J.; Martin, J.L.; Harrod, C.G.; Heald, J.L.; Carden, K.A. Use of Actigraphy for the Evaluation of Sleep Disorders and Circadian Rhythm Sleep-Wake Disorders: An American Academy of Sleep Medicine Clinical Practice Guideline. *J. Clin. Sleep Med.* **2018**, *14*, 1231–1237. [CrossRef]

22. Yuan, H.; Hill, E.A.; Kyle, S.D.; Doherty, A. A systematic review of the performance of actigraphy in measuring sleep stages. *J. Sleep Res.* **2024**, *33*, e14143. [CrossRef] [PubMed]
23. Redline, S.; Purcell, S.M. Sleep and Big Data: Harnessing data, technology, and analytics for monitoring sleep and improving diagnostics, prediction, and interventions—An era for Sleep-Omics? *Sleep* **2021**, *44*, zsab107. [CrossRef] [PubMed]
24. LeBlanc, R.G.; Czarnecki, P.; Howard, J.; Jacelon, C.S.; Marquard, J. Usability Experience of a Personal Sleep Monitoring Device to Self-manage Sleep Among Persons 65 Years or Older With Self-reported Sleep Disturbances. *CIN Comput. Inform. Nurs.* **2022**, *40*, 598–605. [CrossRef]
25. Brückner, S.; Sadare, O.; Fesl, S.; Scheibe, M.; Lang, C.; Gilbert, S. Attitudes of healthcare professionals and researchers toward wearable and app derived patient generated health data. *npj Digit. Med.* **2025**, *8*, 186. [CrossRef]
26. Garbarino, S.; Bragazzi, N.L. Revolutionizing sleep health: The emergence and impact of personalized sleep medicine. *J. Pers. Med.* **2024**, *14*, 598. [CrossRef]
27. Slitzky, M.; Yong, R.J.; Bianco, G.L.; Emerick, T.; Schatman, M.E.; Robinson, C.L. The future of pain medicine: Emerging technologies, treatments, and education. *J. Pain Res.* **2024**, *17*, 2833–2836. [CrossRef]
28. Scott, H.; Lechat, B.; Manners, J.; Lovato, N.; Vakulin, A.; Catcheside, P.; Eckert, D.J.; Reynolds, A.C. Emerging applications of objective sleep assessments towards the improved management of insomnia. *Sleep Med.* **2023**, *101*, 138–145. [CrossRef] [PubMed]
29. Menghini, L.; Cellini, N.; Goldstone, A.; Baker, F.C.; De Zambotti, M. A standardized framework for testing the performance of sleep-tracking technology: Step-by-step guidelines and open-source code. *Sleep* **2021**, *44*, zsaa170. [CrossRef]
30. Schyvens, A.-M.; Peters, B.; Van Oost, N.C.; Aerts, J.-M.; Masci, F.; Neven, A.; Dirix, H.; Wets, G.; Ross, V.; Verbraecken, J. A performance validation of six commercial wrist-worn wearable sleep-tracking devices for sleep stage scoring compared to polysomnography. *Sleep Adv.* **2025**, *6*, zpaf021. [CrossRef]
31. Chinoy, E.D.; Cuellar, J.A.; Huwa, K.E.; Jameson, J.T.; Watson, C.H.; Bessman, S.C.; Hirsch, D.A.; Cooper, A.D.; Drummond, S.P.; Markwald, R.R. Performance of seven consumer sleep-tracking devices compared with polysomnography. *Sleep* **2021**, *44*, zsaa291. [CrossRef]
32. Kahawage, P.; Jumabhoy, R.; Hamill, K.; de Zambotti, M.; Drummond, S.P. Validity, potential clinical utility, and comparison of consumer and research-grade activity trackers in insomnia disorder I: In-lab validation against polysomnography. *J. Sleep Res.* **2020**, *29*, e12931. [CrossRef]
33. Moreno-Pino, F.; Porras-Segovia, A.; López-Esteban, P.; Artés, A.; Baca-García, E. Validation of Fitbit Charge 2 and Fitbit Alta HR Against Polysomnography for Assessing Sleep in Adults With Obstructive Sleep Apnea. *J. Clin. Sleep Med.* **2019**, *15*, 1645–1653. [CrossRef]
34. Doheny, E.P.; Renerts, K.; Braun, A.; Werth, E.; Baumann, C.; Baumgartner, P.; Morgan-Jones, P.; Busse, M.; Lowery, M.M.; Jung, H.H. Assessment of Fitbit Charge 4 for sleep stage and heart rate monitoring against polysomnography and during home monitoring in Huntington's disease. *J. Clin. Sleep Med.* **2024**, *20*, 1163–1171. [CrossRef]
35. Ogasawara, M.; Takeshima, M.; Kosaka, S.; Imanishi, A.; Itoh, Y.; Fujiwara, D.; Yoshizawa, K.; Ozaki, N.; Nakagome, K.; Mishima, K. Exploratory validation of sleep-tracking devices in patients with psychiatric disorders. *Nat. Sci. Sleep* **2023**, *15*, 301–312. [CrossRef] [PubMed]
36. Cook, J.D.; Prairie, M.L.; Plante, D.T. Utility of the Fitbit Flex to evaluate sleep in major depressive disorder: A comparison against polysomnography and wrist-worn actigraphy. *J. Affect. Disord.* **2017**, *217*, 299–305. [CrossRef]
37. Labie, C.; Runge, N.; Mairesse, O.; Nijs, J.; Malfliet, A.; Verschueren, S.; Van Assche, D.; de Vlam, K.; Luyten, F.; Bilterys, T.; et al. Integration of Cognitive Behavioral Therapy for Insomnia in Best-Practice Care for Patients With Knee Osteoarthritis and Insomnia: A Randomized Controlled Trial Protocol. *Phys. Ther.* **2023**, *104*, pzad181. [CrossRef]
38. Association, W.M. World Medical Association Declaration of Helsinki: Ethical principles for medical research involving human subjects. *JAMA* **2013**, *310*, 2191–2194.
39. Depner, C.M.; Cheng, P.C.; Devine, J.K.; Khosla, S.; De Zambotti, M.; Robillard, R.; Vakulin, A.; Drummond, S.P. Wearable technologies for developing sleep and circadian biomarkers: A summary of workshop discussions. *Sleep* **2020**, *43*, zsz254. [CrossRef]
40. Von Elm, E.; Altman, D.G.; Egger, M.; Pocock, S.J.; Gøtzsche, P.C.; Vandenbroucke, J.P. The Strengthening the Reporting of Observational Studies in Epidemiology (STROBE) statement: Guidelines for reporting observational studies. *Lancet* **2007**, *370*, 1453–1457. [CrossRef] [PubMed]
41. Altman, R.; Asch, E.; Bloch, D.; Bole, G.; Borenstein, D.; Brandt, K.; Christy, W.; Cooke, T.D.; Greenwald, R.; Hochberg, M.; et al. Development of criteria for the classification and reporting of osteoarthritis. Classification of osteoarthritis of the knee. Diagnostic and Therapeutic Criteria Committee of the American Rheumatism Association. *Arthritis Rheum.* **1986**, *29*, 1039–1049. [CrossRef]
42. Sleep-Wake Disorders. In *DSM-5-TR*, 5th ed.; Text Revision (DSM-5-TR) ed.; American Psychiatric Association Publishing: Washington, DC, USA, 2022. [CrossRef]

43. Haghayegh, S.; Khoshnevis, S.; Smolensky, M.H.; Diller, K.R.; Castriotta, R.J. Accuracy of wristband Fitbit models in assessing sleep: Systematic review and meta-analysis. *J. Med. Internet Res.* **2019**, *21*, e16273. [CrossRef]
44. Fitbit, I. Available online: https://www.fitbit.com/sg/sense (accessed on 4 May 2025).
45. Berry, R.B.; Brooks, R.; Gamaldo, C.E.; Harding, S.M.; Marcus, C.; Vaughn, B.V. *The AASM Manual for the Scoring of Sleep and Associated Events*; Rules, Terminology and Technical Specifications; American Academy of Sleep Medicine: Darien, IL, USA, 2012.
46. Lee, T.; Cho, Y.; Cha, K.S.; Jung, J.; Cho, J.; Kim, H.; Kim, D.; Hong, J.; Lee, D.; Keum, M. Accuracy of 11 wearable, nearable, and airable consumer sleep trackers: Prospective multicenter validation study. *JMIR Mhealth Uhealth* **2023**, *11*, e50983. [CrossRef]
47. Ong, J.L.; Golkashani, H.A.; Ghorbani, S.; Wong, K.F.; Chee, N.I.; Willoughby, A.R.; Chee, M.W. Selecting a sleep tracker from EEG-based, iteratively improved, low-cost multisensor, and actigraphy-only devices. *Sleep Health* **2024**, *10*, 9–23. [CrossRef] [PubMed]
48. Lim, S.E.; Kim, H.S.; Lee, S.W.; Bae, K.-H.; Baek, Y.H. Validation of fitbit inspire 2TM against polysomnography in adults considering adaptation for use. *Nat. Sci. Sleep* **2023**, *15*, 59–67. [CrossRef]
49. Haghayegh, S.; Khoshnevis, S.; Smolensky, M.H.; Diller, K.R.; Castriotta, R.J. Performance assessment of new-generation Fitbit technology in deriving sleep parameters and stages. *Chronobiol. Int.* **2020**, *37*, 47–59. [CrossRef]
50. Grandner, M.A.; Bromberg, Z.; Hadley, A.; Morrell, Z.; Graf, A.; Hutchison, S.; Freckleton, D. Performance of a multisensor smart ring to evaluate sleep: In-lab and home-based evaluation of generalized and personalized algorithms. *Sleep* **2023**, *46*, zsac152. [CrossRef] [PubMed]
51. Robbins, R.; Weaver, M.D.; Sullivan, J.P.; Quan, S.F.; Gilmore, K.; Shaw, S.; Benz, A.; Qadri, S.; Barger, L.K.; Czeisler, C.A. Accuracy of Three Commercial Wearable Devices for Sleep Tracking in Healthy Adults. *Sensors* **2024**, *24*, 6532. [CrossRef] [PubMed]
52. Kang, S.-G.; Kang, J.M.; Ko, K.-P.; Park, S.-C.; Mariani, S.; Weng, J. Validity of a commercial wearable sleep tracker in adult insomnia disorder patients and good sleepers. *J. Psychosom. Res.* **2017**, *97*, 38–44. [CrossRef] [PubMed]
53. Feige, B.; Baglioni, C.; Spiegelhalder, K.; Hirscher, V.; Nissen, C.; Riemann, D. The microstructure of sleep in primary insomnia: An overview and extension. *Int. J. Psychophysiol.* **2013**, *89*, 171–180. [CrossRef]
54. Riemann, D.; Espie, C.A.; Altena, E.; Arnardottir, E.S.; Baglioni, C.; Bassetti, C.L.; Bastien, C.; Berzina, N.; Bjorvatn, B.; Dikeos, D. The European Insomnia Guideline: An update on the diagnosis and treatment of insomnia 2023. *J. Sleep Res.* **2023**, *32*, e14035. [CrossRef]
55. Baron, K.G.; Abbott, S.; Jao, N.; Manalo, N.; Mullen, R. Orthosomnia: Are Some Patients Taking the Quantified Self Too Far? *J. Clin. Sleep Med.* **2017**, *13*, 351–354. [CrossRef] [PubMed]
56. The Academy of Medical Sciences. *Multimorbidity: A Priority for Global Health Research*; Academy of Medical Sciences: London, UK, 2018.
57. Swain, S.; Kamps, A.; Runhaar, J.; Dell'Isola, A.; Turkiewicz, A.; Robinson, D.; Strauss, V.; Mallen, C.; Kuo, C.-F.; Coupland, C. Comorbidities in osteoarthritis (ComOA): A combined cross-sectional, case–control and cohort study using large electronic health records in four European countries. *BMJ Open* **2022**, *12*, e052816. [CrossRef] [PubMed]
58. Lujan, M.R.; Perez-Pozuelo, I.; Grandner, M.A. Past, present, and future of multisensory wearable technology to monitor sleep and circadian rhythms. *Front. Digit. Health* **2021**, *3*, 721919. [CrossRef]
59. Koerber, D.; Khan, S.; Shamsheri, T.; Kirubarajan, A.; Mehta, S. Accuracy of heart rate measurement with wrist-worn wearable devices in various skin tones: A systematic review. *J. Racial Ethn. Health Disparities* **2023**, *10*, 2676–2684. [CrossRef] [PubMed]

Disclaimer/Publisher's Note: The statements, opinions and data contained in all publications are solely those of the individual author(s) and contributor(s) and not of MDPI and/or the editor(s). MDPI and/or the editor(s) disclaim responsibility for any injury to people or property resulting from any ideas, methods, instructions or products referred to in the content.

sensors

MDPI

Article

A Study on Autonomous Driving Motion Sickness from the Perspective of Multimodal Human Signals

Su Young Kim [1] and Yoon Sang Kim [1,2,*]

[1] BioComputing Lab, Department of Computer Engineering, Korea University of Technology and Education (KOREATECH), Cheonan 31253, Republic of Korea; sgkim6326@gmail.com
[2] Institute for Bioengineering Application Technology, Korea University of Technology and Education (KOREATECH), Cheonan 31253, Republic of Korea
* Correspondence: yoonsang@koreatech.ac.kr

Abstract

In autonomous driving, motion sickness (MS) arises from physical or visual stimuli, or a combination of both. However, objective quantification of MS level (MSL) remains limited beyond questionnaire-based assessments. Using multimodal human signals (physiological and behavioral) collected in an autonomous driving simulator, this study addresses the association between these signals and MSL, across these MS types, by (i) screening and curating a decade of human-signal MS studies (HS-Set) to establish a data-driven foundation for selecting target sensor domains and features, (ii) constructing a dataset with subjective measures of MSL (fast motion sickness scale and simulator sickness questionnaire (SSQ)), alongside human signals (electroencephalogram (EEG), photoplethysmogram (PPG), electrodermal activity (EDA), skin temperature, and head/eye movement), (iii) conducting a correlation analysis between MSL and the identified features from HS-Set, and (iv) quantifying multivariable contributions at the feature and sensor domains through an explainable boosting machine (EBM). Key correlations include head amplitude/energy (pitch/surge) with SSQ total/oculomotor, eye entropy with nausea/oculomotor (positive), and EDA with nausea (negative). The EBM-based contribution analysis highlights EEG connectivity and head kinematics as dominant contributors; excluding EEG, the interpretability of single-domain models remains limited. Additionally, a combination of Head, PPG, and EDA domains retains over 80% of the full model's interpretability.

Keywords: autonomous driving; motion sickness; unified sickness; multimodal sensors; signal processing; systematic feature extraction; physiological and behavioral signals

1. Introduction

An autonomous vehicle (AV) is a vehicle type in which onboard systems perform driving-related tasks (DRTs), such as steering, accelerating, and braking, thereby transforming drivers into passengers. SAE International has defined six levels of driving automation (level 0–5) in its J3016 standard [1], where higher levels indicate greater automation and reduced driver involvement. For example, a level 2 (L2) vehicle assists with both steering and speed control, but requires constant supervision by the driver, whereas level 4 (L4) and higher vehicles handle all DRTs within their operational design domain. Currently, Tesla provides L2 AVs through its Full Self-Driving software, whereas Waymo operates Waymo One, an L4 ride-hailing service, in Los Angeles, USA. As AV commercialization progresses, drivers are expected to spend more time and engage more frequently in non-driving-related tasks (NDRTs) with increasing levels of automation [2–6].

In their new role as passengers, these tasks will primarily be display-based, such as using handheld displays [6–9] or watching integrated displays [5,6]. In recent years, these display devices have evolved into see-through head-mounted displays (HMDs) such as Apple Vision Pro and Meta-Orion, a trend that has led to a corresponding increase in research on in-vehicle HMD usage [5,6,10,11].

However, engaging in display-based NDRTs within moving vehicles has the serious side effect of inducing motion sickness (MS) [12–15]. MS can manifest as classical MS (CMS) from physical stimuli, such as vehicle movement (some studies refer to it as motion-induced MS), visually induced MS (VIMS) from visual stimuli, such as displayed content, or composite MS (Co-MS), resulting from a sensory conflict caused by both stimuli. Although various approaches have attempted to identify the causes of MS and develop mitigation strategies by addressing CMS and VIMS separately or concurrently, the evidence remains fragmentary, and the mechanisms of MS have not been fully elucidated yet.

Quantification of the motion sickness level (MSL) is essential for MS reduction. Conventionally, research has predominantly quantified MSL using subjective questionnaires, including the simulator sickness questionnaire (SSQ) [16], fast motion sickness scale (FMS) [17], and misery scale (MISC) [18]. While these subjective questionnaires have the advantage of being applicable to a wide range of MS types, they have limitations in terms of reliability owing to their dependence on the subjective judgments of the respondents. Some studies on CMS have used the motion sickness dose value (MSDV) (ISO 2631-1) [19], which objectively quantifies ride discomfort that is indicative of MS from vehicle or head acceleration [20–23]. However, MSDV is limited to representing only CMS, not VIMS or Co-MS.

To overcome these limitations, research has been conducted to quantify MSL using objective human signals, including physiological and behavioral data. However, among these MS studies, those that consider Co-MS arising from display-based NDRTs in moving vehicles are insufficient (see Section 2). Furthermore, machine-learning (ML) and deep-learning (DL) models, which are commonly used to quantify MSL, have a critical limitation: their black-box nature makes them difficult to interpret.

Therefore, in this study, multimodal human signals (physiological and behavioral) collected in an autonomous driving simulator have been used to address the quantitative association between these signals and MSL, a relationship that generalizes across CMS, VIMS, and Co-MS. This has been achieved by (i) systematizing human-signal MS studies from the last decade (2015–2024) to organize handcrafted features by sensor domain, (ii) constructing a dataset by simultaneously collecting subjective MSL measures, four physiological signals, and two behavioral signals, (iii) conducting correlation analysis between MSL and the identified features from the HS-Set, and (iv) quantifying multivariable contributions at the feature and sensor-domain levels using an explainable boosting machine (EBM), to identify interpretable multimodal markers of MS.

Four physiological signals, namely, electroencephalogram (EEG), photoplethysmogram (PPG), electrodermal activity (EDA), and skin temperature (SKT), and two behavioral signals, namely, head and eye movements, were considered for building the dataset. Subjective MSLs were assessed using SSQ and FMS, along with the motion sickness susceptibility questionnaire (MSSQ) [24] to estimate the individual MS characteristics based on prior experience and the film immersive experience questionnaire (Film IEQ) [25] to measure content immersion levels. CMS was induced using a six-degree-of-freedom (6-DOF) simu-

 https://doi.org/10.3390/s26051675

lator (sway, surge, heave, pitch, yaw, and roll), whereas VIMS was induced by presenting high-optical-flow movies through two types of displays: a see-through HMD and a tablet.

The remainder of this paper is organized as follows: Section 2 screens and curates MS studies using human signals. Section 3 describes the experimental apparatus, consisting of an autonomous driving simulator and an acquisition system for multimodal human signals, and defines MS-inducing factors. Section 4 details the experiments based on these factors. Section 5 analyzes the results of the subjective questionnaire data and the objective human-signal data, and examines the contributions of sensor domains and features to MSL. Section 6 discusses the findings of this study, followed by conclusions and future research directions in Section 7.

2. Related Work

2.1. MS and Human Signals

Research attempting to explain MS using objective human signals has made steady progress. To establish a data-driven foundation for objective MS evaluation and to identify valid parameters that inform our experimental design, recent studies from the last decade were screened and curated (search period: 1 January 2015, to 31 December 2024). The academic databases used were IEEE Xplore, Web of Science, and ACM Digital Library, with the search keywords "(Motion Sickness OR Cybersickness OR VIMS OR VR sickness) AND (Evaluation OR Assessment OR Quantification) AND (Physiological OR Physiology OR Biometric OR Biosignal OR Objective)," where VR refers to virtual reality. The keyword behavioral was intentionally excluded to minimize search noise from studies with low direct relevance, such as cognitive-behavioral studies. Instead, the methodological core keyword objective was included to comprehensively capture the objective measurement studies, such as MS research based on behavioral signals. The specific search queries used for each academic database are listed in Table S1 in the Supplementary Materials.

The literature search was limited to publications from journals and conferences. Studies that did not attempt to explain MS using objective human signals were excluded. Studies that explained MS based solely on stimulus characteristics (e.g., videos and vehicle acceleration) were also excluded. When a collected publication was a review paper, it was excluded unless an objective MS quantification methodology was suggested. Of the 968 records retrieved through the search, 43 duplicates were removed, and 843 publications were excluded, resulting in 82 studies (last search date: 27 April 2025). Hereafter, the human-signal MS study set (HS-Set) refers to the collection of 82 studies curated in this section.

In the HS-Set studies, CMS cases were classified as simulator-based when physical motion was stimulated, but not in actual transportation. Co-MS cases were limited to studies with separate visual stimuli that were independent of physical motion (e.g., reading a book or watching a display in a moving vehicle). The yearly trends of the classified studies, as shown in Figure 1, were predominantly centered on VIMS, with studies considering CMS and Co-MS gradually emerging since 2018. This trend most likely reflects the rise in data-driven MS quantification approaches, facilitated by advancements in ML/DL technologies in the 2010s and the increasing interest in NDRTs following AV commercialization in the 2020s.

The HS-Set included eight Co-MS studies, four in actual vehicles [11,26–28], and four in laboratory or simulator settings [29–32]. In each study, Co-MS was analyzed in conjunction with various human signals or functional test outcomes. Some studies focused on functional test outcomes to diagnose or predict MS susceptibility. For example, functional head impulse tests (fHIT) and dynamic visual acuity were reported to reflect MS susceptibility and abnormalities in visual-vestibular integration outside transportation

https://doi.org/10.3390/s26051675

simulators [30,31]. In contrast, the vestibulo-ocular reflex (VOR) measured in an actual vehicle showed no significant differences between the susceptible and non-susceptible groups, indicating its limitation as a standalone predictor [26]. Crucially, these approaches focused on diagnostic testing rather than proposing quantitative metrics derived directly from continuous human signals.

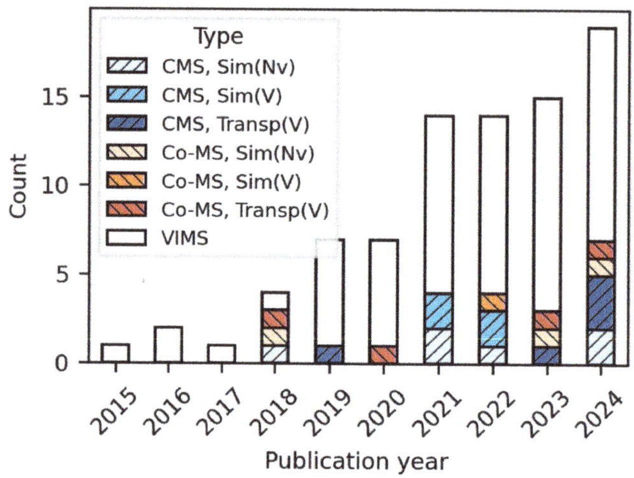

Figure 1. Yearly trends in the HS-Set (Transp = actual transport, Sim = laboratory or simulator settings, V = vehicle, and Nv = non-vehicle).

Other studies collected continuous physiological signals alongside subjective MS questionnaires to evaluate specific experimental conditions. These studies measured signals such as PPG, EDA, electrocardiogram (ECG), and electrogastrogram (EGG) alongside the SSQ under various conditions in actual vehicles (e.g., different types of VR content, scent conditions, or synchronization levels) [11,27,28]; nevertheless, a common limitation of these studies is that the subjective scores and physiological signals were only compared across conditions without performing a direct joint analysis between them. Even when associations were examined, as in the study by Tamura et al. [32], who analyzed EDA and disorientation criteria in a flight simulator, quantitative metrics were not proposed.

While some prior research has explored Co-MS, it remains relatively insufficient compared to CMS and VIMS. In particular, only five studies were conducted in driving environments (including actual vehicles and simulators), and among these, only Kojima et al. [29] conducted a joint analysis of the measured MSL and collected human signals. However, this study did not propose a quantitative metric for MS using human signals. Therefore, an integrated MS quantification approach based on human signals that can be applied to Co-MS is necessary.

2.2. Sensor Domains and Research Trends in the HS-Set

The human signals utilized in the HS-Set can be classified into several sensor domains based on the data acquisition methods; in this paper, both small-scale measurement sensors and larger equipment are collectively referred to as "sensors." For example, if a heart rate (HR) feature is collected using a PPG sensor, it can be considered as a part of the PPG domain. Similarly, features such as the breath rate (BR) can be either indirectly calculated from the PPG domain or collected using a dedicated respiration (RSP) sensor. The number of studies classified according to these sensor domains is listed in Table 1.

https://doi.org/10.3390/s26051675

Table 1. Frequencies of human-signal sensor domains in the HS-Set (not mutually exclusive).

Sensor Domain	Signal Type	Count	Included Studies
Electrocardiogram and Photoplethysmogram [1]	physiological	35	[11,27–29,33–63]
Electrodermal activity	physiological	29	[11,27,32,36–47,56–69]
Eye	physiological and behavioral	24	[26,46–48,60–66,70–82]
Electroencephalogram	physiological	20	[35–40,54–56,83–93]
Head	behavioral	17	[42,46,59–66,77–79,94–97]
Respiration	physiological	9	[27,29,33,36,42,47,57,58,60]
Assessment test	indirectly derived human-signals	8	[26,30,31,44,47,65,80,98]
Skin temperature	physiological	7	[33,37,47,57,61,62,65]
Electrogastrogram	physiological	6	[28,45,53,60,65,99]
Center of pressure	physiological	6	[35,65,98,100–102]
Electromyogram	physiological	5	[35,47,55,57,80]
Functional near-infrared spectroscopy	physiological	3	[103–105]
Functional magnetic resonance imaging	physiological	2	[52,53]
Blood pressure	physiological	2	[53,57]
Impedance cardiography	physiological	1	[57]
Gait	behavioral	1	[106]
Body	behavioral	1	[59]
Endocrine	physiological	1	[53]

[1] ECG and PPG are combined into a single sensor domain as they share major features.

Summaries of the primary characteristics and utilized human-signal features for each sensor domain are extensively described in Section S1 of the Supplementary Materials (Tables S2–S8). Beyond identifying the types of sensors utilized, understanding the broader experimental contexts in which these signals were collected is crucial for establishing an experimental design. To examine these contextual trends in the HS-Set, the studies were categorized according to four key aspects: ① interaction type, ② display used, ③ research purpose, and ④ approach method. Figure 2 shows the categorization results.

As can be seen from Figure 2a, most studies used VR HMDs, indicating that MS research using human signals over the last decade has predominantly focused on VIMS in VR environments. Normal and large displays were frequently used in eight and five studies, respectively, whereas other display types were used in only a few cases. Importantly, despite the trend toward see-through HMDs, only one study used an AR HMD [80]. This underscores the strong need for Co-MS research that combines see-through HMDs with NDRTs in the context of autonomous driving.

Interaction types were classified into two categories based on whether users could actively control the experimental content or only passively experience it (general was used for studies applicable to both categories). The most common interaction was passive, which could stem from efforts to minimize behavioral confounds to clearly analyze the

https://doi.org/10.3390/s26051675

stimulus–response relationships. Active interaction was also frequently employed to induce MS-related factors, such as cognitive load, by requiring users to perform specific tasks.

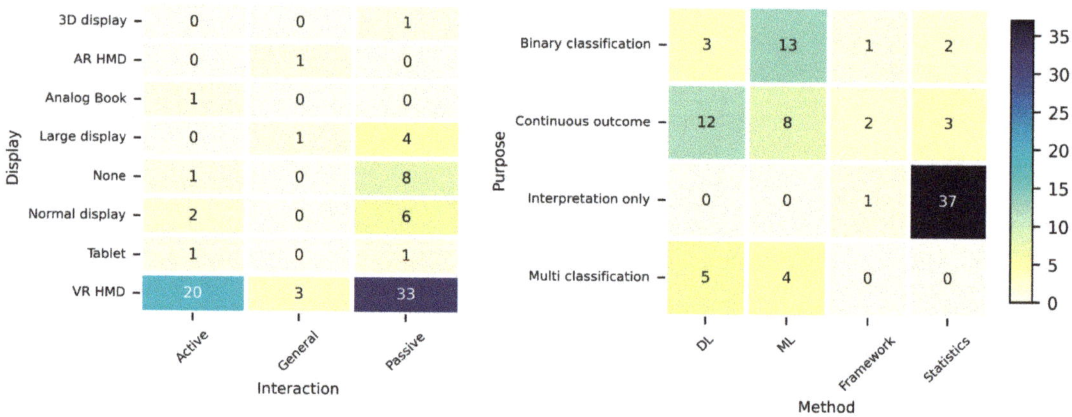

(a) (b)

Figure 2. Results of the four-axis categorization (not mutually exclusive, except for method axis): (**a**) Display × Interaction; (**b**) Purpose × Method. The color scale applies to both subgraphs. (AR = augmented reality; other abbreviations are defined in the main text; other abbreviations are defined in the main text).

Figure 2b shows the approach methods according to the research purpose. Studies that involved calculating continuous values through various models (theoretical, statistical, and neural networks) were classified as continuous outcomes, those that involved distinguishing MS-related states at two levels were classified as binary classification, and those with three or more levels were classified as multi-classification. Studies that did not fit into these categories, such as those that focused solely on interpretation, were classified as interpretation only.

Unlike the other axes, the approach method axis did not allow for overlapping classifications; the studies were classified based on their core approach method. Studies that focused on statistical models or analyses were classified as statistics, whereas those proposing new methodologies (e.g., metrics, theoretical models, and architectures) were classified as frameworks. Among learning-based studies, those employing neural network models were classified as DL, whereas those with shallow networks were classified as ML.

Most studies in the HS-Set were classified as interpretation only × statistics. This indicates that because MS mechanisms have not yet been clearly elucidated and validation frameworks are insufficient, the predominant approach has been to interpret the phenomenon using statistical methods. ML and DL approaches were also frequently used for prediction and classification despite their lower interpretability. ML was primarily used for binary classification, such as determining the presence of MS, whereas DL was mainly used to predict subjective MSL measures, such as SSQ or FMS, using regression. However, ML- and DL-based prediction problems are limited by their low interpretability owing to their predominantly black-box structures.

A Sankey diagram illustrating the relationships between the MS type, research purpose, approach method, and sensor domains used is shown in Figure 3.

Figure 3 reveals an imbalance in MS research using human signals. In terms of MS types, research is excessively biased toward VIMS, whereas Co-MS studies are not only sparse but also limited to interpretation only. Among studies not restricted to interpretation only, some attempted to quantify MS through multi-classification, but these were relatively few compared to continuous outcomes and binary classification. The selection of sensor domains was dispersed without consistent patterns for any specific type-purpose-method combination.

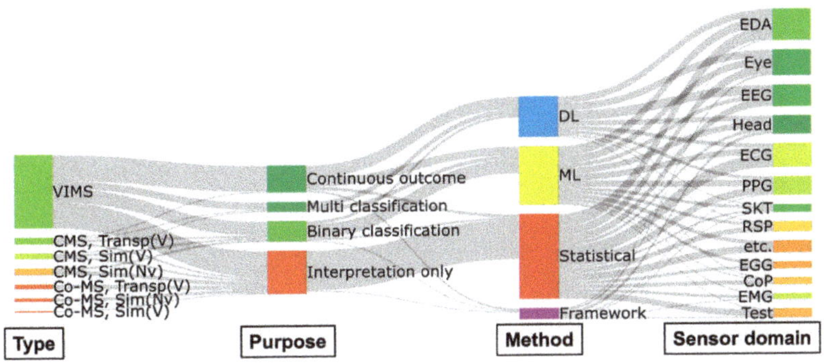

Figure 3. Sankey diagram of the MS type, purpose, method, and sensor domain. (All abbreviations are defined in the main text).

These facts imply that (i) the foundation of MSL research on Co-MS is still weak, (ii) sensor selection was likely driven more by availability or convention than by hypotheses, and (iii) there is a need for an integrated platform to analyze the relationships between complex factors (e.g., MS type and display type) and multimodal sensor domains.

To establish such an integrated approach, it is essential to understand how MS is reflected in human signals. The HS-Set suggests that MS is rarely captured using a single metric. Instead, it corresponds to complex variations across multiple domains, including autonomic nervous system responses (e.g., cardiovascular and electrodermal changes), central nervous system activities (e.g., EEG spectral patterns), and behavioral or oculomotor indicators (e.g., head and eye movements). These cross-domain trends provide an essential rationale for adopting a multimodal measurement strategy. Accordingly, based on the primary domains summarized in this section (Table 1 and Section S1), we configured the targeted sensor setup (Section 3) and established a feature extraction framework (Section 4) to enable joint quantitative analyses with subjective MSL measures across CMS, VIMS, and Co-MS.

3. Materials

3.1. Autonomous Driving Simulator

Collecting data through actual vehicle driving might be the most intuitive approach for quantitatively studying MS in AV environments. However, owing to driving-safety concerns and variable control issues across repeated experiments, it is difficult to provide identical experimental scenarios repeatedly for dozens of participants. Furthermore, because current AVs operate at L4 only in limited regions, there are constraints on the intentional and repeated implementation of various scenarios (e.g., acceleration/deceleration, turning, and slope).

To overcome these limitations, a 6-DOF motion platform was employed to simulate an autonomous driving environment instead of an actual vehicle. The motion platform used was the XTA Pro G6 model from GAMA System in Korea, with maximum motion ranges of sway = ±115 mm, heave = ±120 mm, surge = ±75 mm, pitch = ±33°, yaw = ±20°, and roll = ±28°. The motion platform was installed in a laboratory equipped with window blinds and an air conditioner to minimize environmental factors such as lighting and temperature. It was paired with a large 65-inch display (Samsung Signage Flip) to enhance the visual immersion of passengers.

https://doi.org/10.3390/s26051675

Additionally, green partitions were installed within the experimental environment to minimize external visual stimuli, and an operation console was positioned next to the platform to enable real-time control of the experiment and monitor the sensors. Figure 4a shows the setup of the autonomous driving simulator used in the laboratory.

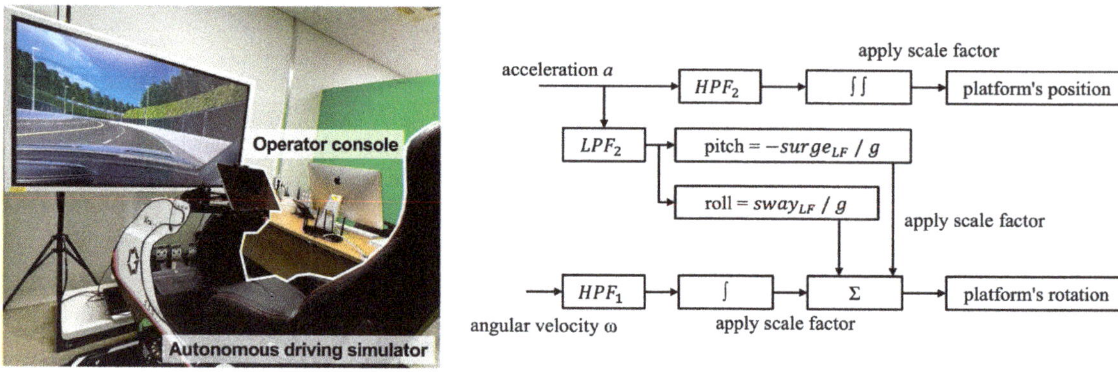

(a) (b)

Figure 4. Implemented autonomous driving simulator: (**a**) Experimental setup; (**b**) Motion cueing algorithm based on the classic washout filter applied to the simulator.

The motion cueing algorithm (MCA) applied to the autonomous driving simulator was the classic washout filter method, as shown in Figure 4b, which is widely used owing to its effective real-time performance [107,108]. Detailed parameter settings of the MCA, including the cutoff frequencies and scale factors for each filter, are provided in Section S2.1 of the Supplementary Materials (Table S9).

The virtual driving content was developed using the Unity engine and EasyRoads3D package, and comprised four driving scenarios: acceleration/deceleration, constant velocity, constant velocity turning, and constant velocity on slopes. The two constant-velocity conditions (turning and slope) refer to a constant linear velocity, not a constant angular velocity, which is similar to the cruise control in actual vehicles. To synchronize with the MCA, each scenario used predefined motion profiles based on acceleration and angular velocity. Detailed procedures for generating these motion profiles and virtual trajectories (e.g., road generation and coordinate transformation) are described in Section S2.2 (Figures S1 and S2).

The virtual driving content excluded surrounding vehicles and traffic signals. The surrounding environment consisted of repeated patterns, namely lamps at regular 25-m intervals and two types of trees at 10-m intervals (with small random offsets to prevent perceptual adaptation to regular patterns), to induce vection related to MS [109]. Based on the 2023 statistics from the Ministry of Land, Infrastructure, and Transport (MOLIT), the road was designed as a four-lane road (two lanes in each direction, with a lane width of 3.5 m), which is the most common configuration for general roads and highways in Korea. In Korea, the first lane is typically the passing lane. Therefore, the second lane was used as the driving lane for the experiment, considering that AVs are expected to travel at a constant speed.

3.2. Multimodal Human-Signal Acquisition System

Although the sensor domains utilized in the HS-Set were diverse, some sensors (e.g., center of pressure (CoP), Gait) were unsuitable for driving environments. Additionally,

 https://doi.org/10.3390/s26051675

ECG and PPG provided largely overlapping feature information, and a few RSP features could be derived from these sensors. Kenward et al. [110] and Wickham [111] reported that nausea, a major symptom of MS, is a higher-level cognitive activity than vomiting. Kim et al. [47] found no significant association between electromyogram (EMG), which belongs to the somatic nervous system rather than the autonomic nervous system, and MS. Considering these factors, the EGG, EMG, CoP, and Gait domains were excluded. PPG was selected for cardiovascular sensing instead of ECG because of the ease of wearability of the PPG sensor, and no dedicated RSP sensor was used because RSP data could be indirectly estimated from the ECG or PPG signals.

The brain activity can usually be measured using various sensor domains such as EEG, functional near-infrared spectroscopy (fNIRS), and functional magnetic resonance imaging (fMRI). However, EEG data were used in 20 studies in the HS-Set, whereas fNIRS and fMRI data were used in only three or fewer studies each. In particular, because fMRI relies on large and fixed equipment, it cannot be applied in driving environments. Moreover, wearing both fNIRS and EEG sensors can cause not only physical interference but also reduced immersion, leading to cumulative fatigue. Thus, only the EEG was selected for measuring the brain activity.

Impedance cardiography (ICG) requires direct electrode attachment to the chest, which can be burdensome for the participants and might cause ethical or psychological discomfort. As these constraints could reduce data reliability, ICG was also excluded. Blood pressure (BP) applies physical pressure to the body, which could divert the attention of the participants from this measurement. This increases the likelihood of bodily responses being distorted by the external factor of pressure rather than by the MS-inducing effects. Therefore, BP was also excluded. The endocrine and assessment test domains were not considered because they were difficult to use as real-time signals. Following this selection process, the human-signal sensors selected for this acquisition system are listed in Table 2.

Table 2. Selected multimodal human-signal sensors and measurement specifications.

Sensor Domain	Sensor	Measurement [1]	Sampling Rate (Hz)
PPG	Shimmer3 GSR+	Voltage (mV)	128
EDA	Shimmer3 GSR+	Skin resistance (kΩ)	128
Eye	HoloLens 2 [2]	Eye position and gaze direction	30
EEG	Enobio 8	Voltage (μV), used channels [3]: F3, F4, C3, C4, P3, P4, O1, O2	500
SKT	Core body temperature	Temperature (°C)	1
Head	HoloLens 2	Head position and rotation	30
Head	IMU (WT901BLECL)	Head acceleration and angular velocity	4
Body	Azure Kinect	Tracked body joints	30

[1] The acquisition system was additionally designed to collect facial expression (iPad), hand tracking (HoloLens 2), viewport image (HoloLens 2–WebRTC), and simulator motion (inertial measurement unit, IMU), although these signals were not analyzed in this study. [2] Eye data was collected from the left, right, and center using an extended eye tracking library. [3] Electrode placement followed the International 10–20 system.

The selected sensors had different sampling rates, making time synchronization essential for simultaneous collection. To address this issue, a multimodal human-signal acquisition system was implemented to collect the sensor data in parallel and align them along a common time axis. As shown in Figure 5, the acquisition system was

 https://doi.org/10.3390/s26051675

designed with multithreading for parallel processing and included two subsystems: a motion hub that distributed GPU-accelerated body-tracking computations, and a peripheral sensor gateway that collected and transmitted data from peripheral sensors lacking communication modules.

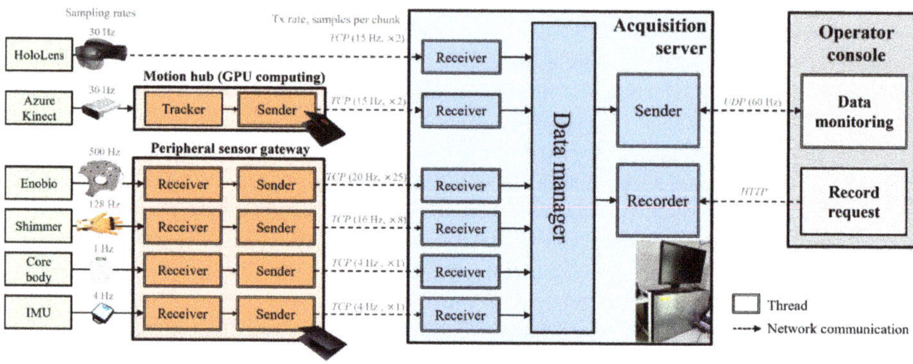

Figure 5. Multimodal human-signal acquisition system. Motion hub = Samsung Odyssey (GPU: GTX 1060), peripheral sensor gateway = Lenovo ThinkPad Gen1, acquisition server = HP Z820 (OS: Windows 10). The core body temperature sensor was sampled at 1 Hz but transmitted at 4 Hz to avoid dropouts from 1 Hz server logging.

To ensure stable parallel recording of multimodal signals with varying sampling rates (up to 500 Hz), the data from each sensor were buffered and time-aligned to a common clock referenced to the EEG stream every second. Further engineering details regarding data synchronization and system I/O optimization are described in Section S3, and the operator console interface is illustrated in Figure S3.

3.3. Design of MS-Inducing Scenarios

For both CMS and VIMS, the MS symptoms can intensify with longer stimulus exposure times [112–117]. Although MS can be induced with just 10–15 min of stimulus exposure [113,116,118], it can also be reduced by adaptation to prolonged repeated stimuli [115,119]. Accordingly, the scenarios were designed with a 20 min stimulus exposure time to induce MS while providing nonrepetitive stimuli.

CMS can be induced by rotation, acceleration/deceleration, and vertical movements [120–124]. According to the sensory conflict theory, the vestibular system acts as a biological accelerometer that detects angular acceleration (semicircular canals) and linear acceleration (otoliths). Therefore, velocity changes amplify sensory conflict and induce MS. CMS can also be caused by vehicle motion that produces optic flow expansion or contraction in the field of view of a passenger [125,126]. Additionally, although not a physical motion, MS in virtual driving environments has been reported to be maximized at a perceived speed of 10 m/s [127]. Considering these CMS-inducing factors, the basic driving speed was set to 40 km/h (rounded up from 36 km/h ≈10 m/s for convenience), and CMS-inducing driving scenarios were designed using the implemented simulator (Table 3).

The designed scenarios consisted of four steps (excluding the start and end steps) to prevent participants from adapting to repetitive stimuli. The intensities of the stimuli were varied within each step to further prevent this adaptation. Figure 6 shows the actual motion of the simulator under the designed driving scenarios.

https://doi.org/10.3390/s26051675

Table 3. Design of driving scenarios for inducing CMS.

Step	Scenario	Duration	Description
Start	straight acceleration	2 min	0 → 40 km/h
1	straight constant velocity	4 min	40 km/h
2	straight acceleration or deceleration	4 min	(1) [2 times] 40 → −40 km/h (30 s), −40 → 40 km/h (30 s)
			(2) 40 → 20 km/h (10 s)
			(3) [2 times] 20 → 60 km/h (20 s), 60 → 20 km/h (20 s)
			(4) 20 → 60 km/h (20 s)
			(5) 60 → 40 km/h (10 s)
3	constant velocity curve (velocity = 40 km/h)	4 min	(1) [2 times] min radius = 135 m (≈total 90°), left curve (30 s) → right curve (30 s)
			(2) [6 times] min radius = 60 m (≈total 67.5°), left curve (10 s) → right curve (10 s)
4	constant velocity slope	4 min	(1) [2 times] slope = 9% (≈max pitch 5.1°), up slope (30 s) → down slope (30 s)
			(2) [6 times] slope = 9% (≈max pitch 5.1°), up slope (10 s) → down slope (10 s)
End	straight deceleration	2 min	40 → 0 km/h

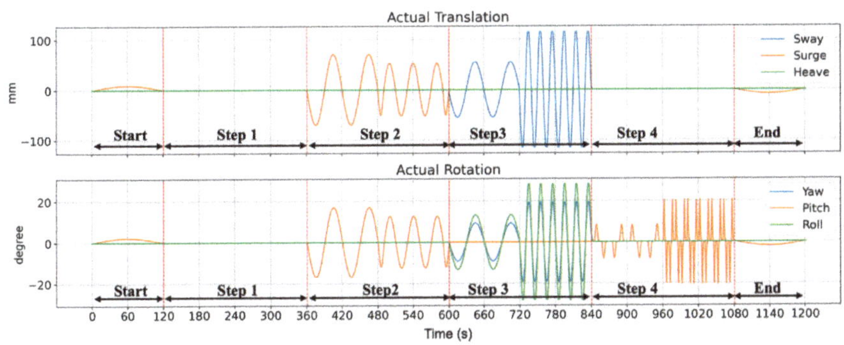

Figure 6. Actual motion of the autonomous driving simulator under the designed CMS-inducing scenarios. The vertical red dashed lines denote the boundaries between different scenario steps.

Modern VIMS can be categorized into two types: MS induced by content on traditional flat displays (e.g., tablets and laptops) and MS induced by content on HMDs. For these categories, an Apple iPad Pro (12.9-in 6th generation) and a Microsoft HoloLens 2 were used in this study. The reason for adopting a see-through HMD rather than a VR HMD was to reflect the trend of next-generation displays evolving toward see-through HMDs and to address the significant research gap in AR HMD studies using human signals. Furthermore, the HoloLens 2 provides native head and eye sensor data, allowing it to be used with the implemented acquisition system without additional hardware changes. Hereafter, the VIMS conditions corresponding to flat-panel displays and see-through HMDs are referred to as F-VIMS and S-VIMS, respectively.

The causes of VIMS have primarily been studied from the perspective of content, with reports indicating that it occurs when the content has a high optical flow (OF) or induces fast vection [127–130]. Additionally, although immersion increases when fidelity, the degree

 https://doi.org/10.3390/s26051675

to which reality is well reproduced, is higher, it can also induce VIMS more strongly than lower fidelity [128,131,132]. Accordingly, to provide high OF and fast vection, films from the live-action category that employed subjective camera technique, which is characterized by high fidelity owing to the use of real-world content, were selected in this study. Since this camera technique is predominantly used in the found footage genre, the initial pool of candidate films was selected from the Internet Movie Database (IMDb) based on the criteria of being a movie in this genre, having more than 100,000 votes, and being released after 2000.

From the initial pool of candidates, films that did not use the subjective camera technique were excluded, and films rated as severe for sex and nudity according to IMDb's parental guidance were also excluded to adhere to research ethics. To ensure that the experimental results were not dependent on a specific video, two films with high OF were selected from the filtered candidates. For this, the DVD version of each film was converted to grayscale, and the mean magnitude of the OF vectors over 20 min was calculated using the OpenCV Farnebäck method (parameters: pyramid scale = 0.5, number of levels = 3, averaging window size = 15 px, iterations per level = 3, neighborhood = 5 px, Gaussian smoothing $\sigma = 1.2$, no additional flags). The results showed that End of Watch (2012, OF = 2.924) and Cloverfield (2008, OF = 2.597) had high mean OF values and were selected as VIMS-inducing videos. Although Chronicle (2012, OF = 2.759) had a higher overall OF than Cloverfield, it was not used because its mean OF in the 10–20 min interval was lower ($OF^{cloverfield}_{10-20\ min} = 2.590$ and $OF^{chronicle}_{10-20\ min} = 1.884$). The first 20 min of the color versions of these two films were used as VIMS-inducing stimuli in the experiment. Copyright compliance for nonprofit experimental use was confirmed through consultation with the Korea Copyright Commission and legal counsel.

In the Co-MS scenarios, the CMS-inducing driving motions and VIMS-inducing videos were presented simultaneously. By superimposing the visual stimuli of the VIMS scenarios onto the physical stimuli defined in the CMS scenarios, the participants experienced sensory conflict from both sources. This condition was intended to replicate the combination of vehicle motion and exposure to visual content that can occur in actual driving environments.

4. Experimental Methods

All research procedures described in this section were conducted in accordance with the guidelines of the Declaration of Helsinki and approved by the Institutional Review Board of KOREATECH in advance (approval on 23 April 2024).

4.1. Participants

A total of 100 participants (70 males and 30 females) were recruited through the intranet of the Korea University of Technology and Education, and a small amount of compensation was provided to encourage active engagement. The eligibility of participants was restricted to non-computer science students or first-year computer science students, who were expected to have limited experience with IT equipment, such as simulators and HMDs. The experiment was conducted from 16 December 2024, to 23 January 2025, with a maximum of four participants per day.

To minimize the effects of sleep inertia and sleep pressure on the analysis, the wake-up time of each participant was confirmed. Eight participants were excluded because they had been awake for less than 1 h or more than 11 h. Furthermore, two participants who had consumed alcohol on the day before the experiment were also excluded. Consequently, a total of 90 participants (64 males and 26 females; mean age = 22.73, SD = 1.67) were included in the final experiment. The participant cohort consisted of 82 undergraduate

https://doi.org/10.3390/s26051675

and 8 graduate students, all with normal or corrected-to-normal vision. Their baseline susceptibility to motion sickness averaged 2.600 (SD = 0.909) on a four-point scale.

4.2. Experimental Procedure

The experiment was designed to measure the CMS using the implemented simulator, VIMS using displays, and Co-MS, where both stimuli were applied simultaneously. The experimental procedure was designed as shown in Figure 7 and was conducted by dividing the participants into two groups according to VIMS type (F-VIMS and S-VIMS). Each participant took part in the experiment only once (no crossover participation).

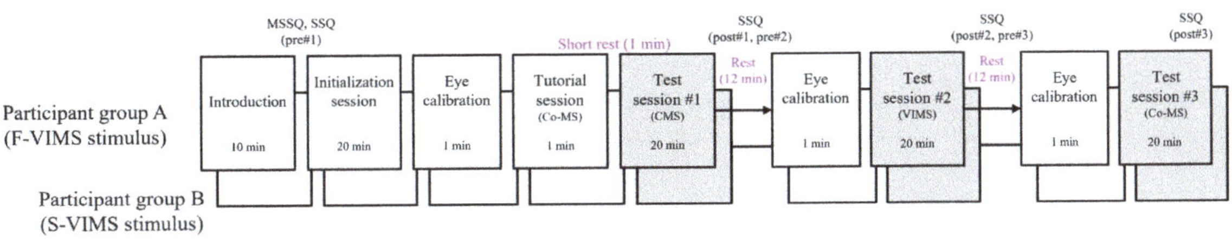

Figure 7. Schematic showing the common experimental procedure for the F-VIMS and S-VIMS groups (approximately 2 h). The two groups follow the identical procedure, differing only in the type of VIMS stimulus. The # symbol denotes the specific ID for each test session (detailed in Section 4.2.2).

The participants were informed about the experiment, the human signals that would be collected, and the subjective questionnaires for MSL measurements before the experiment. We requested that the participants fill out the subjective questionnaires objectively, mentioning the importance of MSL data in the experiment. Additionally, the characteristics of the found footage genre were briefly introduced to prevent interruptions due to aversion while watching VIMS-inducing content.

In the initialization session, the participants were instructed to take a seat in the simulator and adjust their seats for a comfortable posture. All the sensors of the acquisition system were then attached, and a conductive gel was applied to the EEG sensor electrodes. After confirming that all the sensors were operating correctly through the operator console, the tutorial and test sessions began. Figure 8a shows the implemented simulator and acquisition system, and Figure 8b shows the interface for inputting the FMS for subjective MSL measurement.

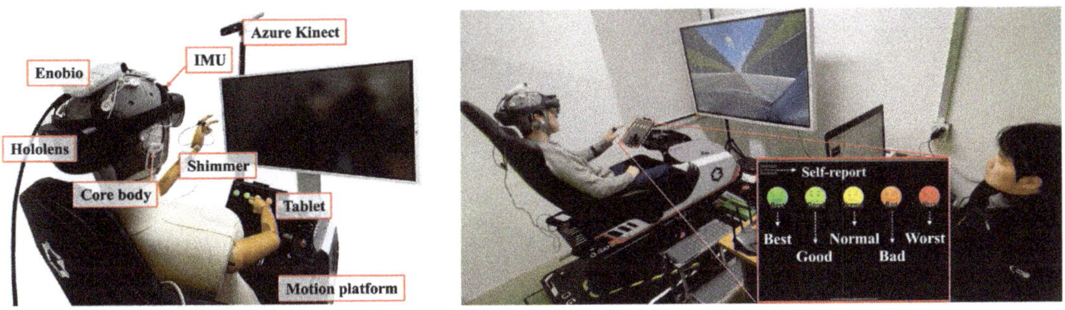

(a) (b)

Figure 8. Experimental environment and MSL measurement interface: (**a**) Autonomous driving simulator and multimodal human-signal acquisition system; (**b**) FMS-based measurement interface for MSL input.

https://doi.org/10.3390/s26051675

4.2.1. Subjective Questionnaires

Subjective MSL was measured using the SSQ and FMS. Additionally, MS susceptibility was assessed using the MSSQ, and content immersion was assessed using the Film IEQ. The SSQ is an MSL assessment tool with 16 items on a four-point scale (0–3), and its results are calculated as subscale scores for three symptom categories: nausea (SSQ_N), oculomotor (SSQ_O), and disorientation (SSQ_D). Weighting factors (SSQ_N: $\times 9.54$, SSQ_O: $\times 7.58$, SSQ_D: $\times 13.92$) are applied to each subscale score, and the total score (SSQ_T) is calculated as 3.74 times the sum of the unweighted raw subscale scores. The SSQ was administered before eye calibration (pre) and after the test session (post), and the ΔSSQ (post-SSQ score − pre-SSQ score) was used for analysis.

The FMS is an MSL assessment tool that tracks the temporal changes by assessing the current MSL state at regular intervals on a single continuous scale (0–20). In this experiment, the FMS was collected during the test session and modified to use a one-to-five-point scale, as performed in the studies by Wen et al. [133] and Won et al. [119]. The FMS was measured by displaying an evaluation layout on the F-VIMS-inducing device (tablet) at 30-s intervals, as shown in Figure 8b. A notification sound was also played to prevent the S-VIMS participants from failing to recognize it. Furthermore, even if 30 s had not elapsed, the participants were allowed to re-enter their FMS immediately via the self-report button if they experienced a change in their MSL.

The MSSQ estimates the MS susceptibility by assessing the frequency of MS experienced on various transportation and amusement rides during childhood and adulthood. The MSSQ was used to analyze the effect of MS susceptibility on MSL and was administered during the pre-briefing stage.

The Film IEQ is a modified version of the IEQ [134] that assesses the depth of immersion in video games. This modified questionnaire is a tool for evaluating the degree of immersion in a video-watching context, with 24 items rated on a seven-point scale (1–7). The evaluation results are calculated as subscale scores for the four areas through a simple summation of the survey items: captivation (IEQ_{cap}), real-world dissociation (IEQ_{dis}), comprehension (IEQ_{com}), and transportation (IEQ_{tra}). In this experiment, the scale was adjusted to a five-point scale to reduce the burden on the participants, and the questionnaire was administered after the test session in which participants viewed the VIMS-inducing videos.

4.2.2. Tutorial and Test Session

The tutorial session was designed to allow participants to preview the stimuli and the FMS measurement method. Participants experienced Co-MS stimuli for 1 min and practiced two types of FMS measurement methods based on 30-s intervals and the self-report button. The CMS stimulus consisted of straight-line driving with 30 s of gradual acceleration followed by 30 s of gradual deceleration (maximum velocity = 40 km/h), and the open-source sample video Big Buck Bunny from Blender was used as the VIMS stimulus. Immediately after the tutorial session, the participants were given a 1 min rest while keeping the sensors on.

The core of the experiment consisted of three test sessions (#1–3). Before each session, the participants wore HoloLens 2 for eye tracking across all conditions (including the F-VIMS condition) and performed a 1 min eye calibration (for Test Session #1, which was performed before the tutorial session). The sessions were conducted in a fixed order, each lasting 20 min: a single CMS stimulus (#1), a single VIMS stimulus (#2), and a Co-MS stimulus (#3).

However, counterbalancing was applied to minimize the adaptation effects of the physical and visual stimuli from #1 and #2, respectively, when conducting #3. A predefined counterbalancing matrix was used across two VIMS types, four driving scenario versions, and

https://doi.org/10.3390/s26051675

two movie orders, yielding 16 combinations (2 × 4 × 2). To maintain balance, participants were sequentially allocated to the next combination in this matrix according to the enrollment order.

The order of the four driving scenario steps (defined in Table 3) was cyclically shifted by one step to create four versions, which were then used to provide different driving scenarios for #1 and #3. For example, participants were assigned to one of four versions based on the starting step of the driving scenario; the first version was driven in the order "Start–1–2–3–4–End" for #1 and "Start–2–3–4–1–End" for #3 (the start and end steps were fixed). In addition, the videos presented in #2 and #3 were counterbalanced by crossing the type and order of the films to control the adaptation effects. For instance, participants were assigned to one of two viewing orders; the first order presented the videos in the sequence "Cloverfield" (#2)– "End of Watch" (#3), while the second order presented them as "End of Watch" (#2)–"Cloverfield" (#3).

After each test session, a 12-min rest period was given to the participants to recover from the MS symptoms [88]. During the rest period, considering the properties of the conductive gel, all sensors except the EEG were temporarily detached and reattached 1 min before the next session. The participants were informed that they could request to stop the experiment at any time if their MS symptoms became severe. After all the experiments were completed, cleaning supplies were provided, and the location of the shower facilities was indicated.

4.3. Preprocessing

Although each test session for collecting sensor data lasted 20 min (1200 s), some data were occasionally saved as fragments due to communication and I/O issues during the experiment. For example, if an issue occurred at t = 500 s, the recording was stopped, and the test session was restarted at t = 490 s to resume data collection immediately. These fragmented data were rearranged chronologically and merged, with the overlapping sections trimmed by at least 5 s on each side to reconstruct a continuous 1200 s timeline (e.g., (0–500 s → 0–495 s) + (490–1200 s → 495–1200 s) = 1200 s timeline). This reconstruction process was applied identically, even when three or more data fragments were present. The reconstructed timeline contained unintended noise and outliers at the beginning and end. Therefore, 30 s were trimmed from the start and end of each session, and the remaining 1140 s segment was used as the dataset.

The acquisition system used in the experiment received data in parallel from the sensors with different sampling rates in chunks at approximately 15–20 Hz. All the received data were extracted after every second, synchronized to the EEG (500 Hz) reference time, and saved. During this process, limitations in network and thread scheduling or OS time resolution can cause the actual number of samples in a 1-s interval to deviate from the expected value by approximately ±1 chunk. Consequently, the sensor data in the dataset may contain slight time warping effects.

To correct for this, an expected sample grid was constructed for each sensor based on its official sampling rate multiplied by 1140 s (e.g., for Shimmer3 GSR+, 128 Hz × 1140 s = 145,920 samples). The recorded samples were then mapped to the nearest index on the expected grid. If two or more samples were mapped to the same expected index, the closest sample was assigned, and any empty indices were filled using piecewise cubic Hermite interpolation. Finally, the dataset was corrected by aligning all sensor time series to a common time axis. Although there could be a loss of approximately one chunk (25 samples) of the EEG stream during the storage process of the data manager, this correction procedure reduced the potential for analytical distortion.

https://doi.org/10.3390/s26051675

The PPG, EDA, EEG, and SKT data in the dataset are raw physiological signals that are significantly affected by noise and artifacts. Therefore, they were preprocessed following the procedure summarized in Figure 9. The notch and band-pass filters used for preprocessing were applied using the MNE package 1.9.0 in Python 3.12, and any parameters not explicitly specified were set to the default values of the package.

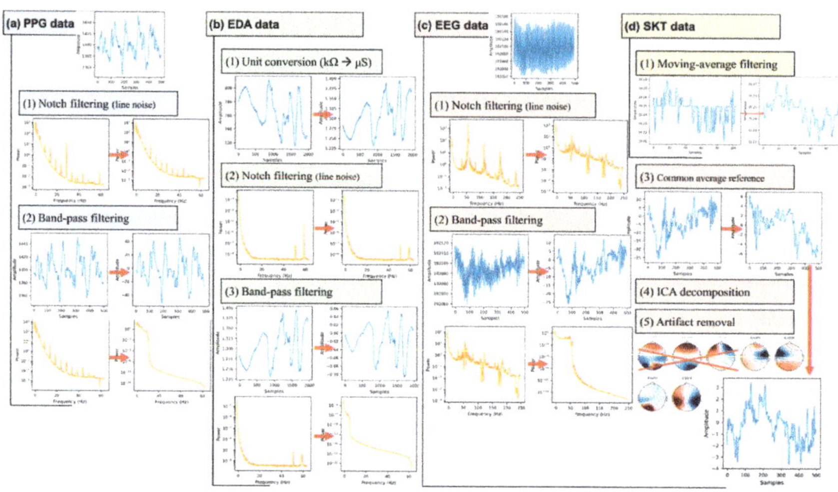

Figure 9. Physiological signal preprocessing pipeline: (**a**) PPG, (**b**) EDA, (**c**) EEG, and (**d**) SKT. The red arrows indicate the sequence of the data processing flow, and the different background colors distinguish the individual sensor domains. The blue and orange line plots represent the signals in the time and frequency domains, respectively.

In the raw PPG data, interference from the Korean power line frequency (60 Hz) and a distinct spectral peak at 30 Hz were observed (Figure 9a, step 1). These two frequency components were suppressed in the frequency domain using a notch filter based on the MNE spectrum-fitting (spectrum_fit) function (all subsequent notch filters were applied using the same method). In addition, the out-of-band noise was suppressed using a Blackman-window-based band-pass filter (all subsequent band-pass filters used the same window), referencing the PPG band used in the previous studies. In the HS-Set, only three studies explicitly reported the PPG band: Dennison et al. [60] used 0.1–10 Hz, Martin et al. [43] used 0.66–3.33 Hz, and Sameri et al. [39] used 0.5–8 Hz. Although these bands partially overlapped, none of the studies used identical bands, and the number of studies was small. Separately, Lapitan et al. [135] reported that a band of 0.1–10 Hz minimizes the pulse waveform distortion in PPG signals, which is identical to the band used by Dennison et al. [60]. Based on this result and the observation of periodic spectral peaks above 10 Hz, as shown in step 2 of Figure 9a, the band-pass range for the PPG preprocessing pipeline was set to 0.1–10 Hz.

While EDA sensors typically measure skin electrical conductance (G) in μS, the Shimmer3 GSR+ used in the acquisition system records the electrical resistance (R) of the skin in kΩ. As the two physical quantities are related by $G = 1/R$, each recorded resistance sample, $R_{k\Omega}$, was converted to the conductance, $G_{\mu S}$ (Figure 9b, step 1). When the EDA data were examined in the frequency domain, unlike the PPG data, no spectral peak was observed at 30 Hz. Therefore, only the 60 Hz component was suppressed using a notch filter (Figure 9b, step 2). In the HS-Set, only the study by Sameri et al. [95] explicitly reported an EDA band of 0.05–4 Hz, and two other studies reported only cutoff frequencies (Lee

 https://doi.org/10.3390/s26051675

et al. [38]: upper cutoff 50 Hz and Rahimzadeh et al. [33]: upper cutoff 1 Hz). Therefore, the evidence for setting a band was limited. Separately, Privratsky et al. [136] fixed the upper cutoff frequency of a band-pass filter at 5 Hz and compared various lower cutoff frequency candidates (0.01–0.10 Hz), reporting that 0.02 and 0.03 Hz were effective. Considering that Sameri et al. [39] used an upper cutoff of 4 Hz and Privratsky et al. [136] used 5 Hz, an upper cutoff frequency of 4–5 Hz can be considered a relatively common choice. The band-pass range in the EDA preprocessing pipeline was set to 0.03–5 Hz, which is one of the configurations reported as effective by Privratsky et al. [136] (Figure 9b, step 3). The choice of the lower cutoff frequency was based on the value (0.03 Hz) closer to the 0.05 Hz used by Sameri et al. [39], as opposed to 0.02 Hz.

In the raw EEG data, distinct spectral peaks were observed at the Korean power line frequency of 60 Hz and its harmonics (120, 180, and 240 Hz) (Figure 9c, step 1). These line noise peaks appeared broad owing to the interpolation of lost chunks during the dataset correction process. Therefore, each component was suppressed by setting the notch filter width to 10 Hz (±5 Hz) at 60, 120, 180, and 240 Hz. Most EEG studies in the HS-Set explicitly specified the bands they used, and the reported cutoff frequencies are listed in Table S8.

In the EEG preprocessing pipeline, the lower cutoff frequency for the band-pass filter was set to 1 Hz, which is the most frequently reported value. The upper cutoff frequency was set to 50 Hz. Although 40 and 50 Hz were the most frequently reported values, the former was selected to retain a broader band of information. Step 2 in Figure 9c shows the result of applying the band-pass filter. In addition, a common average reference was applied to suppress the noise common across the entire scalp (Figure 9c, step 3). Subsequently, an independent component analysis (ICA) based on the MNE extended-infomax algorithm was performed (random state = 97), and the MNE-ICALabel model was used to classify and remove non-brain components (e.g., electrooculogram) as artifacts (Figure 9c, steps 4–5).

The raw SKT data had a very low sampling rate of 1 Hz (Nyquist frequency of 0.5 Hz), indicating that the line noise and high-frequency components were not present as resolvable components in the digital domain. Accordingly, no notch or band-pass filters were applied, and only a three-point moving-average filter was used to suppress sensor-level noise. Step 1 in Figure 9d shows the SKT data obtained after applying the moving-average filter.

4.4. Feature Extraction

As the FMS score was collected at least every 30 s, handcrafted features were calculated using a 30-s window. For a session with an effective length of 1140 s, 38 aligned 30-s windows were generated. The FMS score for each window was mapped as the time-weighted mean of the FMS scores reported in the interval.

Most of the features from the HS-Set were adopted where possible. However, features that were impossible to calculate (e.g., the dataset did not include the pupil diameter), difficult to reproduce (e.g., three intrinsic mode functions of EDA), or had undisclosed calculation methods (e.g., the excitement feature from EEG) were excluded from the analysis. Furthermore, in the cases where features were classified under the same category in the HS-Set but had differing definitions or extraction procedures across studies or were overly granular, they were consolidated into representative features to reduce redundancy and ensure interpretative consistency. For example, the HR variability (HRV) feature group in PPG (SDSD, SDNN, RMSSD, pNN20, pNN50, etc.) was condensed to SDNN and RMSSD, the frequency-domain bands in EEG (simple frequency, standard frequency, and expanded frequency) were condensed to simple frequency and standard frequency, and the cardiac vagal tone (CVT) in PPG was replaced with the cardiac vagal index (CVI).

https://doi.org/10.3390/s26051675

Each feature was classified as either a signal-wise feature (SwF), if it could be calculated as a time-series signal, or a window-derived feature (WdF), if it was summarized as a single scalar value within a given interval. For instance, the head movement features are SwFs, whereas the power ratio features of EEG are WdFs. Time- and frequency-domain templates, which are commonly applied to most sensors in the HS-Set, were used for SwFs. The mean, variance, kurtosis, skewness, sample entropy (embedding dimension $m = 2$, tolerance $r = 0.2 \times$ standard deviation (SD)) [137], and peak-to-peak amplitude were calculated in the time domain. In the frequency domain, the power spectral density (PSD) was obtained using Welch's method. The DC component (0 Hz) was excluded, and any negative PSD values resulting from numerical noise were clipped to zero. Subsequently, the band power and Shannon entropy of the normalized PSD (hereafter referred to as PSD entropy) were calculated. However, if a window contained discontinuous segments owing to preprocessing, features were calculated for each segment and then integrated by averaging. The features extracted from each sensor are listed in Table 4. As an exception, body data were excluded from feature extraction because the tracking performance of the body joints of the Azure Kinect was unstable in the dynamic environment.

Table 4. Extracted features by sensor modality: signal-wise (SwF) and window-derived (WdF).

Sensor Modality	SwF [1]	WdF [1]
PPG	whole, LF, and HF band signals	HR, IBI, LF/HF, HRV, SDNN, RMSSD, SD1, SD2, SD1/SD2, CSI, CVI, CSI/CVI, BR
EDA	whole, SCL, SCR, and Posada-Quintero LF [2] band signals	–
EEG	whole and standard frequency band signals ($\delta, \theta, \alpha, \beta, \gamma$) by brain region (global, frontal, central, parietal, occipital)	β/α, $(\alpha+\theta)/\beta$, θ/α, θ/β, $(\alpha+\theta)/(\alpha+\beta)$, $F\theta/P\alpha$, FBN metrics (coherence, PLV [3]; 8 ch $\times$ 8 ch adjacency = 28 edges) computed from whole and standard frequency band signals
SKT	whole band signal	–
Eye [4]	eye closure signals (L + R, L $\times$ R), convergence distance, gaze direction (yaw, pitch), gaze velocity (°/s)	blink rate, vergence loss ratio, saccade ratio, fixation ratio, VOR, path length, heatmap entropy (visual entropy)
Head (HoloLens 2)	positions (sway, heave, surge), rotations (pitch, yaw, roll), direction (yaw, pitch)	–
Head (IMU)	accelerations (sway, heave, roll), angular velocities (pitch, yaw, roll)	–

[1] Most abbreviations used in this table are defined in Section S1 (Supplementary Materials). Abbreviations not defined in Section S1 are expanded at their first appearance in this table. [2] Posada-Quintero et al. [138]. [3] Phase locking value. [4] L = left, R = right.

In the HS-Set, the only study that explicitly specified the low frequency (LF) and high frequency (HF) component bands for PPG was by Martin et al. [43], which used 0.04–0.15 Hz for the LF component and 0.15–0.4 Hz for the HF component. These are identical to the recommended standard bands for HRV analysis [139]. Accordingly, the HF component band for SwF extraction was set to the standard 0.15–0.4 Hz, because components below 0.1 Hz in the PPG data were suppressed during preprocessing, and the LF component band was set to 0.1–0.15 Hz. The time-domain SwFs for the LF and HF band signals were extracted by decomposition using an MNE band-pass filter, whereas the frequency-domain SwFs were extracted by decomposing the bands from the Welch PSD of the full-band signal

https://doi.org/10.3390/s26051675

(all subsequent band signal decompositions followed the same process). To stabilize the PSD entropy calculation for narrow bands, such as LF, the PSD was calculated with a fast Fourier transform (FFT) length of 8192 and a segment length of 20 s (n_per_segment = 2560) was used to account for the discontinuous data segments.

The WdFs for the PPG, such as HR, were calculated using the HeartPy package 1.2.7 in Python. Before the WdF calculation, a 0.66–3.33 Hz band-pass filter was applied to further suppress noise in the full-band signal. This range was set considering a typical heart rate of 40–200 bpm ($\approx$ 0.66–3.33 Hz), a band also used by Martin et al. [43]. However, the package occasionally failed to compute SD1 and SD2. The missing SD1 values were calculated using the approximation formula $SD1 \approx RMSSD/\sqrt{2}$ [140], and SD2 was calculated using the formula $SD2^2 \approx 2 \times SDNN^2 - 1/2 \times SDSD^2$ [141]. The CSI and CVI were calculated using the approximation formulae given by Toichi et al. [142]: CSI = L/T and CVI = log10(L $\times$ T), where T = 4 $\times$ SD1 and L = 4 $\times$ SD2.

While several studies involving the HS-Set used skin conductance level (SCL) and skin conductance response (SCR) components, none explicitly specified their frequency bands. Separately, Ishchenko and Shev'ev [143] and Greco et al. [144] reported that the band below 0.05 Hz corresponds to the SCL component and the 0.05–2 Hz band corresponds to the SCR component. Based on this, the SCL component band for extracting SwFs was set to 0–0.05 Hz, and the SCR component band was set to 0.05–2 Hz. In addition, an LF component with a 0.045–0.25 Hz band from Posada-Quintero et al. [138] was reported to be influenced by the sympathetic nervous system [144], and SwFs were also extracted from this component. Because the EDA data were collected using the same sensing equipment as the PPG data (Shimmer3 GSR+), they shared the same sampling rate (128 Hz) and the common issue of a narrow band for the SCL component, similar to the LF component of the PPG data. Accordingly, the frequency-domain SwF extraction process for the EDA data was identical to that for the PPG data.

Although the EEG studies in the HS-Set generally used similar bands to decompose the standard components, slight differences were observed, as shown in Table S9. The standard component bands for SwF extraction were decomposed using the most frequently used bands from the HS-Set studies. The upper cutoff frequency for the γ band was set to 50 Hz, chosen over the other frequently used value of 45 Hz, to encompass a broader range of information.

The time- and frequency-domain SwFs were integrated by averaging them for the global head (total channels), frontal lobe (F3, F4), central region (C3, C4), parietal lobe (P3, P4), and occipital lobe (O1, O2) (5 areas $\times$ 6 bands (whole + 5 standard bands) = 30 signals). Sample entropy, a time-frequency domain SwF, was calculated after downsampling to 125 Hz for computational convenience. The frequency-domain SwFs were extracted by decomposing the bands from the Welch-PSD of the full-band signal (1–50 Hz). As the EEG data were sampled at 500 Hz, the PSD was computed with an FFT length of 2048 (segment length = FFT length), which resulted in a frequency resolution of approximately 0.24 Hz/bin. Because the standard EEG bands are relatively wide (several Hz), this resolution can sufficiently distinguish the boundaries. Among the WdFs, the power ratio features were based on the seven features used by Sameri et al. [39]: β/α, $(\alpha+\theta)/\beta$, θ/α, θ/β, $(\alpha+\theta)/(\alpha+\beta)$, $\beta/(\alpha+\theta)$, and $F\theta/P\alpha$. For the functional brain network (FBN), coherence and phase locking value (PLV) were used for each channel pair and were calculated for each of the six band signals (whole, 5 standard bands) (6 bands $\times$ ${}_8C_2$ $\times$ 2 metrics = 336 WdFs).

For the SKT data, only the SwFs from the full-band signal were used. As no separate band-pass filter was applied during preprocessing, the Welch-PSD was calculated with an FFT length of 20, corresponding to a sampling rate of 1 Hz $\times$ a 20-s segment length to account for discontinuous data. For the remaining sensor modalities, excluding PPG, EDA,

https://doi.org/10.3390/s26051675

EEG, and SKT, features were extracted only from the full-band signal because they had no sub-band components. The Welch-PSD for these signals was calculated using an FFT length corresponding to their sampling rate × 20 s, similar to the SKT data.

The data collected from the eye-tracking sensor of HoloLens 2 are limited to the eye position and gaze direction. However, the left/right eye data have null values or the same values as the central gaze (position and direction) when the eye is closed, and it is possible to perform a binary classification of the blink state. The closure signal, which classifies a closed state as 1 and an open state as 0, was calculated as

$$CS_e = \left\{1\left(\text{isnull}(P_e[k]) \vee \|\boldsymbol{P}_e[k] - \boldsymbol{P}_c[k]\|_\infty < 10^{-3}\right)\right\}_{k=1}^{T}, \quad e \in \{L,\ R\} \tag{1}$$

where $\boldsymbol{P}$ is the eye position vector (x, y, z) in meters, L is the left eye, R is the right eye, T is the number of frames in the analysis interval, and $\boldsymbol{P}_c$ is the origin of the HMD gaze. The threshold of 10^{-3} is the tolerance based on the L∞-norm, which is set conservatively to absorb slight rounding errors from floating-point operations and the potential for momentary omissions of actual eye closure periods due to the 30 Hz sampling.

Intermittent single-frame inversions were observed in each closure signal owing to sensor noise and sampling characteristics. To address this, isolated ones (0–1–0) were suppressed to 0, and isolated zeros (1–0–1) were suppressed to 1, and this process was repeated until convergence.

Spontaneous and reflex blinks in humans are fundamentally binocular, meaning that the left and right eyes are usually closed concurrently. Therefore, SwFs were extracted using a frame-by-frame logical AND signal ($CS_L \wedge CS_R$), reflecting the state in which both eyes were closed. Further, to account for minor asynchronous events, a frame-by-frame logical OR signal ($CS_L \vee CS_R$), which allows only one eye to be closed, was also used for SwF extraction. The blink rate WdF was calculated by averaging the values computed for the left and right eyes.

In the HS-Set, studies that used the convergence distance mostly referred to the metric provided by the VIVE Pro Eye SDK. We defined the convergence distance as the minimum Euclidean distance between the closest points on the binocular gaze lines, which was calculated for frame k as follows:

$$d_{conv}[k] = \begin{cases} \frac{|(\boldsymbol{P}_L - \boldsymbol{P}_R) \cdot (\boldsymbol{D}_L[k] \times \boldsymbol{D}_R[k])|}{\|\boldsymbol{D}_L[k] \times \boldsymbol{D}_R[k]\|_2} & if\ \|\boldsymbol{D}_L[k] \times \boldsymbol{D}_R[k]\|_2 \geq 10^{-3} \\ \text{NaN} & \text{otherwise} \end{cases} \tag{2}$$

where $\boldsymbol{P}$ is the eye position vector (x, y, z) in meters, $\boldsymbol{D}$ is the gaze direction vector (x, y, z), L is the left eye, and R is the right eye. The threshold of 10^{-3} is the tolerance for $\|\boldsymbol{D}_L \times \boldsymbol{D}_R\|_2$, which is used to prevent the denominator from approaching zero when the left and right eye gazes are virtually parallel. The d_{conv} data across all frames formed a time-series signal that was used for SwF extraction. Furthermore, as the likelihood of focus convergence failure increased with drowsiness or lack of concentration, the ratio of NaN values to the total number of frames was used as the WdF.

The gaze direction was collected as a unit vector in the head coordinate system (e.g., forward = (0, 0, 1)). Because the L2-norm of this vector is always 1, it has only two DOF, and its Cartesian components (x, y, z) are not independent, and thus the component-wise statistics are difficult to interpret. Therefore, the center gaze (g_x, g_y, g_z) was converted to the spherical coordinates as gaze angles: $\theta_{yaw} = 180/\pi \times \arctan2(g_x, g_z)$ and $\theta_{pitch} = 180/\pi \times \arcsin(\min\{1, \max\{-1, g_y\}\})$. The SwFs for the gaze direction signal were extracted from the θ_{yaw} and θ_{pitch} signals. Berton et al. [80] mentioned the concept of eye-head coordination, the degree of alignment between the gaze and head direction. Since

https://doi.org/10.3390/s26051675

the gaze direction used in this feature extraction is relative to the head coordinate system, it can indirectly represent this concept (as the degree of alignment is expressed by the amount of variance). The gaze distance-to-center feature, which is a measure of the distance from a specific reference, can also be indirectly represented (as the distance from the center is expressed by the amount of variance).

The gaze velocity was calculated by extending the dot product angle between consecutive direction vectors from degrees per frame to degrees per second, and SwFs were extracted from the velocity signal. Referring to Salvucci and Goldberg [145] and Tobii's identification by velocity threshold classification guidelines [146], samples in the gaze velocity signal exceeding 100°/s were classified as saccades and the rest as fixations. The ratios of saccade and fixation frames to the total number of frames were used as the WdFs. The path length WdF was defined as the cumulative sum of the degrees per frame and calculated using the process described above.

In the HS-Set, Lee et al. [76] converted the gaze distribution of 360° videos into a cubemap and corrected for distortion in the spherical projection by defining the visual entropy as the entropy of the normalized heatmap distribution on each face. In contrast, this study involved F-VIMS/S-VIMS environments with the collected gaze directions as unit vectors, making the direct application of the prior method inappropriate. Therefore, a 2D heatmap (64 × 64 bins, range = $[-90°, 90°] \times [-90°, 90°]$) was constructed from the converted θ_{yaw} and θ_{pitch} signals, and its Shannon entropy was calculated and used as the heatmap entropy.

In the Head data, SwFs were extracted for each of the three components of position/acceleration (2 × (sway, heave, surge)) and rotation/angular velocity (2 × (pitch, yaw, roll)). Additionally, considering the use of geographic coordinates in the HS-Set, the rotation vectors were converted to direction vectors, and then spherical coordinate signals (ϕ_{yaw}, ϕ_{pitch} signals) were calculated using the same procedure as the gaze direction. However, it is difficult to directly convert the head rotation into a direction vector because there is no central reference. Thus, the relative rotational difference between the frames was calculated based on the average rotation of the window, which was then applied to the forward direction vector (0, 0, 1), and the result was defined as the direction vector. SwFs were also extracted from the ϕ_{yaw} and ϕ_{pitch} signals and calculated using this process. The VOR represents the degree to which the gaze remains fixed in space as the head moves and is generally calculated as the ratio of the gaze velocity to head velocity. Therefore, the head velocity was calculated from the direction vector defined through the same procedure as the gaze velocity, and the mean of (gaze velocity/head velocity) over the window was used as the VOR WdF.

5. Results

To evaluate the factors contributing to MS across different stimulus conditions, the statistical analysis followed a structured roadmap. First, Section 5.1 profiles the participants' demographic characteristics and baseline susceptibility to MS. Section 5.2 examines the main and interaction effects of MS stimulus type (within-subjects factor: CMS, VIMS, and Co-MS) and user characteristics (between-subjects factors: gender and susceptibility) on ΔSSQ using three-way mixed repeated measures analysis of variances (RM ANOVAs). In addition, because the display mode (F-VIMS vs. S-VIMS) is only applicable when visual stimuli are present, separate three-way ANOVAs were performed on the single-stimulus VIMS and Co-MS conditions to evaluate the main effects of the mode and its interactions with the user characteristics. Section 5.3 investigates the association between psychological immersion (Film IEQ) and MS symptoms, and Section 5.4 explores the correlations between multimodal human-signal features and subjective MSLs (ΔSSQ and FMS) at a nominal significance level. Finally, Section 5.5 evaluates the multivariable

 https://doi.org/10.3390/s26051675

contributions of sensor domains and features to the FMS using an EBM and identifies practical lightweight combinations for actual vehicular environments.

5.1. Participant Demographics

The MS susceptibility of the participants was calculated using the MSSQ-Short percentile formulae [147]:

$$\text{MSSQ}_{\text{percent}} = A\,x - B\,x^2 - C\,x^3 + D\,x^4 \quad (3)$$

$$\text{MSSQ}_{\text{score}} = \frac{9 \times \sum S_{\text{child}}}{9 - N_{\text{child}}^{\text{miss}}} + \frac{9 \times \sum S_{\text{adult}}}{9 - N_{\text{adult}}^{\text{miss}}} \quad (4)$$

where $A = 5.1160923$, $B = -0.055169904$, $C = 0.00067784495$, $D = 0.000010714752$, $x = \text{MSSQ}_{\text{score}}$, S is the score for each survey item, and N^{miss} is the number of types not experiencing MS. Although the actual susceptibility measurement used the MSSQ (0 = never, 1 = rarely, 2 = sometimes, 3 = frequently, 4 = always), the percentile conversion was performed using the MSSQ-Short scores from Golding [147] (0 = never, 1 = rarely, 2 = sometimes, 3 = frequently), and thus the scores were converted as follows: 4 points (always) → 3 points (frequently), with the remaining 0–3 points unchanged. Additionally, since the MSSQ-Short uses only sick/nausea scores, the vomiting scores were excluded from the calculation.

Participants were classified into tertile-based MS susceptibility groups based on their percentile scores: ≤33% = low susceptibility, >33 to ≤66% = moderate susceptibility, and >66% = high susceptibility. The distribution of the MSSQ scores of the participants is shown in Figure 10 and Table 5.

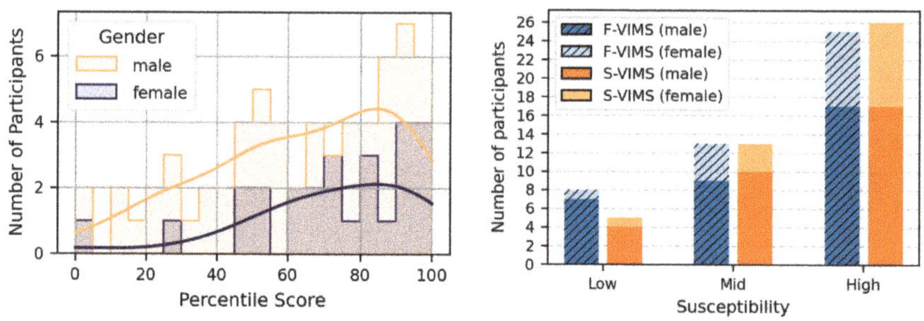

(**a**) (**b**)

Figure 10. Frequency distribution of participants: (**a**) MSSQ percentile scores for the two genders, the lines represent the kernel density curves; (**b**) Distribution according to the VIMS type (F-VIMS, S-VIMS) for different MS susceptibilities and genders.

Table 5. MSSQ-Short percentile scores by susceptibility tertile and gender.

Gender	Low (0–33%)	Moderate (33–66%)	High (66–100%)
Total	(N = 13) Mean = 18.41, SD = 11.58	(N = 26) Mean = 52.54, SD = 7.89	(N = 51) Mean = 85.18, SD = 10.24
Male	(N = 11) Mean = 19.11, SD = 10.73	(N = 19) Mean = 51.15, SD = 7.67	(N = 34) Mean = 84.81, SD = 10.09
Female	(N = 2) Mean = 14.52, SD = 20.54	(N = 7) Mean = 56.32, SD = 7.75	(N = 17) Mean = 85.91, SD = 10.82

N = number of participants.

 https://doi.org/10.3390/s26051675

Figure 10b shows the frequency distribution of the participants by VIMS type according to their gender and susceptibility. Overall, the patterns for both the VIMS conditions were similar, with the high-susceptibility group having the most participants and the low-susceptibility group having the fewest. In addition, most female participants were concentrated in the moderate-to-high susceptibility group. Considering the relatively small number of participants in the low-susceptibility group, the low- and moderate-susceptibility groups were combined into a normal-susceptibility group for subsequent analysis.

5.2. Effects of the MS Types on the SSQ Scores

Table 6 shows the ΔSSQ for each MS type (CMS, VIMS, and Co-MS). Under single-stimulus conditions, the mean values of all ΔSSQs (T, N, O, and D) were the highest for CMS, followed by S-VIMS and F-VIMS. Furthermore, the mean ΔSSQ for Co-MS was generally higher than that for each single-stimulus condition. Among the subscales, there was a tendency for ΔSSQ_O to be relatively high and for ΔSSQ_N to be low, and the SDs indicated large individual differences.

Table 6. Mean and standard deviation of ΔSSQ for the different MS types.

S-VIMS Group	Mean	SD	F-VIMS Group	Mean	SD
CMS	T: 16.065 N: 7.805 O: 17.916 D: 15.502	T: 16.853 N: 13.377 O: 16.025 D: 23.869	CMS	T: 19.107 N: 11.407 O: 20.763 D: 16.341	T: 20.968 N: 18.933 O: 19.004 D: 26.461
VIMS	T: 12.155 N: 6.288 O: 13.437 D: 11.389	T: 16.552 N: 9.855 O: 18.843 D: 21.496	VIMS	T: 8.130 N: 5.600 O: 7.086 D: 9.078	T: 15.893 N: 12.652 O: 16.942 D: 21.806
Co-MS	T: 22.185 N: 13.876 O: 23.429 D: 22.185	T: 27.464 N: 26.624 O: 25.157 D: 29.600	Co-MS	T: 21.139 N: 15.347 O: 18.950 D: 21.485	T: 17.983 N: 17.817 O: 16.433 D: 26.849

T = total, N = nausea, O = oculomotor, D = disorientation.

A three-way mixed RM ANOVA was conducted to examine the effects of the type of MS, gender, and susceptibility level on ΔSSQ. As each participant was repeatedly exposed to the three conditions (CMS, VIMS, Co-MS), the within-subjects factor was the MS type, the dependent variables were the ΔSSQs (T, N, O, D), and the between-subjects factors were gender and susceptibility. Analyses were performed independently for each dependent variable using the afex package in R 4.3.3. The RM ANOVA in this package performs Mauchly's sphericity test for reliability and applies the Greenhouse–Geisser correction if the assumption is violated. Type III sum of squares was used to control the potential imbalances and interaction effects. To reduce the impact of multiple comparisons, the test results were corrected using the Benjamini–Hochberg procedure with the family set for each dependent variable. The results of the three-way RM ANOVA for the S-VIMS participant group are shown in Table 7. None of the main or interaction effects were significant for ΔSSQ, and no trend ($p < 0.1$) was observed.

The results of the three-way mixed RM ANOVA performed on the F-VIMS group using the same procedure are shown in Table 8. Gender had significant effects on the ΔSSQ_T and the ΔSSQ_N and approached significance for the ΔSSQ_O. Susceptibility did not have significant main effects on any SSQ subscales, but when combined with gender, it had a significant interaction effect on ΔSSQ_N. The MS type had significant effects on all ΔSSQs (T, N, O, and D), and the interaction effect was significant only when combined with gender.

 https://doi.org/10.3390/s26051675

Table 7. Three-way mixed RM ANOVA results for the effects of the MS type, gender, and susceptibility on ΔSSQ in the S-VIMS Group.

Effect	ΔSSQ_T			ΔSSQ_N			ΔSSQ_O			ΔSSQ_D		
	F	η^2_G	p	F	η^2_G	p	F	η^2_G	p	F	η^2_G	p
Gender	0.404	0.004	0.711	1.296	0.012	0.826	0.033	<0.001	0.856	0.283	0.003	0.874
Susceptibility	5.345	0.054	0.182	1.995	0.019	0.826	5.781	0.064	0.146	3.890	0.039	0.389
Gender × Susceptibility	0.274	0.003	0.711	0.027	<0.001	0.871	0.558	0.007	0.666	0.379	0.004	0.874
MS type	1.287	0.018	0.711	0.706	0.011	0.870	1.584	0.020	0.531	0.715	0.010	0.874
MS type × Gender	0.308	0.004	0.736	1.052	0.016	0.826	0.165	0.002	0.856	0.093	0.001	0.912
MS type × Susceptibility	0.608	0.009	0.711	0.385	0.006	0.871	0.751	0.010	0.666	0.290	0.004	0.874
MS type × Gender × Susceptibility	0.499	0.007	0.711	0.243	0.004	0.871	1.509	0.019	0.531	0.397	0.006	0.874

T = total, N = nausea, O = oculomotor, D = disorientation. Effect size is reported as η^2_G. p-values were adjusted using the Benjamini–Hochberg procedure.

Table 8. Three-way mixed RM ANOVA results for the effects of the MS type, gender, and susceptibility on ΔSSQ in the F-VIMS Group.

Effect	ΔSSQ_T			ΔSSQ_N			ΔSSQ_O			ΔSSQ_D		
	F	η^2_G	p	F	η^2_G	p	F	η^2_G	p	F	η^2_G	p
Gender	5.078	0.060	† 0.069	9.804	0.090	* 0.011	3.314	0.039	0.177	1.126	0.015	0.412
Susceptibility	0.218	0.003	0.750	2.332	0.023	0.235	0.031	<0.001	0.862	0.063	<0.001	0.804
Gender × Susceptibility	3.109	0.038	0.149	6.510	0.062	* 0.034	0.750	0.009	0.685	2.046	0.027	0.300
MS type	9.923	0.100	* 0.001	6.166	0.078	* 0.011	9.263	0.097	* 0.002	5.748	0.056	* 0.028
MS type × Gender	4.611	0.049	* 0.044	1.619	0.022	0.286	3.463	0.039	0.126	5.102	0.050	* 0.028
MS type × Susceptibility	0.184	0.002	0.832	0.187	0.003	0.829	0.555	0.006	0.806	0.842	0.009	0.507
MS type × Gender × Susceptibility	0.411	0.010	0.576	1.086	0.015	0.399	0.156	0.002	0.862	1.801	0.018	0.300

T = total, N = nausea, O = oculomotor, D = disorientation. Effect size is reported as η^2_G. p-values were adjusted using Benjamini–Hochberg procedure (*: $p < 0.05$, †: $p < 0.1$).

As the mixed RM ANOVA results differed between the S-VIMS and F-VIMS groups, a three-way ANOVA was conducted to examine the effect of the VIMS type on ΔSSQ. The analysis was performed on the single-stimulus VIMS data, excluding the CMS and Co-MS conditions from each VIMS type group, and thus, no repeated measurement factor was included. The dependent variables were the ΔSSQs (T, N, O, and D), and the between-subjects factors were the VIMS type, gender, and susceptibility. The test results were corrected using the same method as that for the mixed RM ANOVA. The results of the three-way ANOVA for the single-stimulus VIMS type are shown in Table 9. All main and interaction effects were nonsignificant, indicating that single-stimulus VIMS was not significantly affected by its type or user characteristics.

https://doi.org/10.3390/s26051675

Table 9. Three-way ANOVA results for the effects of single-stimulus VIMS type, gender, and susceptibility on ΔSSQ.

Effect	ΔSSQ_T			ΔSSQ_N			ΔSSQ_O			ΔSSQ_D		
	F	η^2_G	p	F	η^2_G	p	F	η^2_G	p	F	η^2_G	p
VIMS type	0.147	0.002	0.703	0.232	0.003	0.632	1.118	0.013	0.763	0.034	<0.001	0.855
Gender	1.917	0.023	0.654	2.975	0.035	0.357	1.223	0.015	0.763	0.733	0.009	0.680
Susceptibility	0.534	0.006	0.654	0.293	0.004	0.632	0.912	0.011	0.763	0.979	0.012	0.680
VIMS type × Gender	1.242	0.015	0.654	1.280	0.015	0.457	0.504	0.006	0.763	1.346	0.016	0.680
VIMS type × Susceptibility	0.533	0.006	0.654	2.081	0.025	0.357	0.148	0.002	0.766	0.121	0.001	0.850
Gender × Susceptibility	0.752	0.009	0.654	0.403	0.005	0.632	0.370	0.004	0.763	1.008	0.012	0.680
VIMS type × Gender × Susceptibility	0.241	0.003	0.703	2.183	0.026	0.357	0.089	0.001	0.766	0.490	0.006	0.680

T = total, N = nausea, O = oculomotor, D = disorientation. Effect size is reported as η^2_G. p-values were adjusted using the Benjamini–Hochberg procedure.

To confirm whether the VIMS type also had no effect in a combined-stimulus context, such as Co-MS, the conditions were changed, and another three-way ANOVA was performed. The analysis procedure was identical, but the target of the analysis was changed from the Co-MS condition to the combined-stimulus VIMS data. The results of the three-way ANOVA considering the combined-stimulus VIMS type are shown in Table 10. The interaction effect between gender and the VIMS type had a significant effect on ΔSSQ_N and approached significance for ΔSSQ_T. The other main effects and interaction effects were not significant for any of the subscales. These results indicate that the influence of the VIMS type on Co-MS might be moderated by the user characteristics.

Table 10. Three-way ANOVA results for the effects of the combined-stimulus VIMS type, gender, and susceptibility on ΔSSQ in the Co-MS condition.

Effect	ΔSSQ_T			ΔSSQ_N			ΔSSQ_O			ΔSSQ_D		
	F	η^2_G	p	F	η^2_G	p	F	η^2_G	p	F	η^2_G	p
VIMS type	1.568	0.019	0.266	3.022	0.036	0.182	0.090	0.001	0.872	1.746	0.021	0.270
Gender	1.742	0.021	0.266	0.484	0.006	0.570	1.760	0.021	0.660	1.723	0.021	0.270
Susceptibility	1.286	0.015	0.266	0.053	0.001	0.818	2.838	0.033	0.232	0.647	0.008	0.424
VIMS type × Gender	6.313	0.071	† 0.098	7.692	0.086	* 0.048	4.021	0.047	0.232	2.420	0.029	0.270
VIMS type × Susceptibility	3.740	0.044	0.198	2.701	0.032	0.182	2.780	0.033	0.232	2.524	0.030	0.270
Gender × Susceptibility	1.253	0.015	0.266	3.140	0.037	0.182	0.026	<0.001	0.872	1.158	0.014	0.332
VIMS type × Gender × Susceptibility	1.777	0.021	0.266	1.136	0.014	0.406	0.724	0.009	0.556	2.489	0.029	0.270

T = total, N = nausea, O = oculomotor, D = disorientation. Effect size is reported as η^2_G. p-values were adjusted using Benjamini–Hochberg procedure (*: $p < 0.05$, †: $p < 0.1$).

5.3. MS and Immersion

The subscales of the Film IEQ were calculated by a simple summation of item scores, and thus no separate score-conversion process was performed. Table 11 shows the Film

https://doi.org/10.3390/s26051675

IEQ subscale scores (IEQ scores) for the different VIMS types under single- and combined-stimulus conditions.

Table 11. Mean and standard deviation of IEQ scores by VIMS type.

S-VIMS Group	Mean	SD	F-VIMS Group	Mean	SD
VIMS	Cap: 39.523 Dis: 7.523 Com: 10.977 Tra: 14.682	Cap: 4.810 Dis: 2.246 Com: 2.881 Tra: 3.010	VIMS	Cap: 38.659 Dis: 7.705 Com: 0.795 Tra: 14.409	Cap: 6.911 Dis: 2.041 Com: 3.062 Tra: 3.090
Co-MS	Cap: 39.543 Dis: 6.761 Com: 11.304 Tra: 14.978	Cap: 7.384 Dis: 2.068 Com: 3.926 Tra: 3.201	Co-MS	Cap: 38.978 Dis: 6.804 Com: 11.196 Tra: 15.522	Cap: 8.068 Dis: 2.400 Com: 3.816 Tra: 3.160

Cap = captivation, Dis = real-world dissociation, Com = comprehension, Tra = transportation.

Significance was tested using a two-way mixed ANOVA. As each participant was repeatedly exposed to two conditions (VIMS and Co-MS), the within-subjects factor was set to the MS type. The dependent variables were the IEQ scores (Cap, Dis, Com, and Tra), and the between-subjects factor was the VIMS type (F-VIMS or S-VIMS). As the sample sizes were nearly balanced (the ratio between the VIMS and Co-MS conditions was the same, with 44 S-VIMS participants and 46 F-VIMS participants), a type II sum of squares was used. To reduce the impact of multiple comparisons, the test results were corrected using the Benjamini–Hochberg procedure with the family set for each dependent variable. The analysis was performed using the Pingouin package 0.5.5 in Python, the results of which are listed in Table 12. All effects on immersion were nonsignificant, indicating that the visual stimulus-inducing display modality and MS type had negligible effects on immersion.

Table 12. Two-way mixed ANOVA results for the effects of the MS type and VIMS type on the IEQ scores.

Effect	IEQ_{Cap}			IEQ_{Dis}			IEQ_{Com}			IEQ_{Tra}		
	F	η^2_G	p	F	η^2_G	p	F	η^2_G	p	F	η^2_G	p
MS type	1.219	0.003	0.897	0.254	<0.001	0.132	0.186	<0.001	0.913	0.219	<0.001	0.342
VIMS type	0.017	<0.001	0.818	4.170	0.035	0.754	0.314	0.003	0.913	1.473	0.013	0.641
MS type × VIMS type	0.054	<0.001	0.897	0.099	<0.001	0.754	0.012	<0.001	0.913	1.748	0.004	0.342

Cap = captivation, Dis = real-world dissociation, Com = comprehension, Tra = transportation. Effect size is reported as η^2_G. p-values were adjusted using the Benjamini–Hochberg procedure.

The association between immersion and symptom-specific MSL was examined using a repeated measures correlation (rmcorr) between the ΔSSQs (T, N, O, and D) and the IEQ scores (Cap, Dis, Com, and Tra), as shown in Figure 11. To reduce the impact of multiple comparisons, the test results were corrected using the Benjamini–Hochberg procedure, with the family set as all 16 correlation tests (Film IEQ subscales × SSQ subscales). IEQ_{Cap} exhibited significantly negative correlations with all ΔSSQs, and IEQ_{Com} exhibited significantly negative correlations with ΔSSQ_O and ΔSSQ_T. In contrast, IEQ_{Dis} exhibited significantly positive correlations with the ΔSSQ_O and the ΔSSQ_T, whereas IEQ_{Tra} was not significantly correlated with any of the ΔSSQs.

 https://doi.org/10.3390/s26051675

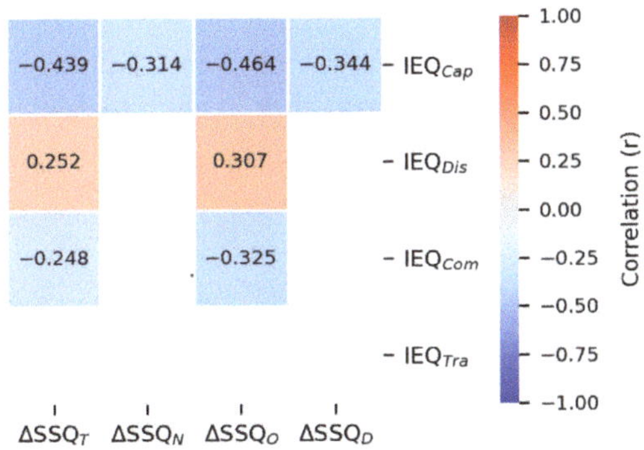

Figure 11. Repeated measures correlations between the IEQ scores and ΔSSQ. Blank cells indicate nonsignificant correlations ($\alpha = 0.05$).

5.4. Correlation Between Human-Signal Sensing Data and MSL

This section presents the results of the correlation analyses performed between the two MSL measures (ΔSSQ and FMS) and the extracted features to explore nominally significant associations. The FMS was analyzed using features on a window-by-window basis, whereas ΔSSQ was analyzed using the mean features of the windows within a session. For the analysis, a baseline sample of 84 participants was used after excluding six participants due to the measurement errors from the acquisition system. Data from some of the participants were further excluded based on the signal quality of each sensor. EDA data were analyzed after excluding two participants with thin fingers, which caused unstable electrode-skin contact across all conditions (CMS, VIMS, and Co-MS) ($N = 82$). EEG data were analyzed after excluding four, three, and two participants in the CMS, VIMS, and Co-MS conditions, respectively, due to measurement errors and poor signal quality (N = 80/81/82).

Although various MS types were used in the experiment, core features with a common association could be identified regardless of the MS type. To examine this, the dataset was analyzed as a single set without partitioning by condition, and rmcorr was calculated to reflect the within-participant variance, considering that the same participant was repeatedly exposed to the three conditions (CMS, VIMS, and Co-MS). Furthermore, rmcorr also reflected the dependency arising from repeated measurements, as a maximum of 38 windows were generated per participant for the FMS. The two continuous variables were defined as pairs of each MSL metric (ΔSSQ_T, ΔSSQ_N, ΔSSQ_O, ΔSSQ_D, and FMS) and extracted features (ΔSSQ on a session basis, FMS on a window basis), and the within-subject correlation coefficient was calculated based on the participant identifiers. Each analysis included only cases with at least two observations per participant, and missing values were excluded pairwise.

As this correlation analysis was exploratory in nature, aiming to broadly identify potential candidates in a multivariate feature space, no correction for multiple comparisons was applied, which could have led to excessive Type II errors. Instead, statistical significance was assessed at $\alpha = 0.05$, but a smallest effect size of interest (SESOI) of $|r| > 0.25$ (medium-large) was set as an additional criterion considering the nature of human-signal features [148]. Hereafter, a conservatively significant correlation refers to a correlation that satisfies both $|r| > 0.25$ and $p < 0.05$. Figure 12 shows the results of the rmcorr between the MSL metrics and the features, indicating that EDA and the Head features have a significant correlation with the overall MSL ($\alpha = 0.05$).

https://doi.org/10.3390/s26051675

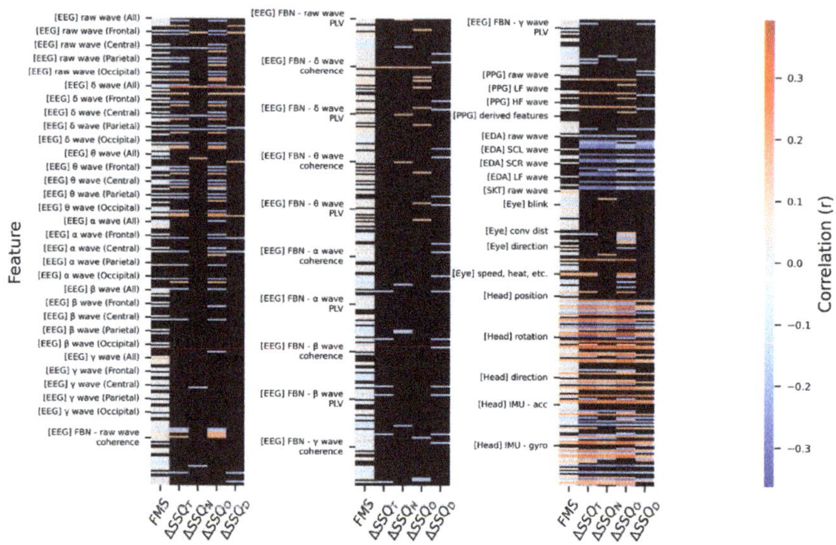

Figure 12. Repeated measures correlations between multimodal features and MSL (FMS, ΔSSQ). Black cells indicate nonsignificant correlations ($\alpha = 0.05$).

While 10 features from the whole and standard band signals of the EEG had a conservatively significant correlation ($p < 0.05$ and $|r| > 0.25$), there were few significant items compared to the extracted features. To reduce candidate omission, the SESOI was relaxed to $|r| > 0.20$. The EEG features with medium-to-large significant correlations are listed in Table 13. Except for the γ-global-mean (band-area-feature), none of the β and γ features met the relaxed criteria. Considering that the significance of the whole band reflects its sub-bands, this indicates that the key bands are δ-α. Except for the parietal lobe, most correlations were confirmed only with the ΔSSQ_O. This could be a result of the visual processing load in the occipital lobe and the visual/ocular control functions of the frontal lobe interacting with the central network, thereby reflecting visual fatigue and oculomotor load. Furthermore, parietal lobe features in the δ-θ signals showed a medium-to-large correlation with ΔSSQ_O and ΔSSQ_N, which could be because the parietal lobe performs spatial cognition through vestibular information, thereby regulating eye-head coordination and postural stability. Among the main SwFs, sample and PSD entropy were negatively correlated with ΔSSQ, whereas kurtosis and skewness were positively correlated. This can be summarized as a common pattern in which the distribution sharpens, and asymmetry is enhanced as irregularity decreases. Among the WdFs, $\beta/(\alpha + \theta)$ and $(\alpha + \theta)/(\alpha + \beta)$ had a medium-to-large significant correlation with ΔSSQ_O. For ΔSSQ_N, the δ-band F3-P3 coherence had a medium-sized correlation, and for ΔSSQ_D, the δ-band O1-O2 PLV had a medium-sized correlation. However, the correlations of these WdFs and those of the frontal-lobe SwFs did not meet the conservative significance criterion, and thus they should be used as auxiliary indicators. In summary, the standard band signals showed significant correlations centered on oculomotor symptoms, with the key signal domains being frequency (δ–α) and space (frontal, central, occipital), and the parietal lobe served as an auxiliary signal.

 https://doi.org/10.3390/s26051675

Table 13. EEG features with significant correlations to the whole and standard band signals ($p < 0.05$ and $|r| > 0.20$). Bold indicates $|r| > 0.25$.

Features			MSL	r	p	Features			MSL	r	p
whole	global	ent	dO	−0.226	0.004	θ	frontal	ent	dT	−0.201	0.010
whole	global	skw	dD	0.212	0.007	θ	frontal	ent	dO	−0.258	<0.001
whole	global	psd_ent	dO	−0.241	0.002	θ	frontal	kut	dO	0.226	0.004
whole	**frontal**	**ent**	**dO**	**−0.250**	**0.001**	θ	frontal	psd_ent	dO	−0.204	0.009
whole	frontal	psd_ent	dO	−0.210	0.007	θ	central	ent	dT	−0.212	0.007
whole	**central**	**ent**	**dO**	**−0.255**	**0.001**	θ	central	ent	dO	−0.248	0.001
whole	central	psd_ent	dO	−0.216	0.006	θ	central	kut	dO	0.230	0.003
whole	parietal	ent	dO	−0.210	0.007	θ	central	psd_ent	dO	−0.210	0.007
whole	parietal	psd_ent	dO	−0.204	0.009	θ	parietal	ent	dO	−0.221	0.005
whole	**occipital**	**ent**	**dO**	**−0.258**	**<0.001**	θ	parietal	kut	dO	0.209	0.008
whole	occipital	kut	dO	0.205	0.009	θ	occipital	ent	dO	−0.241	0.002
whole	occipital	skw	dO	0.218	0.005	**θ**	**occipital**	**kut**	**dO**	**0.260**	**<0.001**
whole	**occipital**	**psd_ent**	**dO**	**−0.256**	**<0.001**	θ	occipital	skw	dT	0.216	0.006
δ	frontal	ent	dO	−0.234	0.003	θ	occipital	skw	dO	0.244	0.002
δ	frontal	kut	dT	0.206	0.009	α	frontal	ent	dT	−0.218	0.005
δ	frontal	kut	dO	0.245	0.002	**α**	**frontal**	**ent**	**dO**	**−0.263**	**<0.001**
δ	central	ent	dT	−0.225	0.004	α	central	ent	dT	−0.224	0.004
δ	**central**	**ent**	**dO**	**−0.257**	**<0.001**	**α**	**central**	**ent**	**dO**	**−0.252**	**0.001**
δ	central	kut	dT	0.207	0.008	α	parietal	ent	dO	−0.203	0.009
δ	**central**	**kut**	**dO**	**0.250**	**0.001**	α	occipital	ent	dT	−0.209	0.008
δ	central	psd_ent	dO	−0.200	0.011	α	occipital	ent	dO	−0.245	0.002
δ	parietal	ent	dT	−0.219	0.005	α	occipital	psd_ent	dO	−0.210	0.007
δ	parietal	ent	dN	−0.204	0.009	γ	global	mean	dD	0.237	0.002
δ	parietal	ent	dO	−0.231	0.003	WdF	$\beta/(\alpha+\theta)$	dT	−0.211	0.007	
δ	parietal	kut	dO	0.217	0.005	WdF	$\beta/(\alpha+\theta)$	dO	−0.243	0.002	
δ	occipital	ent	dT	−0.202	0.010	WdF	$(\alpha+\theta)/(\alpha+\beta)$	dO	0.217	0.006	
δ	occipital	ent	dO	−0.246	0.002	WdF	(δ) Coh	F3-F4	dN	0.205	0.009
δ	occipital	kut	dO	0.223	0.004	WdF	(δ) PLV	O1-O2	dD	−0.205	0.009
δ	occipital	skw	dO	0.227	0.004	WdF	(θ) PLV	O1-O2	dD	−0.201	0.010
θ	global	mean	dN	0.242	0.002						

psd_ent = PSD entropy, ent = sample entropy, kut = kurtosis, skw = skewness, ptp = peak-to-peak amplitude. dT = ΔSSQ_T, dN = ΔSSQ_N, dO = ΔSSQ_O, dD = ΔSSQ_D.

In the PPG signals (whole and sub-band), no features with conservatively significant correlations were identified, and thus the SESOI was relaxed to $|r| > 0.2$ (medium). The PPG features with medium-sized significant correlations are listed in Table 14. Among the SwFs from the whole, LF, and HF band signals, only the variance was significantly correlated with all ΔSSQs (T, N, O, and D); these correlations were positive and medium-sized. This indicates that the variability of the pulse wave increased with the MSL. Among the WdFs, SD1/SD2 exhibited medium-sized negative correlations with ΔSSQ_T, ΔSSQ_O, and ΔSSQ_D, implying that an increase in long-term variability (SD2) relative to short-term variability (SD1) is associated with an increase in oculomotor and disorientation severities. BR exhibited a medium-sized negative correlation with ΔSSQ_N, indicating a tendency for the respiration rate to decrease with an increase in the severity of nausea. However, since the magnitude of all feature correlations was in the range of $0.20 \leq |r| \leq 0.25$, PPG features are better interpreted as auxiliary indicators rather than primary.

In the EDA signals (whole and sub-band), 38 features with conservatively significant correlations were identified (Table 15). In contrast, for the SKT, no features met the criteria even when the SESOI was relaxed to $|r| > 0.2$ (medium). EDA exhibited conservatively significant correlations with ΔSSQs, except ΔSSQ_O, across all bands (whole, SCL, SCR, LF). Unlike the other domains, certain features showed strong correlations ($|r| > 0.3$). In addition, the variance, peak-to-peak amplitude, and band power across all bands showed negative correlations with MSL, with the largest effect in ΔSSQ_N ($|r| \approx 0.30$–0.33). This

https://doi.org/10.3390/s26051675

indicates that EDA can be effectively utilized in environments with physical stimuli such as CMS and Co-MS. Overall, all bands showed similar patterns (symptoms, effect sizes, and features), but the whole-band signal showed conservatively significant correlations in skewness, unlike the sub-bands. The significant appearance of skewness only in the whole band is because the high-amplitude peak characteristics were dispersed through the band decomposition process, alleviating asymmetry.

Table 14. PPG features with significant correlations to the whole band and sub-band signals (medium effect size; $p < 0.05$ and $|r| > 0.20$).

Features		MSL	*r*	*p*
whole	var	dT	0.210	0.006
whole	var	dN	0.226	0.003
whole	var	dO	0.215	0.005
LF	var	dN	0.214	0.005
HF	var	dT	0.206	0.007
HF	var	dN	0.230	0.003
HF	var	dO	0.203	0.008
WdF	SD1/SD2	dT	−0.237	0.002
WdF	SD1/SD2	dO	−0.235	0.002
WdF	SD1/SD2	dD	−0.210	0.006
WdF	BR	dN	−0.201	0.009

var = variance. dT = ΔSSQ_T, dN = ΔSSQ_N, dO = ΔSSQ_O, dD = ΔSSQ_D.

Table 15. EDA features with significant correlations to the whole band and sub-band signals ($p < 0.05$ and $|r| > 0.25$). Bold indicates $|r| > 0.30$.

Features		MSL	*r*	*p*	Features		MSL	*r*	*p*
whole	var	dT	−0.289	<0.001	SCL	band_power	dD	−0.263	<0.001
whole	**var**	**dN**	**−0.321**	**<0.001**	SCR	var	dT	−0.291	<0.001
whole	var	dD	−0.268	<0.001	**SCR**	**var**	**dN**	**−0.323**	**<0.001**
whole	skw	dT	−0.251	0.001	SCR	ptp	dT	−0.285	<0.001
whole	skw	dN	−0.274	<0.001	**SCR**	**ptp**	**dN**	**−0.319**	**<0.001**
whole	ptp	dT	−0.279	<0.001	SCR	ptp	dD	−0.254	<0.001
whole	**ptp**	**dN**	**−0.309**	**<0.001**	SCR	band_power	dT	−0.275	<0.001
whole	ptp	dD	−0.267	<0.001	**SCR**	**band_power**	**dN**	**−0.306**	**<0.001**
whole	band_power	dT	−0.277	<0.001	SCR	band_power	dD	−0.260	<0.001
whole	band_power	dN	−0.309	<0.001	LF	var	dT	−0.295	<0.001
whole	band_power	dD	−0.260	<0.001	**LF**	**var**	**dN**	**−0.327**	**<0.001**
SCL	var	dT	−0.293	<0.001	LF	var	dD	−0.264	<0.001
SCL	**var**	**dN**	**−0.325**	**<0.001**	LF	ptp	dT	−0.284	<0.001
SCL	var	dD	−0.266	<0.001	**LF**	**ptp**	**dN**	**−0.318**	**<0.001**
SCL	ptp	dT	−0.282	<0.001	LF	ptp	dD	−0.258	<0.001
SCL	**ptp**	**dN**	**−0.317**	**<0.001**	LF	band_power	dT	−0.270	<0.001
SCL	ptp	dD	−0.254	<0.001	**LF**	**band_power**	**dN**	**−0.300**	**<0.001**
SCL	band_power	dT	−0.272	<0.001	LF	band_power	dD	−0.260	<0.001
SCL	**band_power**	**dN**	**−0.302**	**<0.001**	LF	band_power	dD	−0.260	<0.001

var = variance, skw = skewness, ptp = peak-to-peak amplitude. dT = ΔSSQ_T, dN = ΔSSQ_N, dO = ΔSSQ_O, dD = ΔSSQ_D.

For the eye sensor domain, only three SwFs had conservatively significant correlations: PSD entropy in the yaw direction with ΔSSQ_T ($r = 0.304$, $p = 0.002$) and ΔSSQ_O ($r = 0.304$, $p = 0.002$), and PSD entropy in the pitch direction with ΔSSQ_O ($r = 0.283$, $p = 0.003$). This indicates that as oculomotor symptoms intensify, the band power of eye movements tends to disperse across multiple frequencies rather than concentrating on a few dominant

https://doi.org/10.3390/s26051675

rhythms. However, as the number of significant items in the eye-tracking data was too small relative to the number of extracted features, the SESOI was relaxed to $|r| > 0.20$, to reduce candidate omission, and features with a significant correlation under this criterion were additionally identified. The eye-tracking features with medium-to-large significant correlations are listed in Table 16. Except for ΔSSQ_T, most features had a medium-to-large significant correlation with ΔSSQ_O. The PSD entropy of the yaw and pitch directions had consistent positive correlations with ΔSSQ_T, ΔSSQ_O, and ΔSSQ_N. Most velocity SwFs were positively correlated with ΔSSQ_T and ΔSSQ_O, except for skewness and PSD entropy, which were negatively correlated. In addition, the cumulative angle WdF was confirmed to have a positive correlation with ΔSSQ_O, and the kurtosis of the convergence distance had a negative correlation. In summary, as the oculomotor symptoms intensified, the gaze direction dispersed across multiple frequencies, leading to an increase in trajectory changes and movement distance. In contrast, the velocity spectrum and distribution became regularized, reducing the number of extreme peaks, representing a dual pattern. This was complementary to the pattern observed in the frontal and occipital lobes of the EEG (entropy↓, kurtosis↑) and was interpreted as the result of the burden on vision-oculomotion-visual processing being reflected in the ΔSSQ_O. However, because these features had medium-sized significant correlations under relaxed conditions, they were interpreted as auxiliary indicators.

Table 16. Eye features with significant correlations (medium effect size; $p < 0.05$ and $|r| > 0.20$). Bold indicates $|r| > 0.25$).

Features		MSL	r	p
Convergence distance	kut	dO	−0.201	0.009
Dir_yaw	mean	dO	−0.212	0.006
Dir_yaw	**psd_ent**	**dT**	**0.304**	**0.002**
Dir_yaw	psd_ent	dN	0.209	0.032
Dir_yaw	**psd_ent**	**dO**	**0.350**	**<0.001**
Dir_pitch	psd_ent	dT	0.210	0.032
Dir_pitch	**psd_ent**	**dO**	**0.283**	**0.003**
Velocity	mean	dT	0.206	0.007
Velocity	mean	dO	0.241	0.002
Velocity	skw	dO	−0.208	0.033
Velocity	psd_ent	dO	−0.206	0.035
WdF	accumulated degree	dT	0.205	0.007
WdF	accumulated degree	dO	0.241	0.002

Dir = direction, psd_ent = Shannon entropy of PSD, var = variance, kut = kurtosis, skew = skewness. dT = ΔSSQ_T, dN = ΔSSQ_N, dO = ΔSSQ_O, dD = ΔSSQ_D.

In the head sensor domain, 94 features were identified to have conservatively significant correlations. To avoid overinterpretation due to redundant features, the SESOI was raised to $|r| > 0.3$ (large). The number of head-tracking features with a large-sized significant correlation was 41, as shown in Table 17. Features with a large-sized significant correlation tended to concentrate on ΔSSQ_O and ΔSSQ_T. Overall, the amplitude/energy features (variance, ptp, and band power) had strong positive correlations with ΔSSQ_O and ΔSSQ_T. In terms of feature metrics, the amplitude/energy features of the pitch/surge exhibited the most consistent positive correlation, and similar tendencies were partially observed for heave and sway. For acceleration/angular velocity, these metrics also showed a predominantly positive correlation with $\Delta SSQ_{O/T}$. This indicates that as oculomotor symptoms intensified, the amplitude and energy of head movements tended to increase and were centered on the pitch and surge axes. In contrast, irregularity features (kurtosis

 https://doi.org/10.3390/s26051675

and sample/PSD entropy) showed strong negative correlations with ΔSSQ_O and ΔSSQ_T. This implies that as the oculomotor symptoms intensified, the position and rotation tended to converge to a specific pattern. Furthermore, the head sensor domain had 94 features with conservatively significant correlations (41 even when retaining only large criteria $|r| > 0.3$), a richer extraction than other domains, and a consistent correlation structure across multiple features. This supports the practicality of head-tracking features as a key, low-burden, and lightweight indicator alongside EDA.

Table 17. Head features with significant correlations (large; $p < 0.05$ and $|r| > 0.3$). Bold indicates $|r| > 0.35$.

Features		MSL	*r*	*p*	Features		MSL	*r*	*p*
Pos_sway	var	dO	0.319	<0.001	Rot_pitch	ptp	dO	0.370	<0.001
Pos_sway	ptp	dT	0.314	<0.001	Rot_pitch	band_power	dT	0.346	<0.001
Pos_sway	ptp	dO	0.336	<0.001	Rot_pitch	band_power	dO	0.331	<0.001
Pos_sway	band_power	dO	0.319	<0.001	Dir_pitch	var	dT	0.347	<0.001
Pos_heav	var	dO	0.306	<0.001	Dir_pitch	var	dO	0.329	<0.001
Pos_heav	ptp	dT	0.319	<0.001	**Dir_pitch**	**ptp**	**dT**	**0.363**	**<0.001**
Pos_heav	ptp	dO	0.337	<0.001	**Dir_pitch**	**ptp**	**dO**	**0.369**	**<0.001**
Pos_heav	band_power	dO	0.301	<0.001	Dir_pitch	band_power	dT	0.345	<0.001
Pos_surg	ent	dT	−0.318	<0.001	Dir_pitch	band_power	dO	0.332	<0.001
Pos_surg	ent	dO	−0.325	<0.001	Acc_heave	ent	dT	−0.345	<0.001
Pos_surg	var	dT	0.331	<0.001	Acc_heave	ent	dO	−0.347	<0.001
Pos_surg	**var**	**dO**	**0.380**	**<0.001**	Acc_surge	ptp	dT	0.345	<0.001
Pos_surg	**ptp**	**dT**	**0.352**	**<0.001**	**Acc_surge**	**ptp**	**dO**	**0.369**	**<0.001**
Pos_surg	**ptp**	**dO**	**0.384**	**<0.001**	AngV_pitch	var	dO	0.306	<0.001
Pos_surg	**band_power**	**dT**	**0.353**	**<0.001**	AngV_pitch	ptp	dT	0.317	<0.001
Pos_surg	**band_power**	**dO**	**0.394**	**<0.001**	AngV_pitch	ptp	dO	0.332	<0.001
Pos_surg	**psd_ent**	**dT**	**−0.359**	**<0.001**	AngV_yaw	kut	dT	−0.307	<0.001
Pos_surg	**psd_ent**	**dO**	**−0.363**	**<0.001**	AngV_yaw	kut	dO	−0.334	<0.001
Rot_pitch	var	dT	0.347	<0.001	AngV_roll	kut	dT	−0.306	<0.001
Rot_pitch	var	dO	0.331	<0.001	AngV_roll	kut	dO	−0.327	<0.001
Rot_pitch	**ptp**	**dT**	**0.364**	**<0.001**					

Pos = position, Rot = rotation, Dir = direction, Acc = acceleration, AngV = angular velocity, psd_ent = PSD entropy, ent = Sample entropy, var = variance, ptp = peak-to-peak amplitude, kut = kurtosis. dT = ΔSSQ_T, dN = ΔSSQ_N, dO = ΔSSQ_O, dD = ΔSSQ_D.

For the FMS, no features from any sensor domain exhibited conservatively significant correlations. Only when the SESOI was relaxed to $|r| > 0.2$ (medium), features with a medium-sized significant correlation were identified (Table 18). All FMS-related features with a medium-sized significant correlation were the head-tracking features, and their correlation structure was consistent with the correlation between these features and ΔSSQ. However, considering that ΔSSQ_T was influenced by ΔSSQ_O, it was difficult to estimate the severity of nausea and oculomotor symptoms in real time using human-signal features that had common correlations across various MS environments.

https://doi.org/10.3390/s26051675

Table 18. Significant correlations between all features and fast motion sickness scale (medium effect size; $p < 0.05$ and $|r| > 0.20$).

Features			r	p
Head	Pos_heav	ptp	0.247	<0.001
Head	Pos_surg	ptp	0.244	<0.001
Head	Pos_surg	psd_ent	−0.210	<0.001
Head	Rot_pitch	ptp	0.226	<0.001
Head	Rot_roll	ptp	0.206	<0.001
Head	Dir_pitch	ptp	0.224	<0.001
Head	Acc_sway	ptp	0.211	<0.001
Head	Acc_surge	ptp	0.206	<0.001
Head	AngV_pitch	ptp	0.214	<0.001

Pos = position, Rot = rotation, Acc = acceleration, AngV = angular velocity, heav = heave, surg = surge, psd_ent = PSD entropy, ptp = peak-to-peak amplitude.

To confirm whether FMS concentrated on ΔSSQ_O in these results, rmcorr was used for examination. To reduce the impact of multiple comparisons, test results were corrected using the Benjamini–Hochberg procedure, with families set as six correlation tests being pairwise combinations of four MSL measures (ΔSSQ_T, ΔSSQ_N, ΔSSQ_O, and FMS). As shown in Figure 13, FMS exhibited significant positive correlations with all ΔSSQs, indicating that FMS is not limited to ΔSSQ_O.

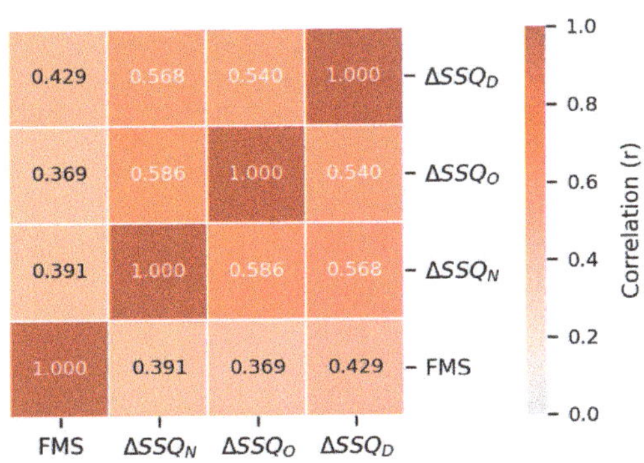

Figure 13. Repeated measures correlations within MSL indices (ΔSSQ_N, ΔSSQ_O, ΔSSQ_D, and FMS). All cells indicate significant correlations ($p < 0.001$).

5.5. Relative Contributions of Multimodal Features to MSL

Statistical analyses effectively identify significant individual factors; however, MSL can involve complex nonlinearity and feature interactions among the multimodal signals. To investigate these complex dynamics and determine the relative importance of each sensor domain, a contribution analysis was performed using an ML approach. In this approach, demographic variables were included as input features because they were confirmed to have significant main or interaction effects (Section 5.2).

An EBM was used as the ML model to ensure interpretability. EBM, which combines a generalized additive model with second-order interactions, can transparently and quantitatively explain nonlinear relationships by estimating the effect curve of each feature and pairwise interaction as separate additive terms. For the regression target, the continuous FMS score was directly employed instead of the session-based ΔSSQ. Although the correlation coefficients between the FMS and SSQ subscales were moderate, they were

https://doi.org/10.3390/s26051675

statistically significant across all symptom categories, indicating that the FMS serves as a valid representative indicator of diverse MS symptoms.

For post hoc explanatory attribution, the EBM was fitted to the entire dataset to obtain global feature importances (seed = 42, interaction = 0.9, validation ratio = 0.2, goodness of fit (R^2) = 0.919, Pearson linear correlation coefficient (PLCC) = 0.961, and Spearman's rank correlation coefficient (SRCC) = 0.877). To rigorously assess the robustness of these feature contributions and ensure that they were not merely artifacts of the dataset, the analysis was supplemented with participant-wise cross-validation. Figure 14 shows the top 20 most important features derived from the full dataset model alongside the mean feature importances maintained across the 10-fold cross-validation.

Among the multimodal features, demographic variables (susceptibility and gender) and skin temperature (SKT_avg) consistently ranked among the strongest multivariable contributors to MSL. For physiological responses, cardiovascular metrics (PPG_Derived_IBI and HR) and electrodermal indices (EDA_raw_avg and SCL_avg) repeatedly appeared in the upper ranks, indicating their stable contributions. Regarding head motion, although the univariate correlation analysis (Section 5.4) highlighted broad linear associations between head-movement features and MS, the EBM emphasized only a small set of representative translational kinematics (e.g., HeadPose_posHeave_avg and HeadPose_posSurge_ptp) among the top-ranked predictors. The most distinct pattern relative to the univariate results was observed for EEG: while individual EEG features showed weaker linear associations, the EBM consistently identified multiple FBN connectivity features (e.g., PLV and coherence-related indices across several frequency bands) as important nonlinear contributors to MSL. Conversely, no gaze features were included in the top-20 multivariable contributors.

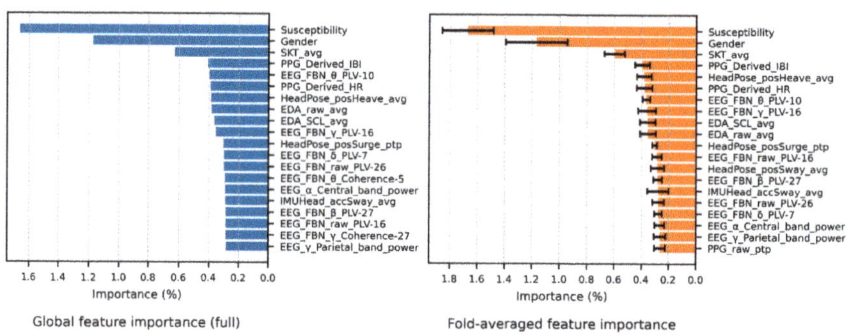

Figure 14. Top-20 global feature importance values (%) from the EBM models. The black error bars in the right panel represent the standard deviation across the 10-fold cross-validation.

The global importance of the model was widely distributed across many features, making it difficult to intuitively identify the sensor domains that contributed the most. Furthermore, owing to the imbalance in the number of features across the domains, the possibility that the model over-relied on a specific domain could not be ruled out. Therefore, an ablation study was conducted by refitting the model after removing all features from each sensor domain while keeping the dataset and hyperparameters the same. The change in explanatory power was defined as the relative degradation rate compared to the baseline model as follows:

$$\Delta m(\%) = \frac{m_{full} - m_{abl}}{m_{full}} \times 100, \quad m \in \{R^2,\ PLCC,\ SRCC\} \tag{5}$$

https://doi.org/10.3390/s26051675

where $\Delta m > 0$ indicates a degradation in explanatory power (decrease), and $\Delta m < 0$ indicates that the model fit improved upon removal. To complement the ablation analysis and address the imbalance in feature counts across domains, single-domain models were also refitted. Figure 15 summarizes the relative degradation in explanatory power when each domain was excluded (w/o) and when only a single domain was used.

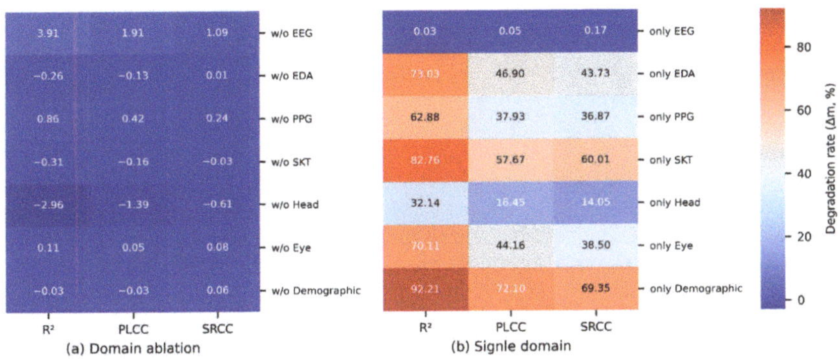

Figure 15. Relative degradation in explanatory power for the sensor domains (w/o = without).

In the domain ablation analysis (Figure 15a), the EEG-ablated model exhibited the largest degradation across all metrics (R^2, PLCC, and SRCC). However, the magnitude of this reduction was relatively modest, not exceeding 5% (e.g., 3.91% for R^2). Furthermore, the ablation of the remaining domains (EDA, PPG, SKT, Head, Eye, and Demographic) resulted in marginal decreases of less than 1% or even slight improvements in the model fit. Notably, the head-ablated model showed a clear improvement in model fit, indicated by a negative degradation rate of -2.96% for R^2 relative to the full model under the same dataset and hyperparameter settings. Although EEG ablation produced the largest degradation among the domains, it should be noted that the EEG features accounted for 583 out of 828 total features ($\approx 70.4\%$), representing a very high proportion. As a complementary analysis, a single-domain evaluation was conducted (Figure 15b) to examine this imbalance and highlight domain contributions without inter-domain redundancy.

The single-domain models provided a domain-wise comparison of explanatory power. The EEG-only model exhibited an exceptionally low degradation rate (0.03% in R^2), maintaining an R^2 almost equivalent to that of the full model. The head-only model recorded the second-lowest degradation rate (32.14% in R^2). Among the physiological signals, the PPG-only model showed the next lowest degradation (62.88% in R^2). The other single-domain models (Eye, EDA, SKT, and Demographic) exhibited substantial degradation in model fit (over 70% in R^2). Although the EDA- and Eye-only models retained partial correlations, with 38.50–46.90% drops in PLCC and SRCC, these substantial losses still quantify their limited standalone modeling capacity.

Given the practical constraints of scalp EEG in actual vehicular environments, lightweight sensor combinations excluding EEG were further evaluated. Based on the domain hierarchy observed in the single-domain evaluation, the Head and PPG were selected as the common baseline modalities. The EDA and Eye domains were conditionally added, resulting in four combinations: Head + PPG, Head + PPG + EDA, Head + PPG + Eye, and Head + PPG + EDA + Eye. As shown in Figure 16a, the Head + PPG + EDA combination exhibited the lowest degradation in model fit across all metrics (17.97% for R^2, 9.13% for PLCC, and 7.15% for SRCC), demonstrating the optimal explanatory capacity among the lightweight models.

https://doi.org/10.3390/s26051675

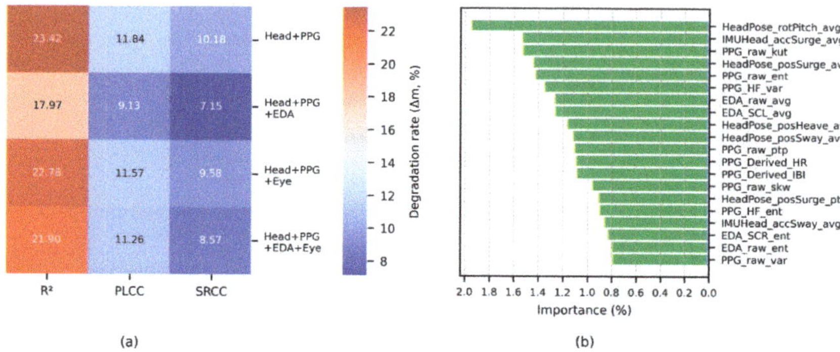

Figure 16. Evaluation of lightweight multimodal combinations: (**a**) Relative degradation rates for the four sensor combinations excluding EEG. (**b**) Top-20 global feature importance values (%) for the optimal combination (Head + PPG + EDA).

The global feature importances for the optimal Head + PPG + EDA combination are presented in Figure 16b. Consistent with the findings from the full multimodal evaluation, head kinematic features maintained the highest ranks, with variables such as average pitch rotation (HeadPose_rotPitch_avg) and surge acceleration (IMUHead_accSurge_avg) serving as the primary explanatory indicators. Furthermore, metrics from the PPG and EDA domains, including raw signal statistics (e.g., PPG_raw_kut and EDA_raw_avg) and derived physiological indices (e.g., PPG_HF_var and EDA_SCL_avg), constituted the remaining top ranks. Overall, head kinematics remained prominent in the lightweight setting, whereas cardiovascular and electrodermal features contributed additional explanatory information.

6. Discussion

MSL was quantitatively examined from a multimodal human-signal perspective under single and composite stimuli in the context of autonomous driving. Considering that the MS quantification studies in the HS-Set were biased towards VR-based VIMS, this study complements the gap in MS quantification research by providing a unified examination of physical stimuli that simulate driving, visual stimuli, and their combination, while also covering see-through HMD and simulator environments.

6.1. Investigation of User Characteristics and Immersion Effects Across MS Types

The absence of significant MSL differences across MS types and user characteristics in the S-VIMS group suggests that, from the perspective of the sensory conflict theory, stimulus variables not treated as independent factors in this study (e.g., stimulus intensity such as optical flow and physical movement of the simulator itself) likely exerted a more dominant influence on MSL than demographic and MS-type variables.

Conversely, the F-VIMS group exhibited significant MSL variations across MS types, generally following an ascending order of VIMS < CMS < Co-MS. The exceptionally low MSL under the single visual stimulus condition may stem directly from the limited field of view (FOV) of the consumer-grade tablet, which likely weakened vection. The subsequent peak in MSL during Co-MS suggests that the combination of a narrowed visual field caused by content immersion and a tilted vestibular axis from a downward posture (neck flexion) amplified the visual-vestibular sensory conflict, a phenomenon absent in the open visual field of S-VIMS. This interpretation is further corroborated by the high predictive importance of the amplitude and energy features of the head domain. Furthermore, the unique deviation for oculomotor symptoms—where CMS induced higher severity than Co-MS—might indicate that physical motion destabilized gaze fixation and tracking, while repetitive and monotonous visual stimuli hindered immersion and exacerbated visual fatigue.

Despite the use of different display technologies (F-VIMS vs. S-VIMS), the display type itself does not appear to be a primary determinant of MSL quantification under single-stimulus conditions. However, the emergence of significant gender effects (a gender effect in F-VIMS and a gender $\times$ VIMS type interaction in Co-MS) specifically impacting nausea suggests that gender acts not as a constant main effect but rather as a conditional amplifier. The absence of this effect in S-VIMS implies that the specific sensory composition, such as display type, posture, and FOV, may modulate the expression threshold of gender-related susceptibilities. Consequently, the design of personalized MS-mitigation strategies requires the distinction of these interaction-driven indicators.

Regarding psychological immersion, the lack of significant effects from either the display modality or the MS type on the Film IEQ subscales implies that user immersion is not directly governed by the technical specifications of the stimulus or the presence of composite stimuli. Consistent with prior research, this finding reaffirms that immersion relies more heavily on content quality and individual cognitive characteristics.

The observed negative correlations of both captivation and comprehension with MS symptoms support the capacity theory, indicating that deep cognitive engagement and higher predictability of the stimulus reduce the resources available to process visuo-vestibular conflicts, thereby facilitating adaptation and alleviating MS. Conversely, real-world dissociation exhibited a positive correlation with oculomotor and total MS symptoms but notably lacked an association with disorientation. This divergence highlights the intrinsic difference between the two concepts: real-world dissociation captures the subjective psychological perception of detachment, whereas disorientation reflects a physiological state of spatial confusion. Thus, despite conceptual similarities, the underlying mechanisms differ, demonstrating that psychological immersion experiences and physiological MS symptoms do not invariably maintain a one-to-one correspondence.

Finally, despite the use of subjective camera techniques, the transportation metric—indicating the sensation of being mentally transported—did not significantly correlate with MSL. This lack of association may stem from the inherent mismatches in the Co-MS condition, where the conflict between the simulator's physical motion and the film's visual motion disrupted the psychological experience of entering the narrative space. Consequently, narrative immersion appears to be fundamentally constrained by sensory conflict in composite-stimulus environments.

6.2. Quantitative Approaches to MS Based on Human Signals

Each human-signal domain exhibited unique characteristics based on the correlation between the extracted features and MSL. In the EEG domain, decreases in entropy and increases in kurtosis and skewness in the δ-α bands showed significant correlations only with oculomotor symptoms, reflecting the phenomenon of brainwave patterns converging on a specific rhythm. The overall correlation, with the oculomotor symptoms observed broadly across the fronto-central-occipital regions, and the partial correlation, with nausea observed in the parietal lobe, are consistent with the perspective of the sensory conflict theory that MS is fundamentally an issue of visuo-vestibular sensory integration. The finding that the FBN features, of which only three had a medium-sized correlation, were numerous in the top ranks of the EBM suggested that nonlinear feature interactions in EEG were more important for MS prediction than simple linear correlations. This implies that

https://doi.org/10.3390/s26051675

complex connectivity patterns between brain networks might reflect MSL changes more sensitively than independent contributions of individual features.

The EDA domain had features with consistently negative correlations across all bands, and features related to nausea exhibited strong correlations ($|r| \geq 0.3$). Considering that Kenward et al. [110] and Wickham [111] reported that nausea is a higher-level cognitive activity than vomiting, this strong correlation suggested that EDA was able to sensitively capture the complex physiological and cognitive responses to nausea symptoms. In contrast, the PPG features showed broad correlations with all symptoms, but were only of medium size, making them more suitable as auxiliary indicators rather than core indicators. The increase in the variance (regardless of the band) exhibited medium-sized significant positive correlations with an increase in MSL, whereas SD1/SD2 exhibited medium-sized significant negative correlations, reflecting an increase in the pulse wave variability and a relative decrease in the parasympathetic tone, respectively. In other words, PPG appears to have captured a general, but relatively insensitive, cardiovascular response to MS.

The Eye and Head domain features exhibited complementary characteristics, capturing different aspects of the oculomotor symptoms. The head domain features showed the richest and most consistent correlation structures, with amplitude/energy features in the pitch and surge axes showing strong positive correlations, which reflected an increase in compensatory head movements in response to anteroposterior visual stimuli. In particular, the strong negative correlation of irregularity features (e.g., entropy and kurtosis) indicated a tendency for head movements to converge to a specific pattern as the oculomotor symptoms intensified, which can be interpreted as an adaptive response of the vestibular system. In the Eye domain, most features with medium-to-large correlations were in the frequency domain (PSD entropy), with positive correlations observed between nausea and oculomotor symptoms. This implies that as the MSL increased, eye movements became more irregular and dispersed across multiple frequencies. However, the VOR (gaze velocity/head velocity) used as an eye feature did not have a medium-to-large significant correlation, suggesting that the traditional VOR concept alone was insufficient to fully explain the complex visuo-vestibular interactions in MS. This result shows that the core mechanism of oculomotor symptoms is a complex response due to visual attention dispersion and visuo-vestibular conflict, rather than a simple VOR gain. Overall, considering that previous studies mainly dealt with eye data in the time domain, this result demonstrates the need to consider the frequency domain as well.

The significant positive correlation between the FMS and all ΔSSQ subsymptom scales supports the validity of the FMS as a window-level MSL indicator that encompasses various MS symptoms. This means that the FMS can reflect the overall MSL in a balanced manner without being biased towards a specific symptom. However, the relatively small correlations between the FMS and individual human-signal features reveal the inherent difficulty of real-time MSL quantification. This could be due to temporal delays or complex nonlinear relationships between the momentary subjective experience captured by the FMS and the physiological changes reflected by the human signals.

Multimodal Contribution Patterns Revealed by Interpretable Modeling

The interpretable modeling results complement the inferential and correlation analyses by characterizing the multivariable contribution patterns to MSL when nonlinearities and feature interactions are present. A notable observation was that global importance was not concentrated in a small set of predictors; instead, relatively small per-feature contributions were distributed across several multimodal variables. Together with the consistently high ranks of demographic variables and SKT-related indices, this pattern supports the view that

 https://doi.org/10.3390/s26051675

MSL is expressed as an integrated response spanning multiple physiological and behavioral channels, rather than being determined by a few highly dominant factors captured by univariate associations alone.

Domain-level ablation and single-domain evaluations provided complementary perspectives on how explanatory capacity was distributed across modalities under substantial feature-count imbalances. In particular, EEG contributed strongly to the domain-wise comparisons; however, the overall ablation magnitudes were modest, and the model fit occasionally improved when specific domains were removed. Such behavior is consistent with substantial redundancy and shared information across modalities in a multivariable setting, where the presence of one domain can partially overlap with the nonlinear information represented by others. Accordingly, it is highlighted that domain relevance in this framework is more appropriately interpreted in terms of incremental changes in model fit, rather than being directly inferred from the number of significant univariate correlations.

From a practical standpoint, the strong contribution of EEG must be weighed against the constraints of in-vehicle deployment. Given the acceptability and setup burden of scalp EEG, EEG-excluded lightweight combinations were evaluated. Among the tested candidates, the Head + PPG + EDA configuration exhibited the most favorable model fit, and its global importance pattern was distributed across multiple features. Head kinematics consistently ranked the highest, while peripheral autonomic indices from PPG and EDA provided additional explanatory contributions, indicating that substantial MSL-related variation can be captured by combining motion-derived behavioral cues with peripheral physiological responses under constrained sensing conditions.

Finally, the limited degradation observed when demographic variables were excluded suggests that a meaningful portion of MSL-related patterns can be characterized using human signals alone, which has practical value for privacy-preserving monitoring when personal information is unavailable. Nevertheless, further validation remains necessary under operational driving conditions, where motion artifacts, sensor stability, and environmental noise may alter the signal quality and relative contribution patterns across MS-inducing factors.

6.3. Limitations

The participants for this study consisted of young adults in their early 20s from a single institution (male/female = 64:26). Therefore, the results of this experiment might be limited to young adults. Susceptibility was also managed as a two-level variable, normal (combining low and moderate) and high, based on MSSQ-Short. This dichotomization could have led to a conservative estimation of interaction effects (e.g., gender × susceptibility and display × susceptibility) and might not fully explain the heterogeneity within the normal group (low versus moderate). In addition, because detailed histories regarding the participants' current pharmacological status and prior experience with see-through HMDs were not collected, this study has limitations in fully accounting for their potential effects on the results. Certain medications may alter baseline physiological signals, and prior HMD familiarity can lead to habituation, which naturally affects the VIMS threshold.

Although Co-MS was intentionally designed to induce visuo-vestibular conflict, it differs from the composite stimuli of actual road environments (e.g., road surface irregularities, cross slopes, traffic situations, and sudden lane changes). The use of heave/roll in the motion platform was limited, and the classic washout filter could not perfectly replicate the low-frequency vibration components of actual vehicles.

The VIMS stimulus was induced in two subjective camera films (found footage). However, for ecological validity, covariates dependent on directorial choices, such as scene

 https://doi.org/10.3390/s26051675

transitions and luminance, were not held constant or removed. Therefore, the estimated effects are closer to the net effect under realistic conditions where content components coexist. However, there are limitations in isolating the contribution of each factor. In addition, F-VIMS (tablet) and S-VIMS (see-through HMD) have structural differences (e.g., FOV, luminance, and viewing distance), which limit a pure comparison of the display effects.

The session order was fixed (CMS → VIMS → Co-MS), which may have introduced carryover/late-session effects (e.g., fatigue, habituation/adaptation, or sensitization). Because Co-MS was always administered last, some portion of the observed Co-MS responses may reflect the accumulated session effects rather than the combined stimulation per se. To mitigate this, the rest was set to 12 min based on prior evidence of EEG recovery following MS exposure [88], and participants could stop at any time if the symptoms became severe. Nevertheless, since the session-type order was not randomized, residual order confounding cannot be fully ruled out, and Co-MS findings should be interpreted cautiously.

Subjective MSL was measured using SSQ (pre/post) and FMS (at 30-s intervals), making it difficult to perfectly capture sudden symptoms. Repeated responses to the FMS itself might have also induced attentional distraction. Although the time axis of the human signals was corrected through synchronization and interpolation, the possibility of slight time warping owing to multithreaded collection and differing sampling rates (30–500 Hz) remains.

The eight-channel resolution of the EEG constrained the precision of interpretation, and the 30 Hz measurement rate of the HoloLens 2 eye tracking resulted in missing data. In addition, PPG was used instead of ECG, but had low precision and artifact resistance, and EDA experienced some data loss owing to unstable contact for participants with thin fingers. Further, some features were indirectly derived indicators, such as BR, which was derived from the PPG instead of a dedicated RSP sensor. Continuous BP monitoring was also omitted to avoid physical distraction, although the extracted PPG features partially compensated for this absence by indirectly capturing cardiovascular dynamics.

While the implemented acquisition system enabled parallel collection across varying sampling rates, minor timing misalignments occurred owing to OS time resolution and thread scheduling limits. Specifically, the actual number of samples within a 1-s interval deviated minimally (e.g., by approximately ±1 chunk) from the expected grid. Although we applied piecewise cubic Hermite interpolation to tightly preserve the original signal morphology and minimize distortion during this time-axis correction, this study did not quantitatively profile the exact proportion of repaired or interpolated segments per sensor modality. Consequently, we cannot completely rule out the possibility that these corrected segments, however minimal, may have partially influenced specific feature calculations. Multiple comparison correction was applied only to the main hypothesis tests (ANOVA, rmcorr), whereas the extensive feature-symptom correlation analysis was treated as exploratory and reported by focusing on the effect sizes instead of adjustments.

The EBM utilized in this study was primarily designed for post hoc explanatory modeling, specifically to assess the relative multivariable contributions of multimodal features, rather than to construct a generalized, real-time prediction system for unseen passengers. Because the modeling objective was targeted at interpretability rather than maximizing predictive performance, extensive comparative benchmarking against predictive baselines was not performed. Furthermore, while participant-wise cross-validation was employed to ensure the robustness of the derived feature importance rankings, the current model focuses on evaluating the concurrent symptom state rather than forecasting future MS onset. Consequently, the reported feature contributions should be strictly interpreted as an internal attribution of multimodal factors within the current sample, as the structural

https://doi.org/10.3390/s26051675

dependencies of the explanatory framework inherently constrain its direct application as a generalized, real-time predictive metric for novel users.

In addition, the multivariable contribution analysis faced a structural limitation due to the substantial imbalance in the number of extracted features across sensor domains, with EEG alone accounting for approximately 70% of the total feature set. Although a single-domain evaluation was explicitly conducted to partially mitigate inter-domain redundancy and isolate individual domain effects, the sheer volume of EEG features intrinsically provided this domain with greater model capacity. Consequently, the observed superiority of EEG in preserving model fit may be partially amplified by its higher degrees of freedom, rather than being solely driven by its inherent informative value.

7. Conclusions

In this work, an integrated framework was designed for CMS-VIMS-Co-MS in the context of autonomous driving, and multifaceted analyses of MSL were performed by simultaneously collecting data from the EEG, EDA, PPG, Eye, Head, and SKT domains in two environments (F-VIMS/S-VIMS) ($N = 90$). Unlike previous studies that were limited to a single device or type, in this study, the display (tablet/see-through HMD), stimulus type, user characteristics (gender and susceptibility), and immersion were compared within a single framework.

The main results obtained from this study are as follows: (i) In the S-VIMS group, there were no significant differences in ΔSSQ according to the MS type, gender, or susceptibility. (ii) In the F-VIMS group, the main effect of the MS type was significant for all sub-symptom scales, with the general order being VIMS < CMS < Co-MS, with the exception of CMS > Co-MS, which was observed only for oculomotor symptoms. (iii) For single-stimulus VIMS, there was no ΔSSQ difference between F-VIMS and S-VIMS, and for Co-MS, the gender × VIMS type interaction was significant for nausea. (iv) IEQ showed no significant difference between F-VIMS and S-VIMS. In rmcorr, patterns of IEQ_{cap} (negative with all ΔSSQ), IEQ_{com} (negative with ΔSSQ_O and ΔSSQ_T), IEQ_{dis} (positive with ΔSSQ_O and ΔSSQ_T), and IEQ_{tra} (non-significant) were confirmed. (v) FMS was positively correlated with all ΔSSQs.

The correlations between human-signal features and MSL revealed distinct patterns across domains. In the EEG domain, a decrease in the entropy and an increase in the skewness and kurtosis in the δ-α bands were primarily linked to oculomotor symptoms. The PPG domain exhibited medium-sized correlations across all symptoms, acting more as an auxiliary indicator than a core indicator. The EDA domain frequently exhibited negative correlations of $|r| \geq 0.3$, particularly with nausea. In the Head domain, the amplitude/energy features of the pitch/surge showed large positive correlations with oculomotor symptoms, whereas the irregularity features (e.g., entropy) showed negative correlations. In the Eye domain, PSD entropy was positively correlated with nausea/oculomotor symptoms, whereas traditional VOR did not have medium-to-large correlations.

The interpretable modeling based on the EBM demonstrated that multivariable contributions to the FMS were distributed across several multimodal features rather than concentrated on a few dominant predictors. The domain ablation and single-domain evaluations confirmed that EEG was the most important explanatory domain, whereas other single modalities (such as Eye, EDA, and SKT) showed substantially lower standalone fit than EEG, Head, and PPG. Nevertheless, the evaluation of lightweight sensor combinations identified the Head + PPG + EDA configuration as the optimal alternative, exhibiting minimal degradation in explanatory power. Notably, this confirmed the feasibility of

 https://doi.org/10.3390/s26051675

characterizing MSL-related patterns using only human signals, even when demographic features were excluded.

The main contributions of this study are as follows: (i) An experimental framework was established to analyze CMS-VIMS-Co-MS within a single framework in the context of autonomous driving, thus complementing the limitations of previous fragmented studies and extending the research scope to next-generation displays. (ii) Through a systematic literature review of the last decade, the usage trends and classification system of features according to the human-signal domain were comprehensively organized, thus providing a reference for enhancing the methodological consistency and reproducibility of future MS-related human-signal studies. (iii) Through multimodal human-signal and MSL correlation analyses, significant features and symptom-specific patterns were derived for each sensor domain. (iv) Through EBM-based domain ablation/single-domain analysis, the core explanatory domain (EEG) and the relative multivariable contributions of multimodal features to the FMS were identified, and the effectiveness of lightweight sensor combinations (Head + PPG + EDA) was examined, thereby demonstrating the practical feasibility of few-sensor-based MSL assessment.

Future research will aim to perform nonlinear quantification of MSL based on DL by combining the stimulus intensity (six-DOF movement of the motion platform and first-person frame of the HMD) with lightweight domains (head, EDA, and PPG-based). Furthermore, despite the correlation structure between the FMS and SSQ, different correlational features were identified for nausea, oculomotor symptoms, and disorientation. Therefore, we intend to explore the latent patterns of MSL from human signals. Subsequently, an expansion to a real-world vehicle environment is planned to examine the generalization and operational feasibility of the MSL prediction model.

Supplementary Materials: The following supporting information can be downloaded at: https://www.mdpi.com/article/10.3390/s26051675/s1, Section S1: Detailed human-signal features by sensor domain in the HS-Set; Section S2: Technical details of the autonomous driving simulator; Section S3: Data synchronization and system I/O optimization; Figure S1: Motion profiles and vehicle trajectories for different driving scenarios; Figure S2: Process of road and surrounding environment generation based on vehicle trajectories; Figure S3: Operator console interface for real-time monitoring and remote recording; Table S1: Specific search queries and results for each database; Table S2: Summary of ECG and PPG features (not mutually exclusive); Table S3: Summary of EDA features (not mutually exclusive); Table S4: Summary of Eye features (not mutually exclusive); Table S5: Summary of EEG features (not mutually exclusive); Table S6: Summary of Head features (not mutually exclusive); Table S7: Summary of EGG features (not mutually exclusive); Table S8: Summary of miscellaneous features (not mutually exclusive); Table S9: Cutoff frequencies and scale factors used in the MCA; Table S10: Lower and upper cutoff frequencies for EEG preprocessing band-pass filters in the HS-Set; Table S11: Lower and upper cutoff frequencies for the EEG decomposition bands (δ-γ) in the HS-Set.

Author Contributions: Conceptualization, Y.S.K.; methodology, S.Y.K.; software, S.Y.K.; validation, Y.S.K.; formal analysis, S.Y.K.; investigation, S.Y.K.; resources, S.Y.K.; data curation, S.Y.K.; writing—original draft preparation, S.Y.K.; writing—review and editing, Y.S.K.; visualization, S.Y.K.; supervision, Y.S.K.; project administration, Y.S.K.; funding acquisition, Y.S.K. All authors have read and agreed to the published version of the manuscript.

Funding: This research received no external funding.

Institutional Review Board Statement: The study was conducted in accordance with the Declaration of Helsinki, and the protocol was approved by the Institutional Review Board of KOREATECH (No. 240423002) on 23 April 2024.

https://doi.org/10.3390/s26051675

Informed Consent Statement: Informed consent was obtained from all subjects involved in the study.

Data Availability Statement: The preprocessing and feature-extraction procedures are fully specified in the manuscript. The feature dataset and the analysis code are publicly available in the GitHub repository at https://github.com/sgkim6326/2025-NRF-Multimodal-MS-Analysis (accessed on 19 February 2026). These datasets are licensed under the Creative Commons Attribution 4.0 International License (CC BY 4.0). The raw multimodal physiological recordings are not publicly available because they contain personal health-related information and are extremely large; however, the full dataset is available on request from the corresponding author.

Acknowledgments: This work was supported by the National Research Foundation of Korea (NRF) grant funded by the Korea government (MSIT) (RS-2023-NR076959).

Conflicts of Interest: The authors declare no conflicts of interest.

References

1. SAE International. *Taxonomy and Definitions for Terms Related to Driving Automation Systems for On-Road Motor Vehicles*; J3016; SAE International: Warrendale, PA, USA, 2021.
2. Ataya, A.; Kim, W.; Elsharkawy, A.; Kim, S. How to interact with a fully autonomous vehicle: Naturalistic ways for drivers to intervene in the vehicle system while performing non-driving related tasks. *Sensors* **2021**, *21*, 2206. [CrossRef] [PubMed]
3. Schartmüller, C.; Weigl, K.; Löcken, A.; Wintersberger, P.; Steinhauser, M.; Riener, A. Displays for productive non-driving related tasks: Visual behavior and its impact in conditionally automated driving. *Multimodal Technol. Interact.* **2021**, *5*, 21. [CrossRef]
4. Yoon, S.H.; Ji, Y.G. Non-driving-related tasks, workload, and takeover performance in highly automated driving contexts. *Transp. Res. Part F Traffic Psychol. Behav.* **2019**, *60*, 620–631. [CrossRef]
5. Müller, A.L.; Fernandes-Estrela, N.; Hetfleisch, R.; Zecha, L.; Abendroth, B. Effects of non-driving related tasks on mental workload and take-over times during conditional automated driving. *Eur. Transp. Res. Rev.* **2021**, *13*, 16. [CrossRef]
6. Suwa, T.; Sato, Y.; Wada, T. Reducing motion sickness when reading with head-mounted displays by using see-through background images. *Front. Virtual Real.* **2022**, *3*, 910434. [CrossRef]
7. Pfleging, B.; Rang, M.; Broy, N. Investigating user needs for non-driving-related activities during automated driving. In Proceedings of the International Conference on Mobile and Ubiquitous Multimedia (MUM), Rovaniemi, Finland, 12–15 December 2016; pp. 91–99. [CrossRef]
8. Detjen, H.; Pfleging, B.; Schneegass, S. A wizard of oz field study to understand non-driving-related activities, trust, and acceptance of automated vehicles. In Proceedings of the International Conference on Automotive User Interfaces and Interactive Vehicular Applications (AutomotiveUI), Virtual Event, 21–22 September 2020; pp. 19–29. [CrossRef]
9. Reed, M.P.; Ebert, S.M.; Jones, M.L. *Naturalistic Passenger Behavior: Posture and Activities*; UMTRI-2020-2; Transportation Research Institute: Ann Arbor, MI, USA, 2020.
10. Li, J.; George, C.; Ngao, A.; Holländer, K.; Mayer, S.; Butz, A. An exploration of users' thoughts on rear-seat productivity in virtual reality. In Proceedings of the International Conference on Automotive User Interfaces and Interactive Vehicular Applications (AutomotiveUI), Virtual Event, 21–22 September 2020; pp. 92–95. [CrossRef]
11. Elsharkawy, A.I.A.M.; Ataya, A.A.S.; Yeo, D.; An, E.; Hwang, S.; Kim, S. SYNC-VR: Synchronizing your senses to conquer motion sickness for enriching in-vehicle virtual reality. In Proceedings of the CHI Conference on Human Factors in Computing Systems, Honolulu, HI, USA, 11–16 May 2024; pp. 1–17. [CrossRef]
12. Diels, C.; Bos, J.E. Self-driving carsickness. *Appl. Ergon.* **2016**, *53*, 374–382. [CrossRef] [PubMed]
13. Metzulat, M.; Metz, B.; Landau, A.; Neukum, A.; Kunde, W. Does the visual input matter? Influence of non-driving related tasks on car sickness in an open road setting. *Transp. Res. Part F Traffic Psychol. Behav.* **2024**, *104*, 234–248. [CrossRef]
14. Turner, M. Motion sickness in public road transport: Passenger behaviour and susceptibility. *Ergonomics* **1999**, *42*, 444–461. [CrossRef]
15. Xie, W.; He, D.; Wu, G. Inducers of motion sickness in vehicles: A systematic review of experimental evidence and meta-analysis. *Transp. Res. Part F Traffic Psychol. Behav.* **2023**, *99*, 167–188. [CrossRef]
16. Kennedy, R.S.; Lane, N.E.; Berbaum, K.S.; Lilienthal, M.G. Simulator sickness questionnaire: An enhanced method for quantifying simulator sickness. *Int. J. Aviat. Psychol.* **1993**, *3*, 203–220. [CrossRef]
17. Keshavarz, B.; Hecht, H. Validating an efficient method to quantify motion sickness. *Hum. Factors* **2011**, *53*, 415–426. [CrossRef]
18. Bos, J.E.; MacKinnon, S.N.; Patterson, A. Motion sickness symptoms in a ship motion simulator: Effects of inside, outside, and no view. *Aviat. Space Environ. Med.* **2005**, *76*, 1111–1118.

 https://doi.org/10.3390/s26051675

19. *ISO 2631-1*; Mechanical Vibration and Shock—Evaluation of Human Exposure to Whole-Body Vibration—Part 1: General Requirements. International Organization for Standardization: Geneva, Switzerland, 1997.
20. Asua, E.; Gutiérrez-Zaballa, J.; Mata-Carballeira, O.; Ruiz, J.A.; del Campo, I. Analysis of the motion sickness and the lack of comfort in car passengers. *Appl. Sci.* **2022**, *12*, 3717. [CrossRef]
21. Htike, Z.; Papaioannou, G.; Siampis, E.; Velenis, E.; Longo, S. Fundamentals of motion planning for mitigating motion sickness in automated vehicles. *IEEE Trans. Veh. Technol.* **2021**, *71*, 2375–2384. [CrossRef]
22. Li, D.; Hu, J. Mitigating motion sickness in automated vehicles with frequency-shaping approach to motion planning. *IEEE Robot. Autom. Lett.* **2021**, *6*, 7714–7720. [CrossRef]
23. Seo, J.W.; Ko, C.H.; Sung, J.H.; Yun, D.G.; Lee, B.; Kim, J.S.; Park, T.; Park, H.S.; Ju, S.P.; Chung, C.C. Data-driven human modeling based on temporal information and nonlinear model predictive control for adaptive cruise control reducing motion sickeness. In Proceedings of the IEEE International Conference on Intelligent Transportation Systems (ITSC), Edmonton, AB, Canada, 24–27 September 2024; pp. 3639–3644. [CrossRef]
24. Golding, J.F. Motion sickness susceptibility questionnaire revised and its relationship to other forms of sickness. *Brain Res. Bull.* **1998**, *47*, 507–516. [CrossRef] [PubMed]
25. Rigby, J.M.; Brumby, D.P.; Gould, S.J.; Cox, A.L. Development of a questionnaire to measure immersion in video media: The Film IEQ. In Proceedings of the ACM International Conference on Interactive Experiences for TV and Online Video (TVX), Salford, UK, 5–7 June 2019; pp. 35–46. [CrossRef]
26. Ramaioli, C.; Steinmetzer, T.; Brietzke, A.; Meyer, P.; Pham Xuan, R.; Schneider, E.; Gorges, M. Assessment of vestibulo-ocular reflex and its adaptation during stop-and-go car rides in motion sickness susceptible passengers. *Exp. Brain Res.* **2023**, *241*, 1523–1531. [CrossRef] [PubMed]
27. Paredes, P.E.; Balters, S.; Qian, K.; Murnane, E.L.; Ordóñez, F.; Ju, W.; Landay, J.A. Driving with the fishes: Towards calming and mindful virtual reality experiences for the car. *Proc. ACM Interact. Mob. Wearable Ubiquitous Technol.* **2018**, *2*, 1–21. [CrossRef]
28. Schartmüller, C.; Riener, A. Sick of scents: Investigating non-invasive olfactory motion sickness mitigation in automated driving. In Proceedings of the International Conference on Automotive User Interfaces and Interactive Vehicular Applications (AutomotiveUI), Virtual Event, 21–22 September 2020; pp. 30–39. [CrossRef]
29. Kojima, T.; Ohsuga, M.; Kamakura, Y.; Hori, J.; Watanabe, S. Assessment of car sickness in passengers using physiological indices. In Proceedings of the IEEE International Conference on Systems, Man, and Cybernetics (SMC), Prague, Czech, 9–12 October 2022; pp. 598–603. [CrossRef]
30. Aydin, K.; Kara, E.; Adatepe, N.U.; Atas, A. Neuro-ophthalmic and neuro-otologic evaluation in individuals with motion sickness susceptibility. *J. Int. Adv. Otol.* **2024**, *20*, 345. [CrossRef]
31. Karababa, E.; Satar, B.; Genç, H. Evaluation of effects of optokinetic and rotational stimuli with functional head impulse test (fHIT) in individuals with motion sickness. *Eur. Arch. Oto-Rhino-Laryngol.* **2023**, *280*, 3149–3156. [CrossRef]
32. Tamura, A.; Iwamoto, T.; Ozaki, H.; Kimura, M.; Tsujimoto, Y.; Wada, Y. Wrist-worn electrodermal activity as a novel neurophysiological biomarker of autonomic symptoms in spatial disorientation. *Front. Neurol.* **2018**, *9*, 1056. [CrossRef]
33. Rahimzadeh, G.; Lacy, K.; Mohamed, S.; Plawiak, P.; Sharifrazi, D.; Toomey, N.G.; Asadi, H. Immediate detection of simulator sickness in virtual environments using integrated subjective feedback and physiological signals. In Proceedings of the IEEE International Conference on E-health Networking, Application & Services (HealthCom), Nara, Japan, 18–20 November 2024; pp. 1–7. [CrossRef]
34. Kobayashi, N.; Yamazaki, M.; Mizutani, R. Impact of visually induced motion sickness from VR depending on viewing patterns, view movement, and background motion. In Proceedings of the Annual International Conference of the IEEE Engineering in Medicine and Biology Society (EMBC), Sydney, Australia, 24–27 July 2023; pp. 1–4. [CrossRef]
35. Recenti, M.; Pescaglia, F.; Guerrini, L.; Maruotto, I.; Gelormini, C.; Jacob, D.; Aubonnet, R.; Petersen, H.; Gargiulo, P. New frontiers in postural control and motion sickness assessment: The BioVRSea paradigm. In Proceedings of the IEEE International Conference on Metrology for eXtended Reality, Artificial Intelligence and Neural Engineering (MetroXRAINE), St Albans, UK, 21–23 October 2024; pp. 1034–1039. [CrossRef]
36. Park, S.-H.; Han, D.-K.; Lee, S.-W. Dynamic multi-modal fusion for biosignal-based motion sickness prediction in vehicles. In Proceedings of the Annual International Conference of the IEEE Engineering in Medicine and Biology Society (EMBC), Orlando, FL, USA, 15–19 July 2024; pp. 1–5. [CrossRef]
37. Su, H.; Jia, Y. Study of human comfort in autonomous vehicles using wearable sensors. *IEEE Trans. Intell. Transp. Syst.* **2021**, *23*, 11490–11504. [CrossRef]
38. Lee, S.; Kim, S.; Kim, H.G.; Ro, Y.M. Assessing individual VR sickness through deep feature fusion of VR video and physiological response. *IEEE Trans. Circuits Syst. Video Technol.* **2021**, *32*, 2895–2907. [CrossRef]

 https://doi.org/10.3390/s26051675

39. Sameri, J.; Coenegracht, H.; Van Damme, S.; De Turck, F.; Torres Vega, M. Physiology-driven cybersickness detection in virtual reality: A machine learning and explainable AI approach. *Virtual Real.* **2024**, *28*, 174. [CrossRef]
40. Oh, H.; Son, W. Cybersickness and its severity arising from virtual reality content: A comprehensive study. *Sensors* **2022**, *22*, 1314. [CrossRef]
41. Reddy, A.; Kim, J.R. Assessment and quantification of virtual reality induced sickness in relation to age and gender: A multimodal approach. In Proceedings of the IEEE MIT Undergraduate Research Technology Conference (URTC), Cambridge, MA, USA, 6–8 October 2023; pp. 1–4. [CrossRef]
42. Kundu, R.K.; Islam, R.; Calyam, P.; Hoque, K.A. TruVR: Trustworthy cybersickness detection using explainable machine learning. In Proceedings of the IEEE International Symposium on Mixed and Augmented Reality (ISMAR), Singapore, 17–21 October 2022; pp. 777–786. [CrossRef]
43. Martin, N.; Mathieu, N.; Pallamin, N.; Ragot, M.; Diverrez, J.-M. Virtual reality sickness detection: An approach based on physiological signals and machine learning. In Proceedings of the IEEE International Symposium on Mixed and Augmented Reality (ISMAR), Porto de Galinhas, Brazil, 9–13 November 2020; pp. 387–399. [CrossRef]
44. Andre, L.; Coutellier, R. Cybersickness and evaluation of a remediation system: A pilot study. In Proceedings of the International Conference on 3D Immersion (IC3D), Brussels, Belgium, 11 December 2019; pp. 1–6. [CrossRef]
45. Gruden, T.; Popović, N.B.; Stojmenova, K.; Jakus, G.; Miljković, N.; Tomažič, S.; Sodnik, J. Electrogastrography in autonomous vehicles—An objective method for assessment of motion sickness in simulated driving environments. *Sensors* **2021**, *21*, 550. [CrossRef]
46. Islam, R.; Desai, K.; Quarles, J. Towards forecasting the onset of cybersickness by fusing physiological, head-tracking and eye-tracking with multimodal deep fusion network. In Proceedings of the IEEE International Symposium on Mixed and Augmented Reality (ISMAR), Singapore, 17–21 October 2022; pp. 121–130. [CrossRef]
47. Kim, H.; Kim, D.J.; Chung, W.H.; Park, K.-A.; Kim, J.D.; Kim, D.; Kim, K.; Jeon, H.J. Clinical predictors of cybersickness in virtual reality (VR) among highly stressed people. *Sci. Rep.* **2021**, *11*, 12139. [CrossRef]
48. Hussain, R.; Chessa, M.; Solari, F. Mitigating cybersickness in virtual reality systems through foveated depth-of-field blur. *Sensors* **2021**, *21*, 4006. [CrossRef]
49. Preciado, C.E.; Starrett, M.J.; Ekstrom, A.D. Assessment of a short, focused training to reduce symptoms of cybersickness. *Presence* **2021**, *27*, 361–377. [CrossRef]
50. Lee, R.; Kim, Y.S. VR sickness evaluation method using recurrence period density entropy. *Appl. Sci.* **2024**, *14*, 4483. [CrossRef]
51. Reyero Lobo, P.; Perez, P. Heart rate variability for non-intrusive cybersickness detection. In Proceedings of the ACM International Conference on Interactive Media Experiences (IMX), Aveiro, Portugal, 22–24 June 2022; pp. 221–228. [CrossRef]
52. Ruffle, J.K.; Patel, A.; Giampietro, V.; Howard, M.A.; Sanger, G.J.; Andrews, P.L.; Williams, S.C.; Aziz, Q.; Farmer, A.D. Functional brain networks and neuroanatomy underpinning nausea severity can predict nausea susceptibility using machine learning. *J. Physiol.* **2019**, *597*, 1517–1529. [CrossRef] [PubMed]
53. Farmer, A.D.; Ban, V.F.; Coen, S.J.; Sanger, G.J.; Barker, G.J.; Gresty, M.A.; Giampietro, V.P.; Williams, S.C.; Webb, D.L.; Hellström, P.M. Visually induced nausea causes characteristic changes in cerebral, autonomic and endocrine function in humans. *J. Physiol.* **2015**, *593*, 1183–1196. [CrossRef] [PubMed]
54. Kim, S.; Lee, S.; Ro, Y.M. Estimating VR sickness caused By camera shake in VR videography. In Proceedings of the IEEE International Conference on Image Processing (ICIP), Abu Dhabi, United Arab Emirates, 25–28 October 2020; pp. 3433–3437. [CrossRef]
55. Recenti, M.; Ricciardi, C.; Aubonnet, R.; Picone, I.; Jacob, D.; Svansson, H.Á.; Agnarsdóttir, S.; Karlsson, G.H.; Baeringsdóttir, V.; Petersen, H. Toward predicting motion sickness using virtual reality and a moving platform assessing brain, muscles, and heart signals. *Front. Bioeng. Biotechnol.* **2021**, *9*, 635661. [CrossRef]
56. Lee, S.; Kim, S.; Kim, H.G.; Kim, M.S.; Yun, S.; Jeong, B.; Ro, Y.M. Physiological fusion net: Quantifying individual VR sickness with content stimulus and physiological response. In Proceedings of the IEEE International Conference on Image Processing (ICIP), Taipei, Taiwan, 22–25 September 2019; pp. 440–444. [CrossRef]
57. Cowings, P.S.; Toscano, W.B.; Reschke, M.F.; Tsehay, A. Psychophysiological assessment and correction of spatial disorientation during simulated Orion spacecraft re-entry. *Int. J. Psychophysiol.* **2018**, *131*, 102–112. [CrossRef]
58. Gavgani, A.M.; Nesbitt, K.V.; Blackmore, K.L.; Nalivaiko, E. Profiling subjective symptoms and autonomic changes associated with cybersickness. *Auton. Neurosci.* **2017**, *203*, 41–50. [CrossRef]
59. Nunes da Silva, W.; Porcino, T.M.; Castanho, C.D.; Jacobi, R.P. Analysis of cybersickness through biosignals: An approach with symbolic machine learning. In Proceedings of the Symposium on Virtual and Augmented Reality (SVR), Manaus, Brazil, 30 September–3 October 2024; pp. 11–20. [CrossRef]
60. Dennison, M.S.; Wisti, A.Z.; D'Zmura, M. Use of physiological signals to predict cybersickness. *Displays* **2016**, *44*, 42–52. [CrossRef]

61. Hadadi, A.; Chardonnet, J.-R.; Guillet, C.; Ovtcharova, J. Machine Learning Application for Real-Time Simulator. In Proceedings of the International Conference on Machine Learning Technologies (ICMLT), Oslo, Norway, 24–26 May 2024; pp. 1–5. [CrossRef]
62. Jeong, D.; Han, K. Precyse: Predicting cybersickness using transformer for multimodal time-series sensor data. *Proc. ACM Interact. Mob. Wearable Ubiquitous Technol.* **2024**, *8*, 1–24. [CrossRef]
63. Kundu, R.K.; Elsaid, O.Y.; Calyam, P.; Hoque, K.A. VR-LENS: Super learning-based cybersickness detection and explainable AI-guided deployment in virtual reality. In Proceedings of the International Conference on Intelligent User Interfaces (IUI), Sydney, NSW, Australia, 27–31 March 2023; pp. 819–834. [CrossRef]
64. Jeong, D.; Paik, S.; Noh, Y.; Han, K. MAC: Multimodal, attention-based cybersickness prediction modeling in virtual reality. *Virtual Real.* **2023**, *27*, 2315–2330. [CrossRef]
65. Li, C.-C.; Zhang, Z.-R.; Liu, Y.-H.; Zhang, T.; Zhang, X.-T.; Wang, H.; Wang, X.-C. Multi-dimensional and objective assessment of motion sickness susceptibility based on machine learning. *Front. Neurol.* **2022**, *13*, 824670. [CrossRef]
66. Jeong, D.; Han, K. Leveraging multimodal sensory information in cybersickness prediction. In Proceedings of the ACM Symposium on Virtual Reality Software and Technology (VRST), Tsukuba Japan, 29 November–1 December 2022; pp. 1–2. [CrossRef]
67. Arafat, I.M.; Ferdous, S.M.S.; Quarles, J. The effects of cybersickness on persons with multiple sclerosis. In Proceedings of the ACM Conference on Virtual Reality Software and Technology (VRST), Munich, Germany, 2–4 November 2016; pp. 51–59. [CrossRef]
68. Tang, Q.; Xiang, H.; Cheng, J.; Xiao, X.; Yu, W.; Zhang, S.; Tang, B.; Guo, G. Study on evaluation method of motion sickness in electric vehicles. In Proceedings of the CAA International Conference on Vehicular Control and Intelligence (CVCI), Chongqing, China, 25–27 October 2024; pp. 1–6. [CrossRef]
69. Deng, Z.; Yuan, K.; Xiao, X. Investigative examination of motion sickness indicators for electric vehicles. *Proc. Inst. Mech. Eng. Part D J. Automob. Eng.* **2024**, 09544070241251521. [CrossRef]
70. Shimada, S.; Pannattee, P.; Ikei, Y.; Nishiuchi, N.; Yem, V. High-frequency cybersickness prediction using deep learning techniques with eye-related indices. *IEEE Access* **2023**, *11*, 95825–95839. [CrossRef]
71. Lopes, P.; Tian, N.; Boulic, R. Exploring blink-rate behaviors for cybersickness detection in VR. In Proceedings of the IEEE Conference on Virtual Reality and 3D User Interfaces Abstracts and Workshops (VRW), Atlanta, GA, USA, 22–26 March 2020; pp. 794–795. [CrossRef]
72. Li, Y.; Pan, L.; Liu, M.; Shao, Z.; Xue, M.; Liao, J.; Zhao, H.; Wu, M.; Yu, S.; Wu, X. Quantitative study on objective indicators for assessing motion sickness susceptibility based on Vestibulo-Ocular Reflex experiments. *Sci. Rep.* **2024**, *14*, 28782. [CrossRef]
73. Josupeit, J.; Greim, L. The optokinetic nystagmus as a physiological indicator of cybersickness–a vergence-based evaluation. In Proceedings of the International Conference on Human-Computer Interaction, Washington, DC, USA, 29 June–4 July 2024; pp. 58–66. [CrossRef]
74. Fujikake, K.; Itadu, Y.; Takada, H. Analyzing gaze data during rest time/driving simulator operation using machine learning. In Proceedings of the International Conference on Human-Computer Interaction, Virtual Event, 26 June–1 July 2022; pp. 420–434. [CrossRef]
75. Park, S.; Mun, S.; Ha, J.; Kim, L. Non-contact measurement of motion sickness using pupillary rhythms from an infrared camera. *Sensors* **2021**, *21*, 4642. [CrossRef]
76. Lee, J.; Kim, W.; Kim, J.; Lee, S. A study on virtual reality sickness and visual attention. In Proceedings of the Asia-Pacific Signal and Information Processing Association Annual Summit and Conference (APSIPA ASC), Tokyo, Japan, 14–17 December 2021; pp. 1465–1469.
77. Islam, R.; Desai, K.; Quarles, J. Cybersickness prediction from integrated HMD's sensors: A multimodal deep fusion approach using eye-tracking and head-tracking data. In Proceedings of the IEEE International Symposium on Mixed and Augmented Reality (ISMAR), Bari, Italy, 4–8 October 2021; pp. 31–40. [CrossRef]
78. Tovar, D.; Wilmott, J.; Wu, X.; Martin, D.; Proulx, M.; Lindberg, D.; Zhao, Y.; Mercier, O.; Guan, P. Identifying behavioral correlates to visual discomfort. *ACM Trans. Graph.* **2024**, *43*, 1–10. [CrossRef]
79. Fan, C.-L.; Hung, T.-H.; Hsu, C.-H. Modeling the user experience of watching 360 videos with head-mounted displays. *ACM Trans. Multimed. Comput. Commun. Appl.* **2022**, *18*, 1–23. [CrossRef]
80. Berton, B.R.; Morel-Targosz, C.; Chaumillon, R.; Wolff, M. Paving the way towards a methodology to faithfully assess physical, physiological, and cognitive impacts of augmented reality under constrained environments: A head-mounted display use case. In Proceedings of the "Ergonomie et Informatique Avancée" Conference, Bidart, France, 10–13 October 2023; pp. 1–8. [CrossRef]
81. Zhu, H.; Li, T.; Wang, C.; Jin, W.; Murali, S.; Xiao, M.; Ye, D.; Li, M. Eyeqoe: A novel qoe assessment model for 360-degree videos using ocular behaviors. *Proc. ACM Interact. Mob. Wearable Ubiquitous Technol.* **2022**, *6*, 1–26. [CrossRef]

https://doi.org/10.3390/s26051675

82. Lopes, P.; Tian, N.; Boulic, R. Eye thought you were sick! exploring eye behaviors for cybersickness detection in VR. In Proceedings of the ACM SIGGRAPH Conference on Motion, Interaction and Games (MIG), Virtual Event, 16–18 October 2020; pp. 1–10. [CrossRef]
83. Hua, C.; Chai, L.; Yan, Y.; Liu, J.; Wang, Q.; Fu, R.; Zhou, Z. Assessment of virtual reality motion sickness severity based on EEG via LSTM/BiLSTM. *IEEE Sens. J.* **2023**, *23*, 24839–24848. [CrossRef]
84. Liu, M.; Yang, B.; Xu, M.; Zan, P.; Chen, L.; Xia, X. Exploring quantitative assessment of cybersickness in virtual reality using EEG signals and a CNN-ECA-LSTM network. *Displays* **2024**, *81*, 102602. [CrossRef]
85. Li, Z.; Zhao, L.; Chang, J.; Li, W.; Yang, M.; Li, C.; Wang, R.; Ji, L. EEG-based evaluation of motion sickness and reducing sensory conflict in a simulated autonomous driving environment. In Proceedings of the Annual International Conference of the IEEE Engineering in Medicine and Biology Society (EMBC), Glasgow, UK, 11–15 July 2022; pp. 4026–4030. [CrossRef]
86. Qin, B.; Wu, B.; Zhou, L.; Chen, Y.; Qian, Z.; Zhu, Q. Study on motion sickness based on EEG power spectrum characteristics. In Proceedings of the IEEE International Conference on Medical Imaging Physics and Engineering (ICMIPE), Hefei, China, 12–14 November 2021; pp. 1–4. [CrossRef]
87. Liu, M.; Yang, B.; Zan, P.; Chen, L.; Wang, B.; Xia, X. Exploring the brain physiological activity and quantified assessment of VR cybersickness using EEG signals. *Displays* **2024**, *85*, 102879. [CrossRef]
88. Woo, Y.S.; Jang, K.-M.; Nam, S.G.; Kwon, M.; Lim, H.K. Recovery time from VR sickness due to susceptibility: Objective and quantitative evaluation using electroencephalography. *Heliyon* **2023**, *9*, e14792. [CrossRef]
89. Hu, H.; Fang, Z.; Qian, Z.; Yao, L.; Tao, L.; Qin, B. Stress assessment of vestibular endurance training for civil aviation flight students based on EEG. *Front. Hum. Neurosci.* **2021**, *15*, 582636. [CrossRef]
90. Lim, H.K.; Ji, K.; Woo, Y.S.; Han, D.-u.; Lee, D.-H.; Nam, S.G.; Jang, K.-M. Test-retest reliability of the virtual reality sickness evaluation using electroencephalography (EEG). *Neurosci. Lett.* **2021**, *743*, 135589. [CrossRef]
91. Xu, M.; Yang, B.; Liu, M.; Huang, Y.; Xia, X.; Zhang, J. SE-CNN attention structure for quantitative EEG-based assessment of VR motion sickness. In Proceedings of the International Conference on Computing and Artificial Intelligence (ICCAI), Tianjin, China, 17–20 March 2023; pp. 606–611. [CrossRef]
92. Jang, K.-M.; Shin Woo, Y.; Kyoon Lim, H. Electrophysiological changes in the virtual reality sickness: EEG in the VR sickness. In Proceedings of the International Conference on 3D Web Technology (Web3D), Virtual Event, 9–13 November 2020; pp. 1–4. [CrossRef]
93. Jeong, D.K.; Yoo, S.; Jang, Y. VR sickness measurement with EEG using DNN algorithm. In Proceedings of the ACM Symposium on Virtual Reality Software and Technology (VRST), Tokyo, Japan, 28 November–1 December 2018; pp. 1–2. [CrossRef]
94. Subudhi, D.; Balaji, P.; Muniyandi, M. Objective quantification of circular vection in immersive environments. In Proceedings of the International Conference on Human-Computer Interaction, Gothenburg, Sweden, 26 June–1 July 2022; pp. 261–274. [CrossRef]
95. Sugiura, T.; Wada, T.; Nagata, T.; Sakai, K.; Sato, Y. Analysing effect of vehicle lean using cybernetic model of motion sickness. *IFAC-PapersOnLine* **2019**, *52*, 311–316. [CrossRef]
96. Bala, P.; Oakley, I.; Nisi, V.; Nunes, N.J. Dynamic field of view restriction in 360 video: Aligning optical flow and visual slam to mitigate VIMS. In Proceedings of the CHI Conference on Human Factors in Computing Systems, Yokohama, Japan, 8–13 May 2021; pp. 1–18. [CrossRef]
97. Bala, P.; Oakley, I.; Nisi, V.; Nunes, N. Staying on track: A comparative study on the use of optical flow in 360 video to mitigate VIMS. In Proceedings of the ACM International Conference on Interactive Media Experiences (IMX), Cornella, Barcelona, Spain, 17–19 June 2020; pp. 82–93. [CrossRef]
98. Laessoe, U.; Abrahamsen, S.; Zepernick, S.; Raunsbaek, A.; Stensen, C. Motion sickness and cybersickness–Sensory mismatch. *Physiol. Behav.* **2023**, *258*, 114015. [CrossRef]
99. Jakus, G.; Sodnik, J.; Miljković, N. Electrogastrogram-derived features for automated sickness detection in driving simulator. *Sensors* **2022**, *22*, 8616. [CrossRef]
100. Xia, Z.; Liu, Y.; Bai, Y.; Zhang, Y.; Cheng, C. Mismatches between 3D content acquisition and perception cause more visually induced motion sickness. *Presence* **2024**, *33*, 481–492. [CrossRef]
101. Wang, Y.; Chardonnet, J.-R.; Merienne, F. VR sickness prediction for navigation in immersive virtual environments using a deep long short term memory model. In Proceedings of the IEEE Conference on Virtual Reality and 3D User Interfaces (VR), Osaka, Japan, 23–27 March 2019; pp. 1874–1881.
102. Choi, C.; Jun, J.; Heo, J.; Kim, K. Effects of virtual-avatar motion-synchrony levels on full-body interaction. In Proceedings of the ACM/SIGAPP Symposium on Applied Computing (SAC), Limassol, Cyprus, 8–12 April 2019; pp. 701–708. [CrossRef]
103. Ren, B.; Guan, W.; Zhou, Q. Study of motion sickness model based on fNIRS multiband features during car rides. *Diagnostics* **2023**, *13*, 1462. [CrossRef]

104. Pöhlmann, K.M.T.; Maior, H.A.; Föcker, J.; O'Hare, L.; Parke, A.; Ladowska, A.; Dickinson, P. I think I don't feel sick: Exploring the relationship between cognitive demand and cybersickness in virtual reality using fNIRS. In Proceedings of the CHI Conference on Human Factors in Computing Systems, Hamburg, Germany, 23–28 April 2023; pp. 1–16. [CrossRef]
105. Yamamura, H.; Baldauf, H.; Kunze, K. HemodynamicVR-adapting the user's field of view during virtual reality locomotion tasks to reduce cybersickness using wearable functional near-infrared spectroscopy. In Proceedings of the Augmented Humans International Conference (AHs), Rovaniemi, Finland, 22–24 February 2021; pp. 223–227. [CrossRef]
106. Feigl, T.; Roth, D.; Gradl, S.; Wirth, M.; Latoschik, M.E.; Eskofier, B.M.; Philippsen, M.; Mutschler, C. Sick moves! motion parameters as indicators of simulator sickness. *IEEE Trans. Vis. Comput. Graph.* **2019**, *25*, 3146–3157. [CrossRef]
107. Cleij, D.; Venrooij, J.; Pretto, P.; Katliar, M.; Bülthoff, H.H.; Steffen, D.; Hoffmeyer, F.W.; Schöner, H.-P. Comparison between filter-and optimization-based motion cueing algorithms for driving simulation. *Transp. Res. Part F Traffic Psychol. Behav.* **2019**, *61*, 53–68. [CrossRef]
108. Stahl, K.; Abdulsamad, G.; Leimbach, K.-D.; Vershinin, Y.A. State of the art and simulation of motion cueing algorithms for a six degree of freedom driving simulator. In Proceedings of the International IEEE Conference on Intelligent Transportation Systems (ITSC), Qingdao, China, 8–11 October 2014; pp. 537–541. [CrossRef]
109. Harm, D.L.; Reschke, M.R.; Parker, D.E. *Visual-Vestibular Integration Motion Perception Reporting*; DSO 604 OI-1; NASA Johnson Space Center: Houston, TX, USA, 1999; pp. 5.2.1–5.2.12.
110. Kenward, H.; Pelligand, L.; Savary-Bataille, K.; Elliott, J. Nausea: Current knowledge of mechanisms, measurement and clinical impact. *Vet. J.* **2015**, *203*, 36–43. [CrossRef]
111. Wickham, R.J. Revisiting the physiology of nausea and vomiting—Challenging the paradigm. *Support. Care Cancer* **2020**, *28*, 13–21. [CrossRef]
112. Irmak, T.; Kotian, V.; Happee, R.; de Winkel, K.N.; Pool, D.M. Amplitude and temporal dynamics of motion sickness. *Front. Syst. Neurosci.* **2022**, *16*, 866503. [CrossRef]
113. Kuiper, O.X.; Bos, J.E.; Diels, C.; Schmidt, E.A. Knowing what's coming: Anticipatory audio cues can mitigate motion sickness. *Appl. Ergon.* **2020**, *85*, 103068. [CrossRef]
114. Lawther, A.; Griffin, M.J. Motion sickness and motion characteristics of vessels at sea. *Ergonomics* **1988**, *31*, 1373–1394. [CrossRef]
115. Palmisano, S.; Constable, R. Reductions in sickness with repeated exposure to HMD-based virtual reality appear to be game-specific. *Virtual Real.* **2022**, *26*, 1373–1389. [CrossRef]
116. Risi, D.; Palmisano, S. Effects of postural stability, active control, exposure duration and repeated exposures on HMD induced cybersickness. *Displays* **2019**, *60*, 9–17. [CrossRef]
117. Stanney, K.M.; Hale, K.S.; Nahmens, I.; Kennedy, R.S. What to expect from immersive virtual environment exposure: Influences of gender, body mass index, and past experience. *Hum. Factors* **2003**, *45*, 504–520. [CrossRef]
118. Caserman, P.; Garcia-Agundez, A.; Gámez Zerban, A.; Göbel, S. Cybersickness in current-generation virtual reality head-mounted displays: Systematic review and outlook. *Virtual Real.* **2021**, *25*, 1153–1170. [CrossRef]
119. Won, J.-h.; Na, H.C.; Kim, Y.S. A new training method for VR sickness reduction. *Appl. Sci.* **2024**, *14*, 3485. [CrossRef]
120. Dam, A.; Jeon, M. A review of motion sickness in automated vehicles. In Proceedings of the International Conference on Automotive User Interfaces and Interactive Vehicular Applications (AutomotiveUI), Leeds, UK, 9–14 September 2021; pp. 39–48. [CrossRef]
121. Henry, E.H.; Bougard, C.; Bourdin, C.; Bringoux, L. Car sickness in real driving conditions: Effect of lateral acceleration and predictability reflected by physiological changes. *Transp. Res. Part F Traffic Psychol. Behav.* **2023**, *97*, 123–139. [CrossRef]
122. Lackner, J.R.; Graybiel, A. The effective intensity of Coriolis, cross-coupling stimulation is gravitoinertial force dependent: Implications for space motion sickness. *Aviat. Space Environ. Med.* **1986**, *57*, 229–235. [PubMed]
123. O'hanlon, J.F.; McCauley, M.E. Motion sickness incidence as a function of the frequency and acceleration of vertical sinusoidal motion. *Aerosp. Med.* **1974**, *45*, 366–369. [PubMed]
124. Wijlens, R.; van Paassen, M.M.; Mulder, M.; Takamatsu, A.; Makita, M.; Wada, T. Reducing motion sickness by manipulating an autonomous vehicle's accelerations. *IFAC-PapersOnLine* **2022**, *55*, 132–137. [CrossRef]
125. Butler, C.A.; Griffin, M.J. Motion sickness during fore-and-aft oscillation: Effect of the visual scene. *Aviat. Space Environ. Med.* **2006**, *77*, 1236–1243.
126. Butler, C.; Griffin, M.J. Motion sickness with combined fore-aft and pitch oscillation: Effect of phase and the visual scene. *Aviat. Space Environ. Med.* **2009**, *80*, 946–954. [CrossRef] [PubMed]
127. So, R.H.; Lo, W.; Ho, A.T. Effects of navigation speed on motion sickness caused by an immersive virtual environment. *Hum. Factors* **2001**, *43*, 452–461. [CrossRef]
128. Chang, E.; Seo, D.; Kim, H.T.; Yoo, B. An integrated model of cybersickness: Understanding user's discomfort in virtual reality. *J. Korean Inst. Inf. Sci. Eng.* **2018**, *45*, 251–279. [CrossRef]

129. Kim, J.; Oh, H.; Kim, W.; Choi, S.; Son, W.; Lee, S. A deep motion sickness predictor induced by visual stimuli in virtual reality. *IEEE Trans. Neural Netw. Learn. Syst.* **2020**, *33*, 554–566. [CrossRef]
130. Tian, N.; Lopes, P.; Boulic, R. A review of cybersickness in head-mounted displays: Raising attention to individual susceptibility. *Virtual Real.* **2022**, *26*, 1409–1441. [CrossRef]
131. McMahan, R.P.; Bowman, D.A.; Zielinski, D.J.; Brady, R.B. Evaluating display fidelity and interaction fidelity in a virtual reality game. *IEEE Trans. Vis. Comput. Graph.* **2012**, *18*, 626–633. [CrossRef]
132. Ragan, E.D.; Bowman, D.A.; Kopper, R.; Stinson, C.; Scerbo, S.; McMahan, R.P. Effects of field of view and visual complexity on virtual reality training effectiveness for a visual scanning task. *IEEE Trans. Vis. Comput. Graph.* **2015**, *21*, 794–807. [CrossRef] [PubMed]
133. Wen, E.; Gupta, C.; Sasikumar, P.; Billinghurst, M.; Wilmott, J.; Skow, E.; Dey, A.; Nanayakkara, S. VR.net: A real-world dataset for virtual reality motion sickness research. *IEEE Trans. Vis. Comput. Graph.* **2024**, *30*, 2330–2336. [CrossRef]
134. Jennett, C.; Cox, A.L.; Cairns, P.; Dhoparee, S.; Epps, A.; Tijs, T.; Walton, A. Measuring and defining the experience of immersion in games. *Int. J. Hum.-Comput. Stud.* **2008**, *66*, 641–661. [CrossRef]
135. Lapitan, D.G.; Rogatkin, D.A.; Molchanova, E.A.; Tarasov, A.P. Estimation of phase distortions of the photoplethysmographic signal in digital IIR filtering. *Sci. Rep.* **2024**, *14*, 6546. [CrossRef]
136. Privratsky, A.A.; Bush, K.A.; Bach, D.R.; Hahn, E.M.; Cisler, J.M. Filtering and model-based analysis independently improve skin-conductance response measures in the fMRI environment: Validation in a sample of women with PTSD. *Int. J. Psychophysiol.* **2020**, *158*, 86–95. [CrossRef] [PubMed]
137. Richman, J.S.; Moorman, J.R. Physiological time-series analysis using approximate entropy and sample entropy. *Am. J. Physiol.-Heart Circ. Physiol.* **2000**, *278*, H2039–H2049. [CrossRef]
138. Posada-Quintero, H.F.; Florian, J.P.; Orjuela-Cañón, Á.D.; Chon, K.H. Highly sensitive index of sympathetic activity based on time-frequency spectral analysis of electrodermal activity. *Am. J. Physiol.-Regul. Integr. Comp. Physiol.* **2016**, *311*, R582–R591. [CrossRef]
139. Malik, M.; Bigger, J.T.; Camm, A.J.; Kleiger, R.E.; Malliani, A.; Moss, A.J.; Schwartz, P.J. Heart rate variability: Standards of measurement, physiological interpretation, and clinical use. *Eur. Heart J.* **1996**, *17*, 354–381. [CrossRef]
140. Ciccone, A.B.; Siedlik, J.A.; Wecht, J.M.; Deckert, J.A.; Nguyen, N.D.; Weir, J.P. Reminder: RMSSD and SD1 are identical heart rate variability metrics. *Muscle Nerve* **2017**, *56*, 674–678. [CrossRef]
141. Brennan, M.; Palaniswami, M.; Kamen, P. Do existing measures of Poincare plot geometry reflect nonlinear features of heart rate variability? *IEEE Trans. Biomed. Eng.* **2002**, *48*, 1342–1347. [CrossRef] [PubMed]
142. Toichi, M.; Kubota, Y.; Murai, T.; Kamio, Y.; Sakihama, M.; Toriuchi, T.; Inakuma, T.; Sengoku, A.; Miyoshi, K. The influence of psychotic states on the autonomic nervous system in schizophrenia. *Int. J. Psychophysiol.* **1999**, *31*, 147–154. [CrossRef]
143. Ishchenko, A.; Shev'ev, P. Automated complex for multiparameter analysis of the galvanic skin response signal. *Biomed. Eng.* **1989**, *23*, 113–117. [CrossRef]
144. Greco, A.; Valenza, G.; Scilingo, E.P. Modeling for the Analysis of the EDA. In *Advances in Electrodermal Activity Processing with Applications for Mental Health*, 1st ed.; Springer International Publishing: Cham, Switzerland, 2016; pp. 19–33. [CrossRef]
145. Salvucci, D.D.; Goldberg, J.H. Identifying fixations and saccades in eye-tracking protocols. In Proceedings of the Symposium on Eye Tracking Research & Applications (ETRA), Palm Beach Gardens, FL, USA, 6–8 November 2000; pp. 71–78. [CrossRef]
146. Tobii. Tobii Pro Lab Gaze Filter. Available online: https://connect.tobii.com/s/article/Gaze-Filter-functions-and-effects?language=en_US (accessed on 25 June 2025).
147. Golding, J.F. Predicting individual differences in motion sickness susceptibility by questionnaire. *Personal. Individ. Differ.* **2006**, *41*, 237–248. [CrossRef]
148. Funder, D.C.; Ozer, D.J. Evaluating effect size in psychological research: Sense and nonsense. *Adv. Methods Pract. Psychol. Sci.* **2019**, *2*, 156–168. [CrossRef]

Disclaimer/Publisher's Note: The statements, opinions and data contained in all publications are solely those of the individual author(s) and contributor(s) and not of MDPI and/or the editor(s). MDPI and/or the editor(s) disclaim responsibility for any injury to people or property resulting from any ideas, methods, instructions or products referred to in the content.

https://doi.org/10.3390/s26051675

MDPI AG
Grosspeteranlage 5
4052 Basel
Switzerland
Tel.: +41 61 683 77 34

Sensors Editorial Office
E-mail: sensors@mdpi.com
www.mdpi.com/journal/sensors

Disclaimer/Publisher's Note: The title and front matter of this reprint are at the discretion of the Guest Editors. The publisher is not responsible for their content or any associated concerns. The statements, opinions and data contained in all individual articles are solely those of the individual Editors and contributors and not of MDPI. MDPI disclaims responsibility for any injury to people or property resulting from any ideas, methods, instructions or products referred to in the content.